Railway Engineering and Systems

Railway Engineering and Systems

Edited by **Marshall Roy**

𝒞ℒ LANRYE
INTERNATIONAL

New Jersey

Published by Clanrye International,
55 Van Reypen Street,
Jersey City, NJ 07306, USA
www.clanryeinternational.com

Railway Engineering and Systems
Edited by Marshall Roy

© 2016 Clanrye International

International Standard Book Number: 978-1-63240-531-9 (Hardback)

Contents

Preface

Railway engineering refers to a dynamic domain of engineering which deals with the design, manufacturing and operation of all kinds of railway networks. It encompasses the elements of civil, mechanical, electrical, production and computer engineering; among many others. This book will unfold the innovative aspects of railway engineering. It has detailed explanations of the various concepts and applications of this field. It is compiled in such a manner, that it will provide in-depth knowledge about this subject. Students, researchers, experts and all associated with this field will benefit alike from this book. It will prove to be a beneficial source of knowledge for readers.

This book is a result of research of several months to collate the most relevant data in the field.

When I was approached with the idea of this book and the proposal to edit it, I was overwhelmed. It gave me an opportunity to reach out to all those who share a common interest with me in this field. I had 3 main parameters for editing this text:

1. Accuracy – The data and information provided in this book should be up-to-date and valuable to the readers.

2. Structure – The data must be presented in a structured format for easy understanding and better grasping of the readers.

3. Universal Approach – This book not only targets students but also experts and innovators in the field, thus my aim was to present topics which are of use to all.

Thus, it took me a couple of months to finish the editing of this book.

I would like to make a special mention of my publisher who considered me worthy of this opportunity and also supported me throughout the editing process. I would also like to thank the editing team at the back-end who extended their help whenever required.

Editor

Why Monorail Systems Provide a Great Solution for Metropolitan Areas

Peter E. Timan

Abstract Faced with the escalating demand for public transportation in metropolitan areas, transportation authorities are challenged to select a technology that will satisfy the often conflicting demands of high capacity and reliable service, urban fit, minimized environmental impact and budget restrictions. There are many technologies available today that can provide medium to high capacity mass transit service, however, in many cases these technologies are costly or are not suited to today's urban environment. High capacity has typically implied costly underground tunneling or obtrusive elevated metro systems that required extensive infrastructure disruption. Although monorail systems have been around for some time, only recent developments such as Bombardier's *INNOVIA* Monorail 300 System have permitted transit authorities to now consider monorail as a mainstream contender to meet their mass transit requirements.

Keywords Elevated · Automated · Driverless · Transit · Monorail · System

1 Introduction

Monorail systems today must enhance city development by exploiting monorail's iconic aesthetics and easy urban integration allowing cost-effective and fast turnkey construction in both greenfield and brownfield mass transit applications.

P. E. Timan (✉)
Bombardier Transportation, Kingston, ON, Canada
e-mail: peter.timan@ca.transport.bombardier.com

Editor: Xihe He

Fully automated operation is an essential ingredient allowing mass transit operators to provide the required frequent, reliable, safe and fast service demanded. Today's automatic train operation technologies must provide reliable short headways that maximize system capacity and permit optimized operations and maintenance regimes.

Today's monorail systems must respect the environment by providing low visual impact of both the guideway and vehicle while offering zero emissions. Notably, Bombardier's *INNOVIA* Monorail 300 solution uses advanced technologies including permanent magnet motors and lightweight construction to provide substantial energy savings resulting in operating cost savings as well as reduced CO_2 emissions.

Monorail systems today must employ international mass transit standards in all aspects of vehicle and system design to provide durable and safe operations. As with any elevated transit system, monorail systems are no exception, requiring emergency walkway to provide assured means of evacuation.

Bombardier's new *INNOVIA* Monorail 300 System enables these key attributes of today's urban mass transit applications.

2 Brief History of Monorails

Early concepts for monorail systems as an alternative to conventional rail began surfacing in the 19th century. However, attempts to develop this technology into a commercial stage did not find success until well into the next century.

Historically, two types of monorail systems emerged from the early stages of development:

- The suspension railway systems on which the vehicle hangs under the fixed track—originally designed as freight transportation. The earliest urban application

Fig. 1 Schwebebahn Wuppertal, Germany in operation since 1901

Fig. 3 INNOVIA Monorail 100—Jacksonville (1998)

Fig. 4 INNOVIA Monorail 200—Las Vegas (2004)

was the Wuppertal Monorail (Fig. 1) that was installed in 1901 and is still in use today.

- The straddle-beam monorail system uses a vehicle that straddles a reinforced beam. In the 1950s, a German company by the name of ALWEG pioneered this technology, and installed its first system at Walt Disneyland in California (Fig. 2). Most successful wheeled monorails today can trace their roots to the straddle beam ALWEG type monorail technology including the extensive Chongqing Monorail as well as Hitachi, Scomi and Bombardier Monorail.

3 Bombardier's History in Monorail Systems

Embarking in a new chapter of turnkey solutions, Bombardier installed its first complete automated *INNOVIA* Monorail 100 System in 1991 in Tampa, Florida; followed by the JTA Skyway Monorail system in 1998 in Jacksonville, Florida (Fig. 3). This was the beginning of extensive development of the technology, further supported by the acquisition of Adtranz, (owners of Von Roll Monorail technology), and a license agreement with Disney.

In 2004, Bombardier launched its second-generation monorail technology in Las Vegas, Nevada (Fig. 4). This new *INNOVIA* Monorail 200 System technology featured many aesthetic and operational improvements including design and testing to mass transit industry standards.

Leveraging the experience gained from earlier monorail projects and from other Bombardier installed automated mass transit systems around the world; Bombardier has now developed a new generation of its monorail, the *INNOVIA* Monorail 300 System (Fig. 5). The *INNOVIA* Monorail 300 System provides a truly robust mass transit solution for today's urban mobility challenges.

Fig. 2 First ALWEG monorail in commercial operation—Disneyland 1959

Fig. 5 INNOVIA Monorail 300—Sao Paulo and KAFD

4 Today's Urban Mobility Challenges

Public planning authorities often face conflicting demands to provide reliable and accessible high capacity public transportation. In the past, city planners have provided this kind of service by specifying metro systems. However, installing metro systems are typically very costly and often involved extensive relocation and destruction of valuable existing infrastructure. In some cases, metro systems are not an option due to existing infrastructure, which results in no high capacity transit system at all.

Congested city streets result in inefficient flow of goods, services and people with the result that commerce is restricted and productivity declines (Fig. 6). Expanding existing streets is often not an option due to already intense development and the severe limitation of available land. In any case, adding traffic lanes to already congested streets only results in added pollution and noise with costly health and social implications (Fig. 7).

Many cities around the world today have well-established infrastructure such as water, hydro and electric services that have been developed and expanded over time making tunneling or surface transportation solutions very costly. In addition, many older cities have valuable or culturally sensitive historical buildings that are often in the path of preferred public transportation routes. Transit systems must accommodate steep grades and sharp curves to avoid destruction of valuable infrastructure and to minimize land acquisition and expropriation.

Elevated transit systems avoid the need for costly tunneling and minimize the need to relocate existing utilities; however, elevated metro systems typically require wide visually obtrusive deck construction and often still require extensive destruction of existing infrastructure.

Mass transit systems must be able to attract people away from existing inefficient modes of transport, providing not only reliable high capacity transportation, but also visually

Fig. 6 Congestion limits commerce

Fig. 7 Pollution results in costly health and social implications

enhancing the community. The key to attracting ridership is the need to provide a transit system with efficient and reliable service through fully automated driverless operation, which not only provides enhanced reliability but also operational flexibility, reduced energy consumption and increased passenger safety. As with any elevated mass transit system, passenger safety must be a key consideration through specified mass transit industry safety standards including specification of emergency evacuation walkway along the entire elevated guideway.

Although monorail systems have been around for many years, only recent developments such as Bombardier's *INNOVIA* Monorail 300 System have permitted transit authorities to now consider monorail as a mainstream contender to meet their mass transit requirements.

5 New Generation Monorail for Metropolitan Areas— Market Requirements

A need for new mass transit system technologies that meets market requirements and cost less within a reduced timeframe prompted the development of Bombardier's new *INNOVIA* Monorail 300. Market studies clearly indicated a need in the medium to high capacity range from 2000 to 20,000 pphpd (passengers per hour per direction) and up to 40,000 pphpd (Figs. 8, 9).

The market studies also concluded that key to market acceptance of any new development was the need to ensure that mass transit industry standards must form the basis for any new technology as well as stringent safety standards and features.

Key requirements for market penetration include the need for the transit system to provide what could be termed "delighter" features, which would attract increased

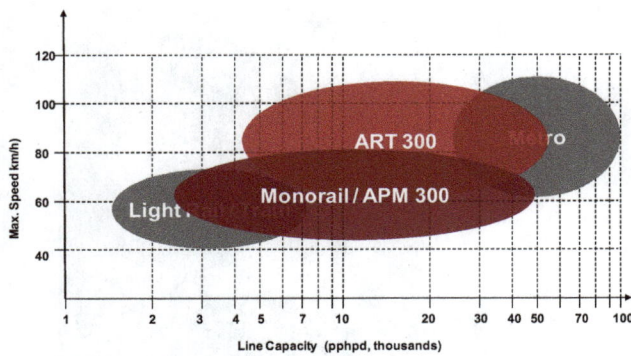

Fig. 8 Typical market segments

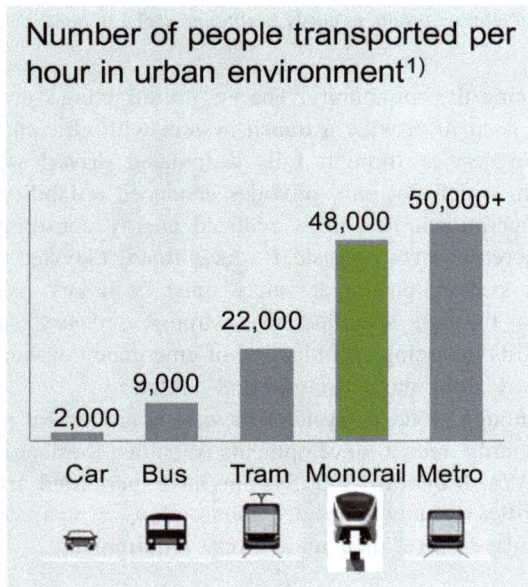

Fig. 9 Transport capacity. Number of people crossing a 3–5 m-wide space in an hour in an urban environment. *Source* International Association of Public Transport (UITP)—Monorail data added by Bombardier

ridership away from existing congested, foot print intensive, polluting automotive modes of transport. The kinds of delighter features that are proven to be effective in other mass transit applications are frequent, efficient and reliable service implemented through fully automated driverless operation. Other delighter features required in modern transportation systems today include attractive aesthetics, energy efficient, non-polluting, quiet and comfortable operation that blend well into existing communities.

Optimization of a monorail system to best fulfill the market requirements can only be done when considering all elements of the transit system over the complete system life cycle from guideway and power distribution systems to the vehicle and operations & maintenance (Fig. 10).

Bombardier Transportation has taken this whole system approach to optimize the new *INNOVIA* Monorail 300 system design resulting in an optimized medium to high capacity mass transit system for today's congested urban environments.

Historically monorail developments have focused on specific single elements of the mass transit system such as the vehicle resulting in reduced overall competitiveness and failing to meet many market requirements. From a public transport authority perspective, the best way to achieve an optimized monorail transit system is to use a full system Turnkey procurement process ideally including a significant operations and maintenance element of at least 10–15 years. Turnkey system procurement allows the best tradeoff between vehicle and wayside elements such as guideway and station construction [1]. When a significant operations and maintenance component is also included as part of the initial contract, then total life cycle and maintenance costs are automatically included in the design optimization thus resulting in the lowest overall life cycle cost of the system. By placing the full system under one consortium there are clear lines of responsibility since critical system integration tasks are the responsibility of the single consortium entity thus ensuring that the transit authority obtains the contracted system performance (Fig. 11).

The full system turnkey procurement process places responsibility for all system elements under one roof thus minimizing transit authority risk while at the same time providing the transit authority with the lowest fixed price and avoiding scope and budget creep. A pre-qualification process may be beneficial to ensure that only qualified and proven suppliers can prepare submissions.

6 Bombardier's *INNOVIA* Monorail 300 System—Answering The Market

Bombardier considered all aspects of the system in the development of its *INNOVIA* Monorail 300 System. As discussed above, key to low cost and minimized disruption of existing infrastructure was the need to optimize for primarily elevated alignments, which then became the key driving force behind many of the design decisions. The *BOMBARDIER INNOVIA* Monorail 300 System is particularly suited for connector distributor and line haul type systems with line capacities in the range of 2000–20,000 pphpd and readily expandable to 48,000 pphpd.

7 Driverless Automated Operation

By taking a full system approach to the design of the *INNOVIA* Monorail 300, it was clear that fully

Fig. 10 Full system integration is critical to success

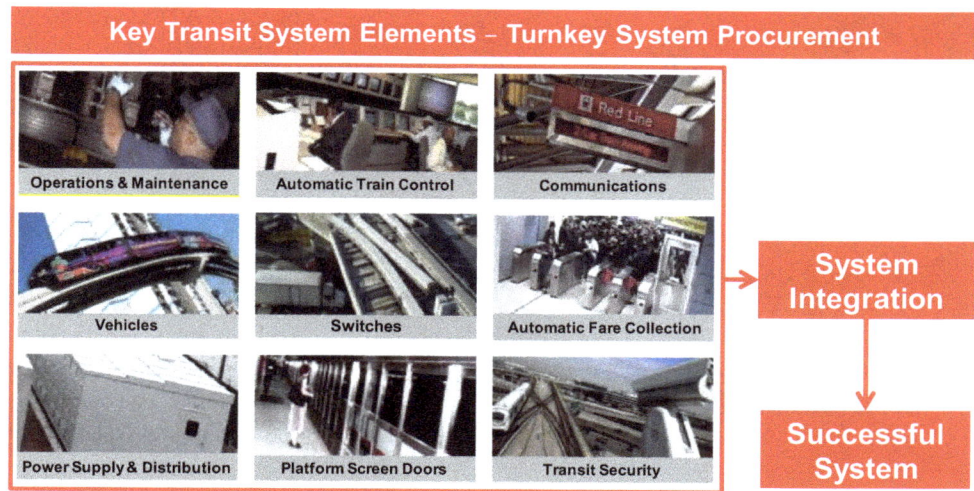

Fig. 11 A procurement structure with full turnkey system responsibility under a single entity provides best opportunity for project success

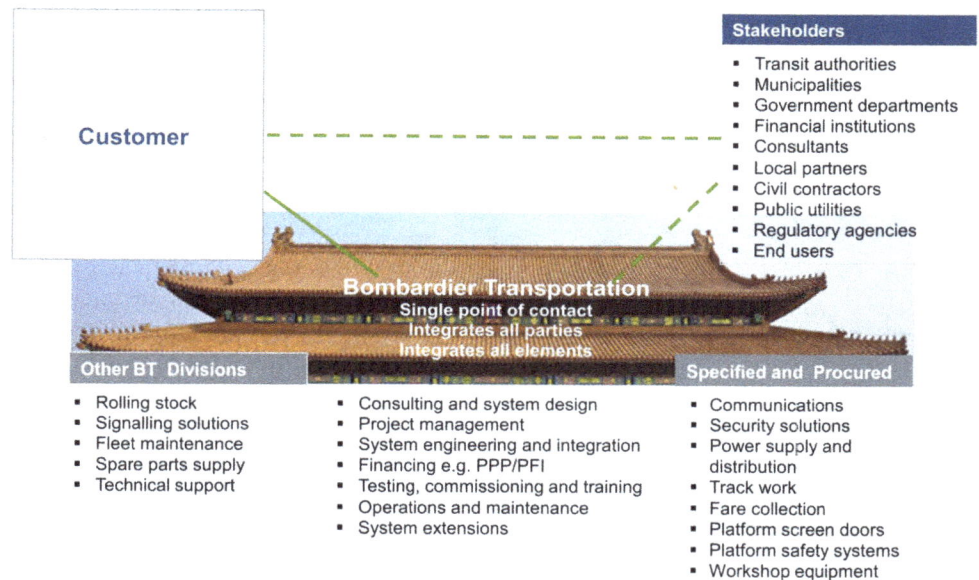

automated driverless operation was required to provide the market requirement for frequent, efficient and reliable service.

Bombardier's *CITYFLO* 650 solution, the latest generation automatic train control (ATC), is a communications-based train control (CBTC) moving block technology which enables full driverless operation of the system. This state-of-the-art technology increases system performance and passenger safety, comfort, and convenience. It is fully compatible for overlay applications at existing sites, and it is inexpensive to install, maintain and expand.

Automated driverless systems are experiencing rapid growth in the mass transit industry as reported by Malla [2] in 2011 (Fig. 12).

Bombardier is the leader in driverless transit systems around the world with over 40 driverless systems installed since 1971 (Fig. 13).

Ridership studies have shown that increasing train frequency has very significant impact on attracting new riders on a transit system. In retrospect, this would seem obvious since the key element for a passenger is to reach his final destination reliably and in the shortest possible time. A rider would typically start his journey on a bus of feeder system, transfer to a line haul system and then perhaps transfer to a second feeder system at the end of his route. As train intervals are increased, the rider must account for this by leaving earlier for his journey to the point that riders will choose other less desirable forms of transportation as the total trip time increases. Studies have shown that train

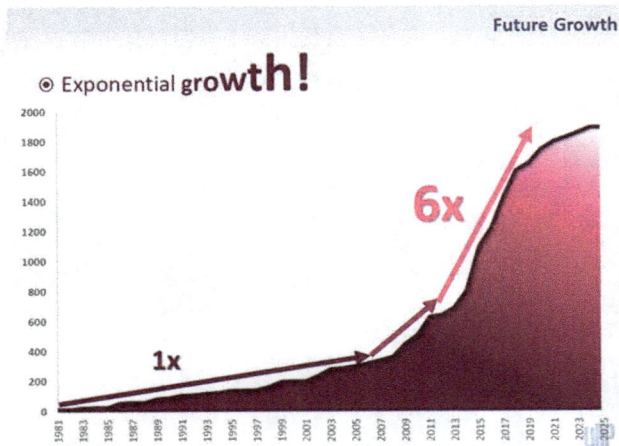

Fig. 12 Rapid growth projected for automated driverless systems [2]

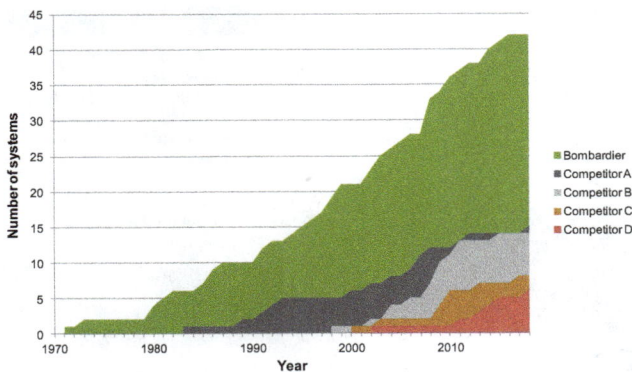

Fig. 13 Bombardier a leader in driverless systems

frequency with headways less 2 min between trains can significantly increase ridership.

The benefits of fully automated driverless operation go beyond increased ridership having significant cost advantages to the overall system. The ATC system is capable of very short headways, which permits the shortest possible trains (Fig. 14) to provide the required system service capacity. Shorter, more frequent trains permit much shorter station platforms, which in turn result in reduced need for land expropriation, reduced land acquisition cost, reduced station cost and reduced quantity of other system elements such as station communications equipment, platform screen doors, maintenance workshop, etc. Manually

operated ATP protected systems today are typically limited to headways greater than 120 s or more and are often not able to reliably sustain even these headways over extended periods. Automated driverless systems on the other hand are proven to sustain reliable headway performance as low as 60 s depending on train length and system configuration although most common minimum headways range from 75 to 90 s for most automated systems.

Fully automated driverless operation results in reliable increased average train speed since acceleration rate, braking rate and dwell times can be consistently maintained independent of any particular driver. Increased average train speed can significantly reduce required fleet size thus reducing system capital cost and maintenance costs (Fig. 15).

Automated operation provides the added benefit of reduced energy consumption by employing consistent acceleration, cruising and braking parameters optimized for the particular alignment and train technology.

8 Elevated Guideway

Bombardier Monorail guideways comprise slender beams that provide flexibility in alignment and ease of construction with minimum visual impact.

Designed to integrate seamlessly into different environments, including through buildings and structures, *INNOVIA* Monorail 300 System infrastructure requires minimal land expropriation and meets the most stringent urban transit, environmental and safety standards. The system technology permits slender contemporary guidebeams, which both guide the vehicle and provide its structural support.

Fig. 14 Frequent short trains result in increased ridership

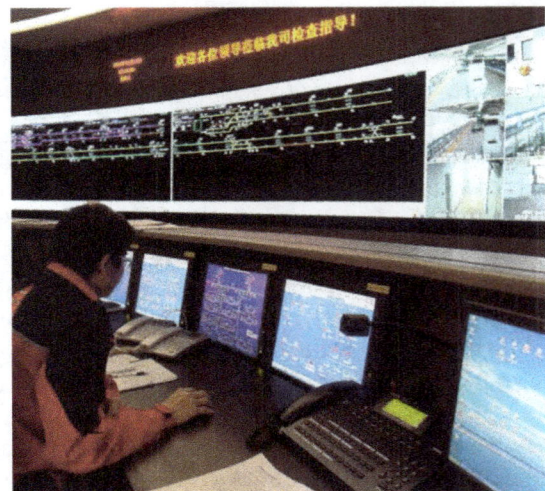

Fig. 15 Automated operation key to optimized performance

Today's fast-paced urban environment creates a high demand for quick implementation of efficient transportation systems. Disruption to everyday city life is simply not an option. Time is short to transform a city's transportation network, and the *INNOVIA* Monorail 300 System permits cost-effective, rapid installation in comparison to other mass transit technologies.

Bombardier developed its *INNOVIA* Monorail 300 System to minimize the costs and disruption of civil construction. The pre-cast, post-tensioned elevated guideway structure is constructed off-site to allow for exceptionally rapid assembly on site (Figs. 16, 17, 18, 19). In addition, the elevated guideways avoid the need for potentially expensive and time-consuming tunneling works, a major advantage when introducing a new transit system in existing dense urban areas.

Bombardier's focus for designing minimized guidebeam size was to optimize civil construction cost. This was primarily accomplished by minimizing the train loading applied to the beam. This led to an important requirement for the vehicle design to minimize weight and center of gravity as well as to minimize the overall vehicle silhouette thus significantly reducing static and dynamic loading as well as wind loading on the guidebeam structure.

The lowest vehicle weight in its class and narrowest guidebeam width requirement of 690 mm, contribute to these cost savings by maximizing pier spacing and reducing substructure requirements.

9 Urban Fit

Transit systems must be able to accommodate steep grades and sharp curves to fit the existing conditions of congested urban environments. The *BOMBARDIER INNOVIA* Monorail 300 System can accommodate grades of up to 6 % and curve radii of 46-m radius permitting the guideway alignment to optimally follow existing rights of way (Fig. 20).

Fig. 17 Beam casting

Fig. 18 Spanning existing infrastructure

Fig. 16 Construction yard

Fig. 19 Beam placement

Fig. 20 Minimize costs by following existing rights of way

The ability to follow existing rights of way such as existing roadways results in minimized land intake which reduces the need for divisive property expropriation as well as reducing the need to destroy existing valuable or culturally sensitive buildings. The long span capability of the multi-span posted tensioned beam design minimizes the need to relocate existing utilities such as water, sewer, communications and hydro.

A key requirement for good urban fit includes unobtrusive stations, which is accomplished by implementing short headways between trains thus permitting shorter and less obtrusive stations.

Good urban fit requires improved conditions for people and city living (Fig. 21). The social aspects of mass transit cannot be overlooked, as they are critical to market acceptance and successful transit system implementation. Bombardier paid special attention to the social implications of mass transit by designing the INNOVIA Monorail 300 System to be an attractive and efficient public transit system. It can be installed close to existing homes and businesses with low visual impact. The Bombardier INNOVIA Monorail 300 uses low noise metro style rubber tires as well as direct drive permanent magnet hub motors

minimizing noise and vibration. The INNOVIA Monorail 300 has zero emissions helping to improve air quality in congested city landscapes.

10 Safety

In addition, using driverless operation and design to mass transit industry standards, other key safety requirements for a mass transit system include the recommended use of platform screen doors. Platform screen doors not only prevent inadvertent falls from the platform edge, but also avoid typical rider induced service interruptions as well as providing the capability to have climate controlled station platforms (Fig. 22).

Platform screen doors fully enclose the platform edge, and are synchronized to open and close with the vehicle doors to allow access and egress to and from the vehicle. It is in fact the consistent stopping accuracy provided by the automatic driverless train control, which allows for the safety benefits of platform doors. Emergency access doors, included in the panels between the automatic doors, allow additional access in case of emergency or failure.

As with any elevated transit system, a safe means of evacuation is required to be available at any point along the alignment. Existing standards such as NFPA 130 recommend the use of an emergency walkway or equally safe means of evacuation; it must also demonstrate a safe evacuate the entire fully loaded train within 15 min. Many independent studies show that the only reliable fail-safe means of evacuation that is able to meet this requirement is an emergency walkway along the entire alignment. All other means of evacuation fail to achieve the required evacuation time in all locations or in all anticipated conditions of failure or hazard.

Emergency walkways should be part of initial contract to ensure aesthetic integration into the guideway design (Fig. 23). Emergency walkways can be implemented in a visually aesthetic way and are cost-effective since they

Fig. 21 Quite operation and iconic aesthetics enhance city living conditions

Fig. 22 Platform doors enhance passenger safety

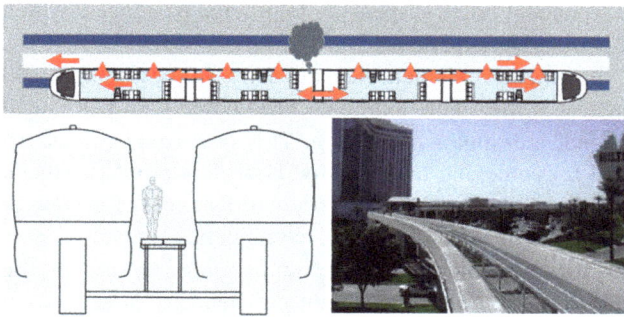

Fig. 23 Emergency walkway permits safe and fast evacuation

provide multiple additional functions of improved maintenance access to guideway elements as well as providing a natural support for the transit system power and communication cables (Fig. 24).

Including requirements for an emergency walkway in the initial contract helps avoid contractual disputes with costly budget overruns and compromise solutions that either are marginally effective (walkway not easily accessible from the vehicle) or visually intrusive.

11 Switching

INNOVIA Monorail 300 switches are either beam replacement or multi-position pivot switches. The beam replacement switches are used on the mainline and multi-position pivot switches are used in storage yard areas. In both cases, they are structural steel assemblies incorporating steel guideway beams that pivot to provide guidance and support, routing the train from one lane to another. Power rails are continuous throughout the switches to maintain vehicle propulsion.

11.1 Beam Replacement Turnouts

The beam replacement switch consists of beams with tangent and spiral sections pinned at one end and supported on carriages at the other to permit rotation of the beams (Figs. 25, 26). These switches permit operation at line speed in tangent direction while spiral geometry in the turnout direction allows maximum curve speed while enhancing ride comfort. Beam replacement switches are used in various configurations to provide single or double crossovers.

11.2 Pivot Switches

The pivot switch is a low speed switch with a tangent steel beam that pivots at one end and capable of rotating into up to four positions is used to provide access into storage and maintenance lanes (Fig. 27).

12 Bombardier's *INNOVIA* Monorail 300—Vehicle Development and Key Features

Bombardier's *INNOVIA* Monorail 300 is designed to comply with stringent international mass transit standards and architecture typically associated with light metro rail technologies. The use of these standards and architecture provides a robust vehicle built on proven metro subsystem technologies providing the operator with reliable performance and 30-year design life (Fig. 28).

The *INNOVIA* Monorail 300 System vehicle allows total flexibility of car configuration in up to 8-car trains and with a newly designed gangway, riders can freely move between cars and enjoy improved visual aesthetics.

INNOVIA Monorail 300 System vehicles operate from transit standard 750 Vdc power supply distribution system.

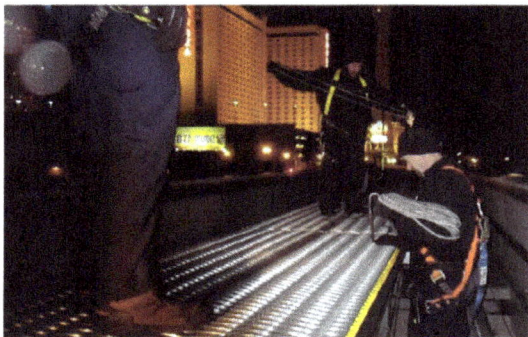

Fig. 24 Emergency walkway provides guideway maintenance access

Fig. 25 Turnout position

Fig. 26 Tangent position

Fig. 27 Multi-position pivot switch

Fig. 28 Bombardier's INNOVIA Monorail 300 System

The *INNOVIA* Monorail 300 System features a sleek and modern vehicle equipped with energy efficient technologies. The combination of lightweight carshell design and other advanced system technologies results in substantial energy savings compared to other monorail and metro technologies. Bombardier designed its monorail vehicles to combine futuristic aesthetics in conjunction with open visual vehicle interior, large side and end windows and

independent bogie suspension to provide passengers with a superior ride experience.

A key development for the *INNOVIA* Monorail 300 vehicle is the single axle dual tired bogie with permanent magnet hub motor propulsion. This development was critical for achieving both the low floor, low profile aesthetic vehicle while simultaneously providing access for intercar walkthrough capability. The combination of single axle bogie and low floor configuration permitted the low profile vehicle with the lowest weight per standard passenger capacity in the industry.

The single axle bogie configuration provides significant performance advantages due to perfect radial alignment of each axle at all times regardless of travel direction or curve condition (Fig. 29). This configuration results in further reduced vehicle weight, reduced rolling resistance as well as reduced tire wear in curves compared to other dual axle bogie configurations.

Creating a comfortable and impressive interior is an important part of drawing passengers to any transportation system. Elegantly designed, the passenger compartments allow flexible seating arrangements to meet a variety of system capacities.

When designing the *INNOVIA* Monorail 300 System vehicles, Bombardier made incorporating an inter-car gangway a top priority. These vehicle gangways greatly increase capacity, but also increase the passenger's sense of safety and security (Fig. 30). The freedom to walk between cars is an appealing feature and contributes to the spacious and open feel of the vehicle.

Large side and front windows add to the feeling of spaciousness while allowing for natural light and a scenic view (Figs. 31, 32). The passenger-friendly interiors are

Fig. 29 Single axle bogie radially aligned in curves

Fig. 30 Gangway permits free passage between cars

Fig. 31 Spacious interior ease passenger flow

Fig. 32 Large windows contribute to passenger-friendly experience

accessible for passengers in wheelchairs, with strollers/prams or even bicycles. These customizable interiors can be equipped with bike or luggage racks.

Large side doors permit fast egress and ingress of passengers at stations thus minimizing required station dwell time and reducing the trip time for the passenger as well as minimizing fleet size for the operator.

Vehicles are outfitted with an advanced information system for station details or emergency evacuation instructions, thus passengers will receive clear and timely instructions. Modern LCD display screens can be included in each vehicle as an additional option for general information or advertising.

The *INNOVIA* Monorail 300 System vehicles have low noise through the effective carshell insulation design, low aggression metro rubber tires, low vibration direct drive hub motor drive and micro-plug side doors, effectively creating a quiet and enjoyable trip.

In accordance with ASCE 21 Automated People Mover Standard [3], the maximum operational load of the *INNOVIA* Monorail 300 System end car is 16-seated passengers and 105 standees at 6 passengers/m^2. The operational load of the *INNOVIA* Monorail 300 System vehicle middle car is 16–18 seated plus 115–117 standees at 6 passengers/m^2.

Engineered for increased efficiency and capacity, these attractive and low profile vehicles allow for flexible arrangement. The standard vehicle configuration permits two, four, six and eight car trains; however, three, five and seven car trains are also an option, if best suited to project requirements.

The maximum operating speed of the *INNOVIA* Monorail 300 is 80 km/h which when combined with reliable high acceleration and braking capability result in high average train speed.

13 Design for the Environment

Design for the environment is a major focus of the *INNOVIA* Monorail 300 vehicle. Bombardier's focus on the environment is seen in all aspects of the system design including minimized consumption of material, minimized consumption of energy, minimized rolling resistance and wear, minimized noise, low visual intrusion system elements and maximized use of recyclable materials.

Material consumption to construct the system guideway and stations is much smaller than similar capacity elevated metro systems and has been designed to be the best in the industry for mass transit monorail applications.

Energy consumption is minimized through best in the industry lightweight vehicle construction, as well as use of energy efficient technologies; such as permanent magnet hub motors, LED lighting, sealed microplug doors, vehicle thermal insulation, low drag coefficient vehicle design, automated operation to optimize train performance. Short frequent trains increase power distribution system receptivity during electric braking regeneration, and permit short stations to minimize lighting and environmental control energy consumption. Intelligent power management shedding of non-essential loads when the vehicle is placed into sleep mode reduces energy consumption.

The environmental benefits of the single axle bogie include reduced energy consumption due to reduced rolling resistance as well as reduced tire dust pollution due to reduced tire wear.

Other aspects of environmental consideration include Bombardier's industry leading position to ensure stringent elimination of prohibited materials.

14 Operations and Maintenance

INNOVIA Monorail 300 Systems includes the capability to provide full life cycle operations and maintenance services, which is a core part of Bombardier's Systems division product offering. Bombardier is a leading operator and maintainer of many automated transit systems around the world.

Direct operation and maintenance involvement, and vast transit system experience enables Bombardier to provide system availability guarantees and to provide optimized system and vehicle maintenance regimes (Fig. 33).

Passenger and employee safety, world-class customer service, high system and equipment reliability and overall cost effectiveness are central to Bombardier's approach to operations and maintenance.

Fig. 33 Fleet maintenance services in Las Vegas

Key to success has been Bombardier's move toward predictive maintenance to optimize operation and maintenance regimes (Fig. 34). This approach is based on collection of key performance data, which is analyzed for data trending and visualization in order to provide advanced predictive maintenance and to permit rapid fault identification and resolution.

The benefits of this focus on predictive maintenance include increased system availability by minimizing the occurrence of service affecting failures through tracking of trends and early mitigation.

Other benefits include improved customer service by extending the system's operating life and extending equipment life through optimal performance of maintenance.

The net benefit to the transit authority is reduced total cost of ownership through extended maintenance intervals, potential elimination of historically identified frequent tasks, automated vehicle inspections, reduced planned maintenance activities and reduced spares holdings.

15 Conclusion

Monorail technology is a very attractive transit technology. Its futuristic look and appeal is a big attraction for riders.

Recent developments such as Bombardier's *INNOVIA* Monorail 300 System have opened opportunities for new applications of medium to high capacity mass transit monorail systems. The benefits of urban fit, light and unobtrusive civil structure, high grade and small curve capabilities, fast project implementation, etc., can be fully leveraged by transit authorities to address transit challenges. To ensure reliable, efficient high capacity service, transit authorities must specify mass transit industry standards for both system and vehicle elements.

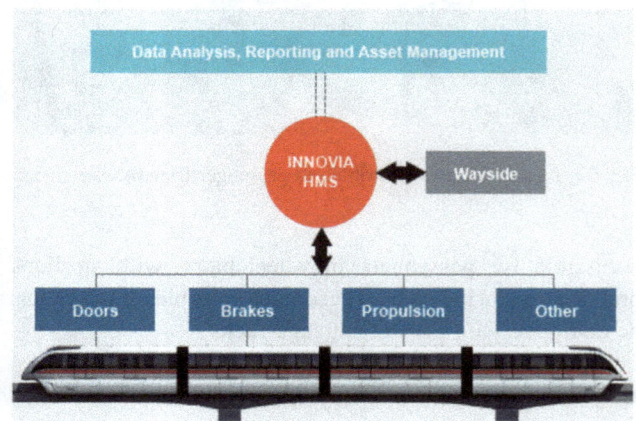

Fig. 34 Predictive maintenance optimizes system performance

Transit authorities now have the opportunity to pursue monorail systems for mass transit applications, which were not previously considered viable alternatives to heavy metro technology when existing infrastructure and high tunnel cost were an issue.

To obtain best overall monorail system cost and performance, transit authorities should use a full turnkey procurement approach or similar based on clearly defined performance parameters and preferably including a significant period of operations and maintenance to capture full life cycle costs. In this way, the transit authority will be able to evaluate submissions with the true full picture in terms of overall system life cycle cost. A pre-qualification process may be beneficial to ensure that only qualified and proven suppliers are permitted to prepare submissions.

Peter E. Timan was born in 1956 in Kingston, Canada. He received a B. Sc in Applied Science, Mechanical Engineering from Queen's University, Canada in 1979. He has been employed by Bombardier Transportation since 1980 and is presently a Senior Expert in Mass Transit and Monorail Systems. He held the position of Monorail System Product Manager for eight years culminating in the development of the Bombardier *INNOVIA* Monorail 300 Mass Transit Product. His current interests are focused in the area of research and development related to Monorail and Metro LIM systems. He is an executive member of the International Monorail Association (IMA) and he has presented at many IMA conferences worldwide since 2010. He has also presented a paper at the Chongquin International Monorail Conference in 2013 where he was presented with an "Outstanding Expert Award".

References

1. Didrikson PV (2008) Specification and procurement of fully automated transit systems. Presented at rail solutions asia conference, Kuala Lumpur, 12 June 2008
2. Malla R (2013) UITP observatory of automated metros: the trend to automation. Presented at 4th automated metro seminar, London, 6 Sept 2013
3. ANSI/ASCE/T&DI 21-13 (2013) Automated people mover standards. American Society of Civil Engineers

2

Application and Prospect of Straddle Monorail Transit System in China

Xihe He[1]

Abstract Straddle monorail, using rubber wheels and precast concrete track beams, is a kind of distinctive urban rail transit system, featured with strong climbing capability, small turning radius, less land occupation, low noise, moderate volume, and low cost. Those unique technical characteristics have played an important role in Chongqing urban rail transit lines. Chongqing provides a typical demonstration project, and straddle monorail transit system will be a favorite urban rail transit system for our other mountain cities, landscape cities, coastal cities, historical and cultural cities, etc. And it has laid a good foundation and favorable conditions for further popularization and application of straddle monorail transit system.

Keywords Urban rail transit · Straddle monorail · Project application

1 Summary

Among the urban rail transit systems, monorail transit system is a typical and popular system with nearly 50-year safe operation in foreign countries especially in Japan, Southeast Asian countries such as Malaysia, Europe, the USA, Canada, and so on.

Monorail is a moderate volume rail transit system with electric vehicles running on a track beam. According to the positional relationship between the vehicle and the track beam, it can be divided into two types of monorail such as straddle monorail and suspend type monorail.

Straddle monorail is a kind of monorail system, in which the vehicles use rubber wheels traveling across on the track/beam. Except the walking wheels, there are guiding wheels and stabilizing wheels on both sides of the bogie, which clamp on both sides of the track beam, to ensure the vehicle safely and smoothly running along the track [1].

Not only is Straddle monorail very suitable for mountain city, coastal city, urban and complex terrain, and road city but also for the urban areas surrounded by a high concentration of buildings as well as suburban areas; at the same time, it is a good choice for a low economic growth and fiscal revenue region to develop urban rail transit.

At present, except for the sightseeing area or domain-specific (such as airports, schools) small monorail line, monorail has been built or under construction in the world as a moderate volume urban rail transit system, in the following countries: Japan, Canada, the USA, Germany, Russia, India, Malaysia, Brazil, and South Korea. In Chongqing, we have successfully operated for about 75 km lines of straddle monorail. Now straddle monorail transit lines in operation and under construction both at home and abroad have exceeded 600 km.

2 The Technical Characteristics of Straddle Monorail

2.1 Unique Technical Advantages of Straddle Monorail

Track beam system: track beam system of monorail is a kind of precast reinforced concrete beam (PC beam),

✉ Xihe He
cq-hexihe@163.com

[1] Chongqing Rail Transit Design and Research Institute, Chongqing 401122, China

Editor: Gary Barber

integrated with some embedded parts, structural parts, power supply, and signal facilities. As a result, it has prominent characteristics of small volume, compact structure, good light transmittance, and good landscape for passengers.

Because the vehicle can travel in a small curve radius, and the track are mostly elevated above road Barrier, so a monorail line, which requires less land use than traditional rail system, can make better use of limited city space, is suitable for city construction and could effectively reduce unneeded urban demolition.

The track structure is simple, and its cost is less than metros. Compared with the traditional metro mode, straddle monorail transit system has the characteristics of low cost and high transport efficiency, with about 1/3-1/2 cost of metro.

Vehicle can travel in line of large slope and small curve radius, bypassing urban buildings, so that straddle monorail system is suitable for mountain city and complex terrain environment with strong environmental adaptability.

Vehicles are walking with air springs, rubber wheels, and electric drive, resulting in low noise, comfortable ride, and less environmental pollution. Generally, its noise is over 10 db lower than metro, so it is a better choice for urban traffic.

Line is elevated over urban area, which makes the passengers fell comfortable with good vision, so it has the dual function of urban transportation and tourism (Table 1).

2.2 Three Main Technical Characteristics

Monorail, as a new urban rail transit form, has some unique system technologies. In general, the construction of monorail requires three main technical features including monorail vehicles, monorail PC beam, and monorail track switch.

At first, the basic and the core technology of monorail technology system is the monorail vehicle technology. On the one hand, vehicle technology itself is the core technology of the entire rail transit projects; on the other hand,

only when the vehicle technology is determined, can we select match switch, PC beams, and other supporting electromechanical equipment systems. The difference of various monorail vehicle technology is mainly reflected by the structure of bogies which is completely different from the subway [2] (Fig. 1).

The second one is monorail PC beam production process and erection technology. Monorail track beam can be divided into two categories of pre-stressed concrete beams and steel track beam according to its different materials: in general, concrete beams use I-shaped sections, and hollow structural and steel beam uses a box-section structure. In a monorail traffic system, track beam is both bearing beam and guide rail, which has the following characteristics:

(1)   Small volume, compact structure, factory production, good light transmission, and good landscape.
(2)   Precast track beam uses pre-stressed concrete structure, with higher quality and precision when manufacturing and erecting.
(3)   Both sides of the track beam have rigid central catenaries.
(4)   Beam bottom or the access is provided with cable tray for traction power supply, communications, and signal
(5)   Embedded parts of communication, signal and the traction power supply system for installation and interface are embedded into track beams during precast.

Therefore, the most prominent problem is how to solve the balance and stability during the vehicle traveling. In order to ensure requirements of safety operation, smooth, comfort, except for technical problem of the vehicle itself, we shall also resolve the security and linear problem of rail. Our rail is track beam instead of the ordinary steel rail, which is very different at present, we used to adopt pre-stressed concrete track beam (referred to PC track beam). However, it is far more difficult to make PC track beam to meet the linear requirements of train while running faster than on steel rail. As a result, our track beam system integrated with many kinds of important functions becomes

Table 1 Metro and straddle monorail comparison

	Metro			Straddle monorail	
	Minimum radius of curve (m)	Maximum longitudinal slope (‰)		Minimum radius of curve (m)	Maximum longitudinal slope (‰)
Main line	250	35	Main line	100	60
Parking line	110	40	Parking line	50	60

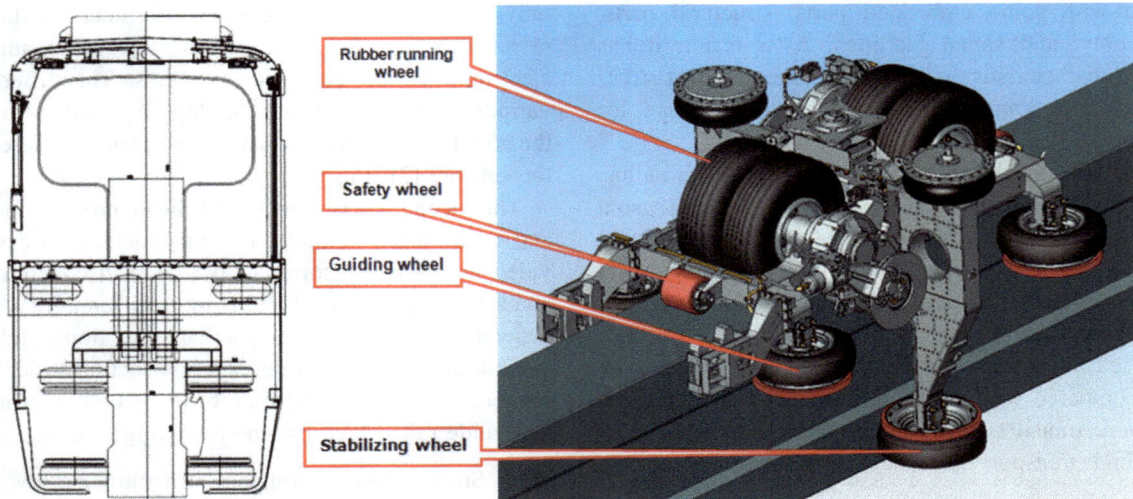

Fig. 1 Typical bogie structure of monorail vehicle

one of the key parts of straddle monorail system. During production and construction, PC track beams are required of not only enough stiffness and strength but also enough surface smoothness. At the same time, the unique erection technology of the track beam is also one of the difficulties (Fig. 2).

75 km straddle monorail line in Chongqing line 2 has been completed, and most of them use PC simply supported beam in standard span of 22–24 m, so that the project was implemented with advantages of easy, reasonable cost, convincible line maintenance, etc. For section of crossing interchange, 40-m-span steel rail beam was adopted, as well as some new simple supported beam techniques such as using "simultaneous construction of pier and beam" instead of "beam after pier," the disk rubber bearing beam, 6 × 30 m continuous beam, etc (Fig. 3).

After 40 min of 5.12 Wenchuan earthquake in 2008, Chongqing monorail was resumed operations, while the whole track beam system is not damaged (Fig. 4).

Thirdly, straddle monorail system adopts beam type switch, its load, linear, switch, locking, and information feedback which need to meet the requirements for train operation, which is one more key technology to study and solve (Fig. 5).

Monorail switch is a special steel-structure switch in the same section shape with PC track beam. Switch beam is switched to another track beam or another switch beam to realize alignment and to form a fork through electric driving, so that we can help vehicle to change its traveling line. According to the shape of switch beam in transition line (it can be broken line or arc line), monorail switch can be divided into joint switch, joint flexible switch, and flexible switch; according to the number of line alignment, the switch can realize; line switch is divided into single switch, single crossover switch, three switch, and five switch (Fig. 6).

Joint switch and sliding switch are controlled by signal system and switched by drive unit. The structure is greatly different from conventional subway system (Fig. 7).

3 The Construction of Straddle Monorail in Chongqing

Chongqing rail transit line no. 2 was constructed as the largest urban infrastructure project after Chongqing became a municipality, which is the first time to introduce monorail transit system in china as an important trial and demonstration project (see Fig. 8).

Fig. 2 Simple track beam structure

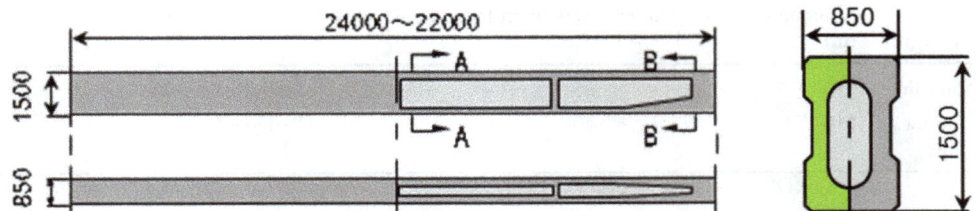

Fig. 3 Continuous steel system of track beam

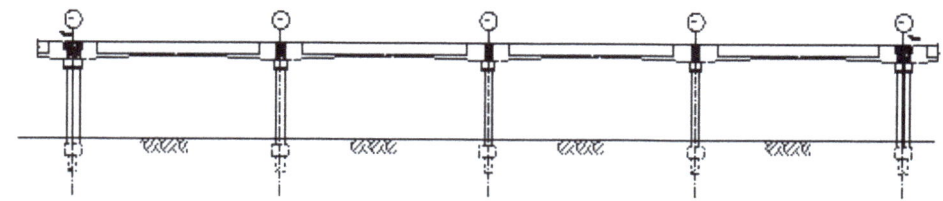

VIEW FROM TYPICAL IPs

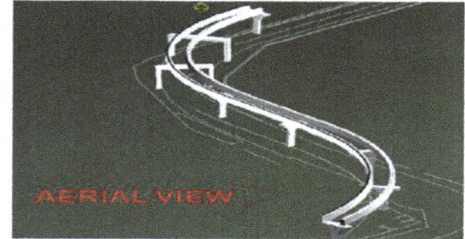

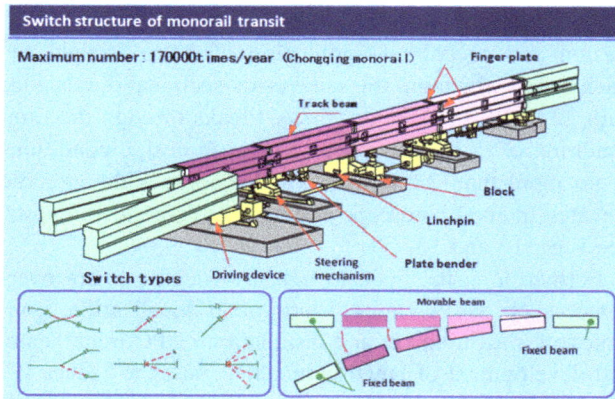

AERIAL VIEW

Fig. 4 Joint switch of monorail

Fig. 6 Switch area of line 3 in Chongqing

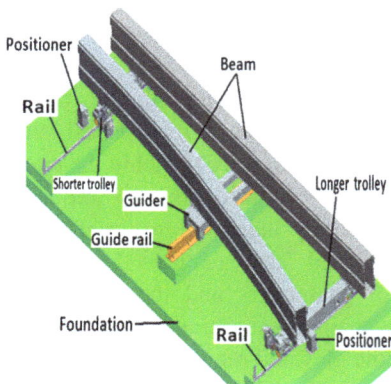

Fig. 5 Monorail translational switch

Fig. 7 Switch area of line 2 depot in Chongqing

In 2005 June, with the contemporary international advanced level, as one of the ten key projects of western development, our firstly introduced straddle monorail line— the first phase of Chongqing rail transit line 2 of 14 km with 14 stations was opened for test operation. After several years of safe and efficient operation, it showed unparalleled features and benefits than other rail transit systems.

According to the features that line runs through a certain area of high mountains and steep, narrow urban roads, and

Fig. 8 Through the downtown area of the city

Fig. 9 Along Jialing River

complex terrain, Chongqing rail transit line 2 adopted straddle monorail system and successfully completed and smoothly opened into operation, so as to obviously improve that the residents traveling rate and the quality of urban environment greatly promote the economic development along the route and show a new scene of Chongqing. See Fig. 9.

Chongqing rail transit line three of 65 km length, among which 56 km has the trial operation of, 39 stations, 6-car marshalling in initial stage, load capacity of 962 staffs: while in recent and future stage of 8 cars and 1292 passengers; 9 km north extension section is under construction now. At present, the largest number of passengers in single day is 751,000 people, 700,000 in daily trip, and 2.9 million in peak hour section.

The characteristics of Chongqing straddle monorail: the advantages of straddle monorail system are vividly showed that Chongqing is a city around the world with the biggest scale monorail, the longest monorail line, the largest monorail day volume, which perfectly shows the characteristics of monorail (see Table 2). Monorail line 3 runs through the Yangtze Caiyuanba dual combination of public

rail bridge (length: 1866 m, maximum span 420 m) and Jialing dedicated monorail bridge (length: 352 m, maximum span 160 m). Monorail line 2 sets plum dam station in residential building. See Figs. 10 and 11.

4 Key Technologies of Straddle Monorail System are Improved and Innovated in Chongqing

In Chongqing, depending on the construction of straddle monorail transit system in line 2 and line 3, we carry out some localization research upon key technologies, including PC track beam system, switch system, and vehicle bogie system.

4.1 The Exploration and Innovation of PC Track Beam Production and Construction Technology

Design and successful application of 24 m straight PC track beam, which has the same cross-sectional dimensions with 22 m PC track beam, firstly breaks through the limit condition of PC simply supported beam design conditions of no more than 22 m span which improves the straddle elevated interval landscape and reduces construction costs. See Figs. 12 and 13.

Furthermore, the main technical difficulties were overcome, such as PC track beam mold localization, construction, measurement and installation of PC track beam, and development of erection machine, so as to reduce the construction investment. See Figs. 14, 15 and 16.

The project won 2008 National Environment friendly Project Award, the Eighth China Zhan Tianyou Civil Engineering Award and science and Technology Innovation Achievements Prize of China, and other national important awards.

As an important measure of the people oriented, Chongqing rail transit line 3, on the basis of the original line 2, added a maintenance channel in the interval between the two PC track beams overhead. Setting maintenance channel substantially increases in the efficiency of maintenance work and also features both the emergency evacuation of passengers and the role of cable laying. Maintenance channel using transparent steel grid form not only retains the elevated range of good permeability effect but also replaces the cable tray under the track beam (see Fig. 17) [3].

4.2 Improve the Standard System to Lay the Foundation for Application

National standard edited mainly by Chongqing Rail Transit (Group) Co, Ltd. "*Code for design of straddle monorail transit*" (GB 50458-2008), "*Code for construction and*

Table 2 Characteristics of Chongqing monorail

Technical parameters	Main content
Topography	City of mountains and rivers, group-style city, road winding down
Route	The length of operation line is 75 Km in Chongqing
	Line under construction: about 12 km south extension section of line 2, and 10 km north extension section of line 3
	The minimum curve radius is 100 m on main line, while 50 m in depot
	The maximum gradient is 50 ‰
Marshaling	Line 2: 4 and/or 6 cars marshaling
	Line 3: 6 and/or 8 cars marshaling
Capacity	The transportation capacity of 6 cars is up to 1300
	Peak hour traffic capacity in one-direction section is over 30,000
	Max. capacity/day of line 3 is nearly 1,000,000
Traffic organization	The minimum operating interval can be 2 min in the future; at present, it is 2 min and 40 s
Operation service	Perfect urban monorail operation services elevated and underground, including air-conditioned waiting rooms on platforms
Security	When vehicles or signal system is in trouble, stereo rescue mode and evacuation maintenance channel can be used
Shortest construction period	Second phase in line 3 is about 19 km; it is planned to be completed in 28 months

Fig. 10 Across Jialing River Bridge

Fig. 12 Fabrication module of track beam

Fig. 11 Through the small curve

acceptance of straddle monorail transit" (GB 50614-2010), and "*General technical specification for straddled monorail vehicle*" (CJ/T 287-2008) were respectively formally promulgated and implemented in 2008 and 2009, which filled the gaps between national and industry standards.

Local standards: as the editor in chief, CRT also completed Chongqing local standard "Standard for construction quality acceptance and evaluation of straddle monorail transit," which covers equipment installation and civil engineering of straddle monorail transit system.

Corporate standard: CRT organized and completed some corporate standards with independent intellectual

Fig. 13 Factory production of track beam

Fig. 14 All terrain, universal PC track beam erecting machine

Fig. 15 Beam transport vehicle and erecting machine

property rights, such as "straddle monorail traffic PC track beam finger board manufacturing and acceptance criteria" and "straddle monorail traffic PC track beam steel tension bearing manufacture and acceptance standard."

Fig. 16 Maintenance channel in section

Regulations and rules for operation: CRT organized to publish hundreds of regulations and rules for operation, check, maintenance, emergency rescue, and operation management of straddle monorail transit, which forms a perfect management standard system of straddle monorail transit, as well as provides strong guarantee for the safety, efficient operation maintenance, and emergency rescue.

The establishment and improvement of national, local, corporate standards and the standardizing operation system laid a foundation for the further promotion and application of straddle monorail transit system at home and abroad.

4.3 Technological Innovation Promote Industrial Upgrading

Based on **the ten key projects of western development—** Chongqing rail transit line 2, combined with the national and provincial science and technology projects, we completed the straddle monorail transportation equipment system integration, straddle monorail vehicle integration and other key technology R&D and industrialization technology research, to realize five major and key technology innovations. The research achievements filled the technical gaps in this field of China, totally reached the international advanced level, in which the vehicle traction, disk brake, a new car ATP, sliding switches and other core technology reached the international leading level.

A thorough research on the core technology of key parts and integrated monorail vehicle, switch, train operation control and monorail transit security, system integration and specification standards, made innovative technology system and a number of independent intellectual property rights results.

After the technical innovation for more than 5 years, we completed the whole vehicle integrated design and development, vehicle bogie development, key technology of

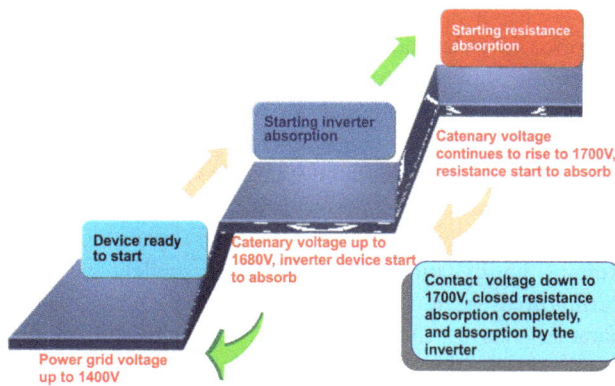

Fig. 17 Inverter absorption device of regeneration energy

vehicle electric traction R&D, vehicle system integration, switch, research, and development of ATP/TD signal system, and development of vehicle mobile real-time video monitoring system, so as to realize the independent development and Industrialization of vehicle and master the core technology of three key components of car body, bogie, and electric traction. Now we have a number of key technologies of straddle monorail switch, automatic overspeed protection and position detection system, mobile real-time video monitoring system, inverter absorption device of regeneration energy (see Fig. 17), security, and emergency rescue, which provides reliable support to technology and equipments, aiming to cultivate and develop straddle monorail equipment emerging industries in china, to popularize straddle monorail technology to the world [4] [5].

Combined with good industrial foundation and strong competitiveness of Chongqing Municipality, relying on the construction of Chongqing Monorail line 2 and line 3 and the R&D of straddle monorail key technologies, Chongqing straddle monorail equipment manufacturing industry base and a industrial chain came into being after a period of planning and cultivation, a lot of R&D institutions, academic exchange, and industrial groups were established, such as Chongqing Urban Monorail Engineering Technology Research Center, International Association of Chongqing Monorail, Chongqing Rail Transit Industry Technology Innovation Strategic Alliance and others. By further polymerization of the talent, technology, and manufacturing equipments from enterprises and scientific research institutes, we formed a model of industrial organization featured with external highly specialized division and internal close collaboration, so as to give full play to their core competencies, therefore, we greatly enhanced the Organizational efficiency of urban monorail equipment R&D and manufacturing resources.

On the other hand, during the promotion and application of straddle monorail system, we need do further research upon safety rescue in elevated section, lightweight of monorail vehicle, and serialization of monorail system, which lead to a more perfect monorail system.

5 Application and Prospect of Straddle Monorail Transit System

At present, with Chinese rapid development of small and medium cities urbanization and rapid growth of urban population, increase of urban volumetric rate cars is surged. Due to limited road resources, road congestion is very serious, which seriously affects economic development. To solve the traffic jam, it is necessary to vigorously develop public transport. In a small or medium city with road congestion, we can use tram, monorail, and light rail in a different marshaling best at a high price, an independent right of way to deal with different levels of one-way traffic along the line in order to solve its traffic pressure, while in a city with one-way traffic over 30,000, we recommend it to choose metro or light rail.

In today's china, both the total operating mileage and the total mileage under construction are more than 2000 km, and the total planning mileage is over 13,000 km. Facing to such scale of Chinese urban rail transit construction, operation, and development, it is an important way to adhere to the principle of "Summary, improvement, innovation," aiming to further improve the city rail transit service quality and promote the city rail traffic health and sustainable development. Therefore, according to urban characteristics (such as mountain city, urban landscapes, coastal cities, historical and cultural city, etc.), how to scientifically select the appropriate urban rail transit system, how to improve its quality and operation service levels through scientific and technological means, how to promote its sustainable and healthy development through technological innovation, will become an important job to play a key role in urban rail transit and to promote its healthy development.

According to preliminary investigation, the global urban rail transit operating mileage of the top ten countries is 19,169 km, of which 12,312 km, or 64 %, is light rail and tram, while MTR systems mileages are 6857 km, or 36 %. Russian is the No.1 with urban rail traffic operation mileage up to 3853 km, of which 3357 km is light rail and tram, accounting for 87 %, MTR 496 km, accounting for 13 %. The second country is German with 3795 total km of which 2,876 km is light rail and tram, accounting for 76 % and MTR 919 km, accounting for 24 %. Thirdly, America is 2607 km in total, of which 1391 km is light rail and tram, accounting for 53 % and MTR 1216 km, accounting for 47 %. Fourth is China inland, with the total mileage of 2077 km, of which 338 km is light rail, tram, and maglev, accounting for 16 % and MTR 1740 km, accounting for 84 %. Here is no longer enumerating. In short, in the ten most populous countries of urban rail transit, China is the only country to adopt metro system primarily; the rest are light rail and tram. Obviously, it is indeed necessary to optimize the constituent ratio of Chinese urban rail transit systems. Generally speaking, large volume metro is suitable for mega city center area; in suburbs, especially outer suburb district, moderate volume light rail including monorail and low volume modern tram is appropriate. To contact the satellite cities, we use city domain fast track, and light rail and tram are more suitable for the center city. Taken Chongqing as an example, monorail can also be adopted in appropriate number in center city. Medium-sized cities are also advised to use lighter rail vehicles and tram. Such multi-system coordinated development is scientific, economic, and sustainable. Therefore, the demand of straddle monorail transit system with medium capacity, low cost, and short construction cycle will gradually increase. In the future urban rail transit construction, straddle monorail transit system has great room for development in China.

6 Conclusion

With the rapid development of small and medium cities urbanization in China and rapid growth of urban population, increases of urban volumetric rate cars are surged at present. Due to limited road resources, road congestion is very serious, which seriously impacts on economic development. To solve the traffic jam, it is necessary to vigorously develop public transport. With the road congestion, we can use tram, or monorail, or light rail in a different marshaling best at a high price, an independent right of way to deal with different levels of one-direction traffic along the line in order to solve its traffic pressure. While in a city with one-way traffic over 30,000, we advise it to choose metro or light rail.

Straddle monorail traffic system has mature technology with superior technical performance of vehicles. what is more it has significant advantages in operating noise, climbing ability, turning radius, circuit model, and engineering investment; thus, it not only meets with the terrain characteristics of Chongqing, a city with mountains and rivers, but also has very broad promotion and application prospects in other cities of China.

References

1. GB50458 (2008) Code for design of straddle monorail transit [S]. China building industry press, Beijing
2. CJ/T287 (2008) General technical specification for straddled monorail [S]. Standards Press of China, Beijing
3. Zhong J (2009) Idea of green light rail reflected and developed in Chongqing rail transit [J]. Urban Rapid Rail Transit 22(1):2–6
4. He X (2014) Equipments' developing and industrialization of straddle monorail transit in Chongqing [J]. Urban Rapid Rail Transit 27(2):6–10
5. Lu M, Yang Q, He X et al (2012) Numerical simulation analysis of the translational turnout in the straddle type monorail. China Railw Sci 33(1):13–18

3

Comparative Assessment of Virtual Track Circuit Based on Image Processing

Florin Codruţ Nemţanu[1] · Dorin Laurenţiu Bureţea[1] · Luigi Gabriel Obreja[1]

Abstract For urban rail track, it is important to detect the presence of the tram or light train in black spots (like urban tunnels, bridges and low visual contact). The classical solution is to use track circuit which is safety oriented designed. The paper proposes a virtual track circuit as an alternative solution. For this proposal a comparative assessment was done to identify the main issues of this solution. For both systems analysed the authors defined and calculated two special functions: one is safety function which is a probability function (together with a distribution function) and the second one is error function which has the same type as previous one.

Keywords Virtual track circuit · Probability function · Comparative assessment · Railway safety

1 Introduction

One of the most important systems in our society is transport system. This system is a very complex one and it is composed by different transport modes. In this paper one important problem is addressed and this problem is related with urban transport and railway transport for urban area. Safety in transport systems is the main issue which is addressed by different systems and the main objective of those systems is to permit the movement of persons and goods without any danger or threat [1]. A system which is able to increase the level of safety in railway transport system (for urban area in this case) is the track circuit which is a sensor installed to collect information about the presence of the rolling stock on a given controlled area [2].

The image processing and communication systems are now very well developed and designed and they are able to support applications for safety and railway safety (the majority of applications are developed for road transport) [3]. Pattern recognition, in terms of image analysis, could be a good solution to detect the presence of the rolling stock on a specific monitored area. The application of pattern recognition could increase the efficiency of the method in terms of providing additional information about an object placed on the rails, the physical obstacle between the camera and the rails. Various methods of pattern recognition were defined and tested and one of them could be selected to be applied in this research (the next stage of the research) [4]. This is the reason to introduce a new type of track circuit which is based on image processing. The main issue of this virtual track circuit is to demonstrate that this solution is able to provide the same level of safety as classical track circuit based on using tracks as wires for an emitter–receiver system. This demonstration is based on comparative assessment of those two different sensing solutions (a model of this assessment came from decision support systems domain) [5].

The authors defined the two main hypotheses of this research as following:

– The virtual track circuit is a safety oriented system and it is able to provide the same level of safety where they installed the system.
– The probability functions and a comparative assessment could provide enough information to support the first hypothesis.

✉ Florin Codruţ Nemţanu
  florin.nemtanu@ieee.org

[1] Transport Faculty, University Politehnica of Bucharest, Bucharest, Romania

Editor: Baoming Han

The method of comparative assessment is based on, firstly, establishing a reference, in the case of this paper the reference is the track circuit and its specific probability functions, and secondly to compare two probability functions associated with those two different types of track circuit.

In Fig. 1 a principle scheme of track circuit is presented with the main components, these components will be defined by their variable. The principle of this track circuit is: the tram which is ocupping the track will have a similar effect as an electrical shunt and the signal emitted by E is no longer received by R, which means Red light is turning on and the next tram is stopped.

An equivalent model (Fig. 2) of this track circuit is proposed by authors in terms of defining probability function. The components of this model are: emitter (E), connectors (Con), tracks as wires (Track), receiver (R) and control subsystem (Ctrl).

The second track circuit, the virtual one, is described in Fig. 3 and the principle is: the tram is detected on specific area by a camera which is able to process the information and to send the information to a control subsystem and the red light is turned on.

An equivalent model is proposed and the structure (presented in Fig. 4) has the following components: Track

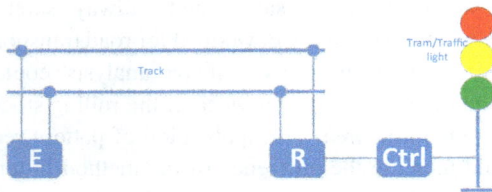

Fig. 1 The principle of track circuit

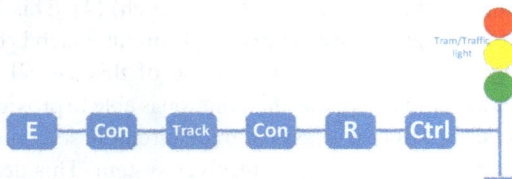

Fig. 2 Equivalent model of track circuit

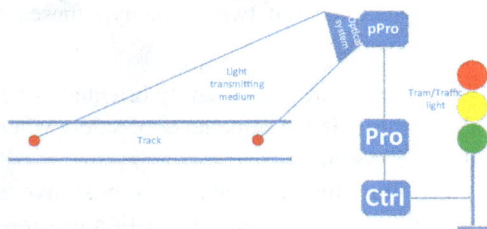

Fig. 3 The principle of virtual track circuit

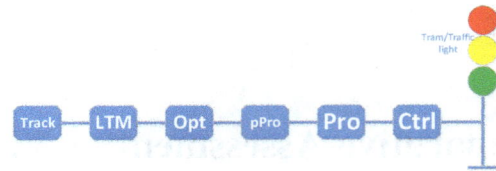

Fig. 4 Equivalent model of virtual track circuit

(Track), light transmission medium (LTM), optical system (Opt), pre-processing (pPro), image processing (Pro) and control subsystem (Ctrl).

2 The Mathematical Instrument for Comparative Assessment

The *probability* is defined in terms of likelihood of a specific event. If X denotes an event, the probability of occurrence of the event X is denoted by $P(X)$ [6].

$$P(X) = \lim_{n \to \infty} \frac{m}{n} \tag{1}$$

where m is the number of the successful occurrences and n is the number of observations.

$$0 \leq P(X) \leq 1 \tag{2}$$

A *probability density function* is a function $f(x)$ defined on interval (a, b) and having the following properties [7, 8]:

$$f(x) \geq 0 \text{ for every value of } x \tag{3}$$

$$\int_a^b f(x) = 1 \tag{4}$$

A continuous random variable X admits a probability function f if for every c and d,

$$P(c \leq X \leq d) = \int_c^d f(x)dx \tag{5}$$

Let X be a $K \times 1$ continuous random vector. The *joint probability density function* of X is a function $f_X : R^K \to [0, \infty)$ such that:

$$P(X \in [a_1, b_1] \times \cdots \times [a_k, b_k])$$
$$= \int_{a_1}^{b_1} \cdots \int_{a_k}^{b_k} f_X(x_1, \ldots, x_k)dx_k \ldots dx_1 \tag{6}$$

A *discrete probability distribution* shall be understood as a probability distribution characterised by a probability mass function [9]. The distribution of a random variable X is discrete, and X is then called a discrete random variable if:

$$\sum_a P(X = a) = 1 \tag{7}$$

where a runs through the range of values of variable X.

The probability that a discrete random variable X takes on a particular value x ($P(X = x)$), is denoted $f(x)$. The function is called probability mass function. The *probability mass function*, $P(X = x) = f(x)$ of a discrete random variable X is a function that satisfies the following properties: [10]

$$P(X = x) = f(x) > 0 \text{ if } x \in \text{ the support } S \quad (8)$$

$$\sum_{x \in S} f(x) = 1 \quad (9)$$

$$P(X \in A) = \sum_{x \in A} f(x) \quad (10)$$

3 Comparative Assessment Based on Probability Functions

The first step in this assessment is to identify the variable which are suitable to be part of this multivariate analysis of probability functions. In the Fig. 2 the authors presented a model with six components and a vector of six variables was defined based on Eq. (6).

$$X = [x_1, x_2, x_3, x_4, x_5, x_6] \quad (11)$$

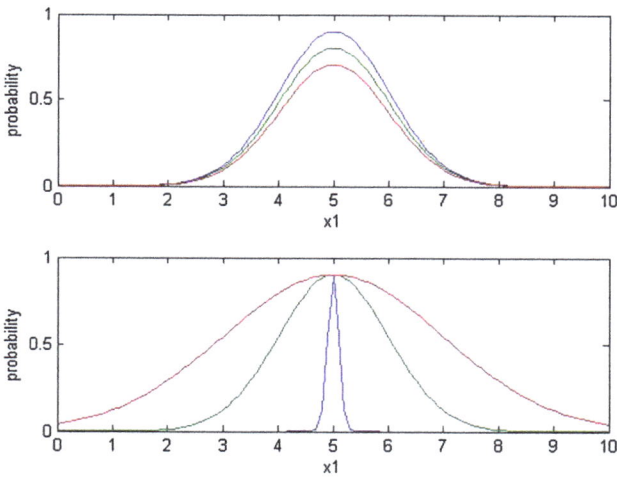

Fig. 5 Gaussian distribution of probability function—variable x_1

The definitions of this variables are available in the table as well as some other characteristics useful for the objectives of the research.

We assumed that the probability functions of all these variables have a Gaussian distribution (initially, after some iteration we can reconsider this distribution) and these functions have the graph represented in Fig. 5 and the graph is directly influenced by σ (if sigma is near zero the probability of that event is near 1) and the probability associated with the variable in discussion.

The *safety function* is defined as the probability function calculated for entire chain of components (defined in equivalent model), that means for all specific variables (defined in Table 1) associated with these components, and this is a joint probability density function and, based on Eq. (6) and Table 1 the expression is:

$$P(X \in [a_1, b_1] \times \cdots \times [a_7, b_7])$$
$$= \int_{a_1}^{b_1} \cdots \int_{a_7}^{b_7} f_X(x_1, \ldots, x_7) dx_1 \ldots dx_7 \quad (12)$$

The authors proposed a reference safety function (which is a probability function of a system to attempt a given level of safety) as a sum of the probability functions of all components of equivalent model. This joint probability density function will be considered as a reference for future comparative assessment of different systems. The safety function (in this model) has 6 components one for every components of the equivalent model and for other application the equivalent model could have more components (in fact, for real systems this number of components is the total number of devices, equipment and any other hardware or software entities). Using Matlab, the authors proposed a graphical representation of this safety function and a procedure to compare this figure with the figure of any other system (this figure is named *foot print* of safety function) [11]. The foot print of existing, accepted track circuit is the reference and the safety function of any other solution will be compared with this and the characteristics, in terms of safety, will be improved based on this comparative assessment. A partial safety function is the probability function for one component which is part of the safety

	Variable	Definition	Component	Range	Probability
Table 1 Definition of variables x_i	x_1	Power of emitted signal	Emitter	0.5–1.5 W	Gaussian distribution
	x_2	Resistance of connector	Connection	0–10 Ω	Gaussian distribution
	x_3	Resistance of the ballast	Track	0.8–2 Ω/Km	Gaussian distribution
	x_4	Resistance of connector	Connection	0–10 Ω	Gaussian distribution
	x_5	Amplification	Receiver	0.1–1.5	Gaussian distribution
	x_6	Command signal (voltage)	Control	10–14 V	Gaussian distribution

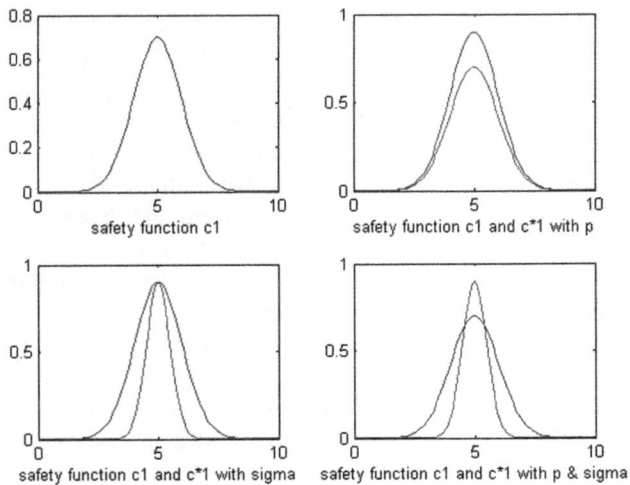

Fig. 6 Partial safety functions for one component (c_1 and c_1^*)

function. In this case c_1 is the component of track circuit used as reference and c_1^* is the safety function for a similar component from the virtual track circuit (Fig. 6).

4 Results

The authors assumed that all variables have a Gaussian distribution of their probability functions or partial safety functions.

In the Fig. 4 the authors presented a model with six components for the virtual track circuit and a vector of six variables was defined based on Eq. (6).

$$Y = [y1, y2, y3, y4, y5, y6] \tag{13}$$

The definitions of this variables are available in the table as well as some other characteristics of the components of the equivalent model for a virtual track circuit (see Fig. 4).

The authors assumed that the weight of every partial safety function in entire safety function is equal. Based on this assumption all six components of the model of virtual track circuit (see Table 2) could replace all six components of the reference model (track circuit).

In Figs. 7 and 8 the partial safety functions of the track circuit defined as reference is revealed to provide an

instrument to determinate the comparative safety function of virtual track circuit. To improve the safety of the track circuit the designer has to propose a component with a better partial safety function that means the shape of partial safety function has to be narrow (with a smaller sigma, in the case of Gaussian distribution of probabilities) and tall (higher probability for the normal state of the component) (Fig. 9).

A partial code in Matlab is presented by authors to show the software tool which is needed to generate safety image of the track circuit:

$x1 = 20{:}0.05{:}100;$—the variable $x1$ is running between 20 and 100
$y1 = 0.8{*}gaussm\,f(x1,[1\ 60]);$—Gaussian distribution of partial safety function
subplot (1,6,1);
plot $(x1/10,\ y1);$—normalised $x1$ (maximum limit was decreased from 100 to 10)
$x$label ('$y1$');
grid on;
axis([$0,10,0,1$])—normalised axis limits

If the partial safety functions for component c_1 are compared, that means x_1 as reference and y_1 which comes from virtual track circuit, the virtual track circuit has a better partial safety function y_1, the shape is narrow and the probability is 0.8 (0.1 more than x_1).

5 Extension of the Solution

In this paper, the authors presented two probability functions generated by a single variable per component of the equivalent model (for each type of track circuit). The authors proposed an extension of this solution through the extension of the number of variables per components of equivalent model. The extended number of variable will generate a modification of Eq. (6) as following:

- A set of vectors for the track circuit on per component (vectors of variables);
- composite variable could be defined to simplify the analysis;

Table 2 Definition of variables y_i

Variable	Definition	Component	Range	Probability
x_1	Physical reference points and areas	Track	20–100 m	Gaussian distribution
x_2	The quality of light transmission medium—the visibility	Light transmission medium	10–100 m	Gaussian distribution
x_3	Opacity of optical system	Optical system	0–30 %	Gaussian distribution
x_4	Time of pre-processing	Pre-processing	0.1–3 s	Gaussian distribution
x_5	Time of image processing	Image processing	0.1–5 s	Gaussian distribution
x_6	Command signal (voltage)	Control	10–14 V	Gaussian distribution

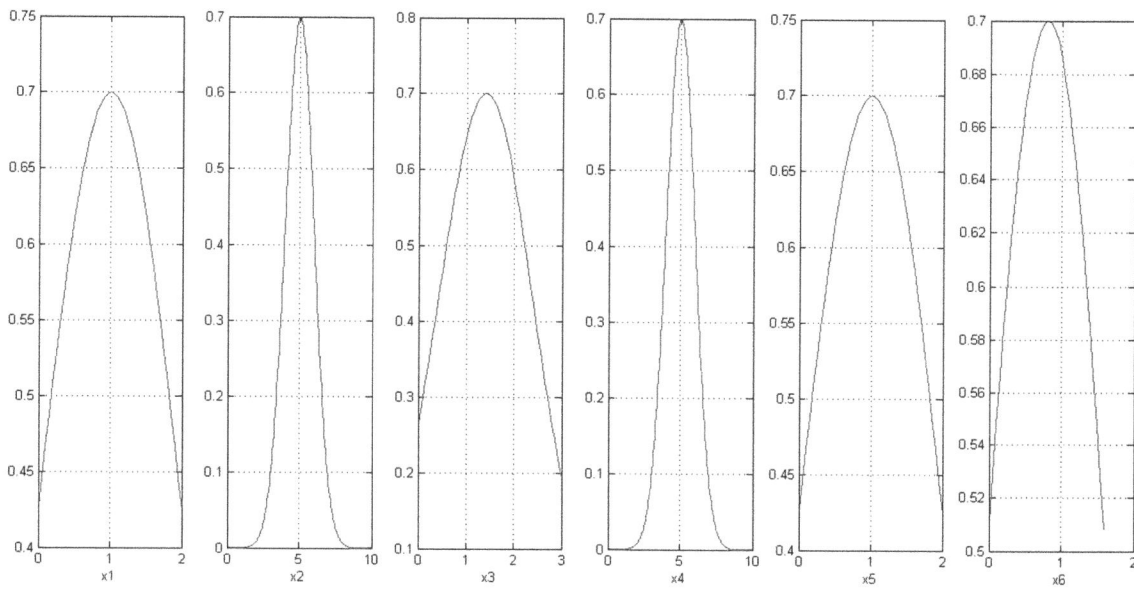

Fig. 7 Partial safety functions as components of safety function

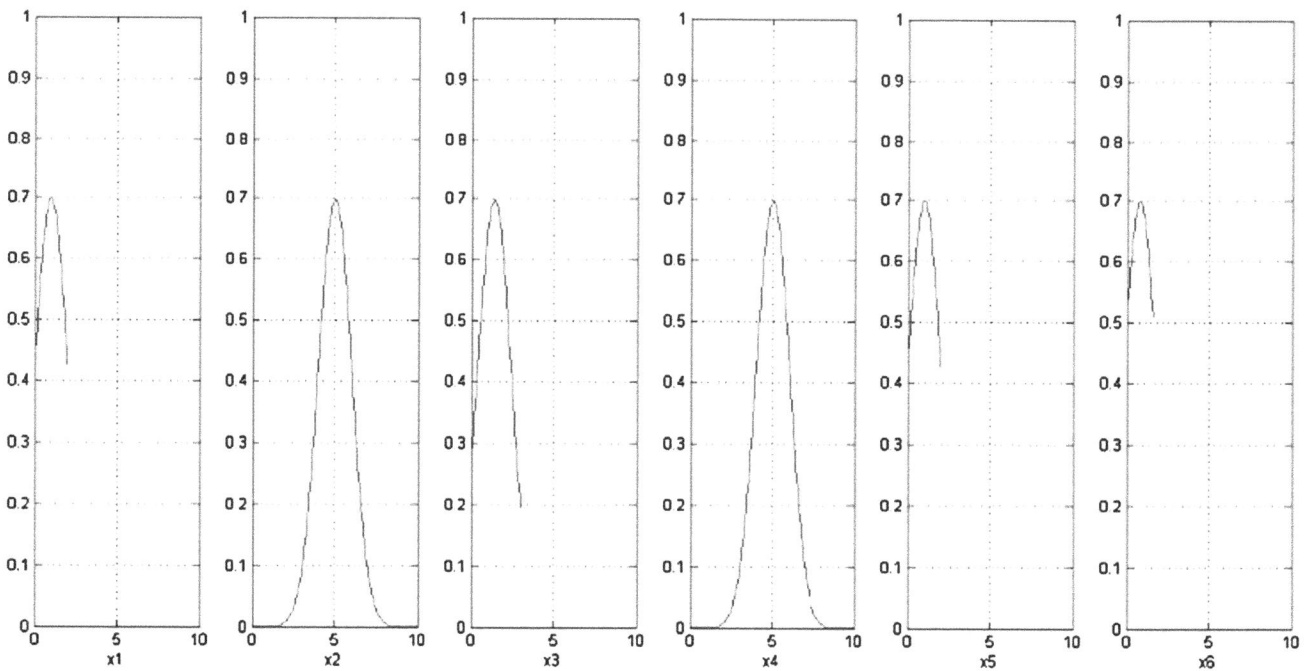

Fig. 8 Partial safety functions—safety image of the reference track circuit

- if the variables are discrete the integral becomes a sum.
- 3D representation (maximum 3 variables per components) is recommended.

In the case of two variable per component of equivalent model the following graph is generated as a representation of probability function as a part of safety function (Fig. 10).

Based on the functional models developed in the laboratory the comparative assessment will be done close to real conditions and the Gaussian distributions assumed in this paper will be replaced with the real distribution of the safety function (probability function). The same problem could be rose for another important analysis which is closed to safety, error function.

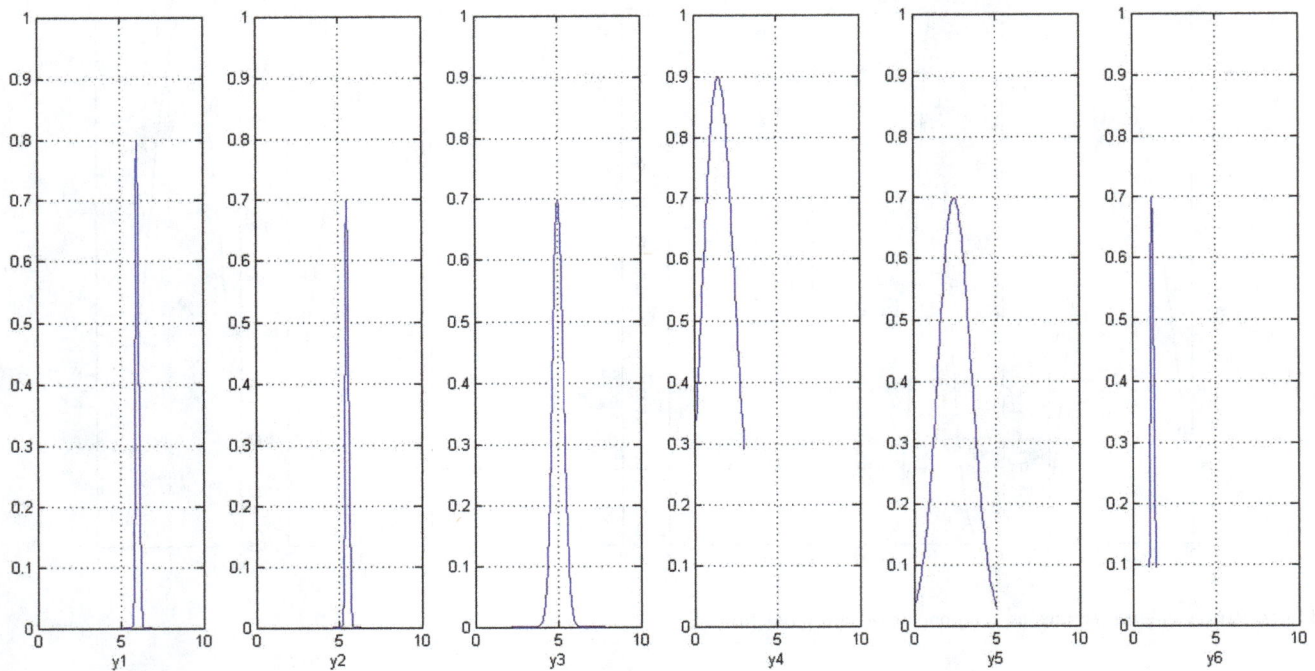

Fig. 9 Partial safety functions—safety image of the virtual track circuit

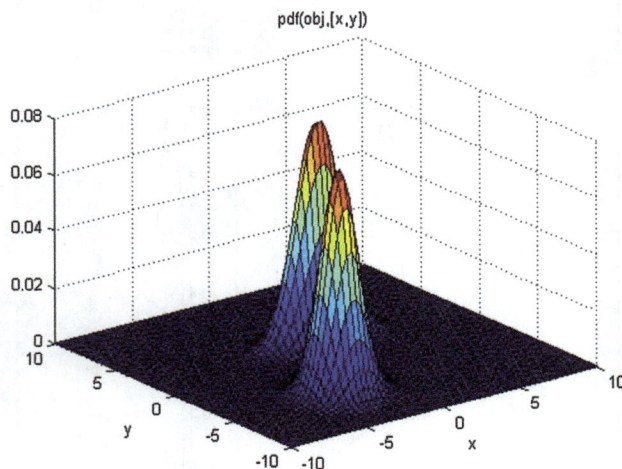

Fig. 10 Partial safety function for one component with two variables

6 Conclusions

The main idea of the paper is to elaborate a mathematical tool for comparative assessment of different technical systems based on probability theory, especially using probability functions.

The main advantage of this comparative assessment is the possibility to identify the weaknesses of the track circuit, in terms of partial safety functions, and to improve these components to obtain a better partial safety function (based on probability and sigma, in case of Gaussian distribution).

The reference will be considered an existing installed track circuit and a new track circuit, in this case a virtual one based on image processing will be compared to find the safety function and to improve partial safety functions. The research will be conducted to demonstrate the equivalence in terms of safety and reliability of virtual track circuit and the first propose is to use this virtual track circuit in urban tram network especially in branches of this network characterised by blind spot (tunnels, bridges etc.). The approach has to steps: the first one is to develop a test platform in laboratory—this platform will be based on an existing track circuit with a known safety function and a virtual lab based circuit with a calculated safety function—and the second one is to move this platform in real condition (the method will be tested in Bucharest, Romania in a tunnel of tram network).

The author presented in this paper the method for comparative assessment and the next step is to develop a functional model in laboratory for both, reference track circuit and the virtual track circuit.

The best safety function is defined also from the context perspective, which means urban railway has another context than interurban railway (in terms of speed and masses).

In these models all partial safety functions have a Gaussian distribution and for real equipment the real distribution has to be considered.

References

1. Flammini F (2012) Railway safety, reliability, and security: technologies and systems engineering. IGI Global
2. Franklin F, Nemtanu F, Teixeira PF (2013) Rail infrastructure, ITS and access charges. Res Transp Econ 41(1):31–42
3. Nagamalai D, Renault E, Dhanuskodi M (2011) Advances in digital image processing and information technology. In: first international conference on digital image processing and pattern recognition, DPPR 2011, Tirunelveli
4. Bezdek J, Keller J, Krisnapuram R, Pal N (2005) Fuzzy models and algorithms for pattern recognition and image processing. Springer, New York
5. Kiang MY (2003) A comparative assessment of classification methods. Decis Support Syst 35(4):441–454
6. Rao S (1996) Engineering optimization: theory and practice. Wiley, New York
7. www.zweigmedia.com (2012) [Online]. Available via http://www.zweigmedia.com/RealWorld/index.html. Accessed 9 Oct 2014
8. Taboga M (2014) Statlect: the digital textbook. [Online]. Available via http://www.statlect.com/. Accessed 9 Oct 2014
9. Silverman B (1986) Density estimation for statistics and data analysis. CRC Press, London
10. STAT 414 (2014) The Pennsylvania State University. [Online]. Available via https://onlinecourses.science.psu.edu/stat414/node/57. Accessed 9 Oct 2014
11. MathWorks Documentation (2014) MathWorks, [Online]. Available via http://www.mathworks.com/help/index.html. Accessed 10 Oct 2014

4

Status Analysis and Development Suggestions on Signaling System of Beijing Rail Transit

Zhili Zhang[1] · ChunQiang Wang[2] · Wenqiang Zhang[2]

Abstract The current statuses on signaling system of Beijing rail transit, including the signaling devices, repair and maintenance, spare parts, and personnel training, were investigated and presented in general. The existing issues were also presented. Considering the development of Beijing rail transit in future, some suggestions are proposed to the local government for Beijing rail transit in future development.

Keywords Beijing subway · Rail transit · Signaling system · Status analysis · Development suggestions · Maintenance · Spare parts · Personnel training

1 The Overview of Beijing Rail Transit

In 2013, the passenger flow of Beijing rail transit network presented a rapid growth. The passenger traffic reached 3.205 billion in the whole year, i.e., 8.78 million per day, with year-on-year growth of 30.54 %. Specifically, over the weekly period, the passenger traffic reached on average 9.489 million per day; the largest number of passenger traffic per day reached 11.055 million. Rail transit plays an important role in easing the pressure of ground traffic jam.

At present, Beijing rail transit systems are operated by Beijing Subway Corporation Limited (Beijing Subway)

✉ Zhili Zhang
zlzhang@bjtu.edu.cn

1 School of Mechanical, Electronic and Control Engineering, Beijing Jiaotong University, Beijing 100044, China

2 Transportation Administration of Beijing Municipal Commission of Transport, Beijing 100053, China

Editor: Tao Tang

and Beijing MTR (Mass Transit Railway) Corporation Limited (BJMTR).

Beijing Subway, with 29,117 employees, is a professional oversized, state-owned operator focusing on managing the operating lines of urban rail transit; its predecessor is Beijing Underground Railway Corporation. At present, lines operated by this company include Line 1, Line 2, Line 5, Line 6, Line 8, Line 9, Line 10, Line 13, Line 15, Ba tong Line, Airport Line, Fang shan Line, Chang ping Line, and Yizhuang Line. The company operates 231 stations, with its total operating distance reaching 395 km.

Mass Transit Railway, established in January 2006, is the first corporative venture which introduced foreign capital in domestic urban rail transit field. The funding contributors of this company are Beijing Infrastructure Investment Corporation Limited, accounting for 2 %; and Beijing Capital Group and MTR Corporation Limited, accounting for 49 % each. The company has 4680 employees, operates Beijing Metro Line 4, Da xing Line, and Line 14; its total operating distance coverage is 62 km. Table 1 illustrates the overall operating statuses of the two companies and the operating situation of each line.

Table 2 illustrates the operating condition of each metro line).

2 The Current Situation of Signaling Systems and Equipments

As a component of technical equipment which ensures safe, punctual, fast, convenient, high-density, ceaseless operation of the train, signaling system plays an important role in rail transit system [1–3]. The signaling system consists of four subsystems: automatic train supervision

Table 1 Overall operation status of Beijing rail transit

Items	Beijing subway	BJMTR	Total
Operating distance/km	403	62	465
Numbers of stations	235	41	276

system (ATS), automatic train protection subsystem (ATP), automatic train operation (ATO), and computer-based interlocking subsystem (CI). The four subsystems form a closed-loop system via message switching network—automatic train control (ATC) that manages to combine ground-control and train-control, local control, and central control together. In order to facilitate safety in operation, an ATC system which integrates the functions of traffic control, operating adjustment, and automatic driving is constructed. The signaling system operates in modes of distributed management and centralized control. Distinguished from the distribution of section, the centralized signaling system is made up of controlling equipments in the center, station equipments, vehicle-based equipments, trackside equipments, and on-board equipments.

ATC has two kinds of classification: (1) According to the classification of block systems, it can be divided into fixed block, quasi-moving block, and moving block; (2) According to the classification of train–ground communication, it can be divided into continuous and point models.

Fixed block ATC system is an automatic block system based on traditional rail circuit. Based on line conditions, the block section is confirmed by traction calculation. Once confirmed, the section would be permanent. The train's tracking intervals are divided into several block sections, which is irrelevant to the train's actual location in the section. The starting point and terminal point are usually considered to be the boundaries of certain section. Train-tracking diagram is shown in Fig. 1.

Quasi-moving block ATC system is also an automatic block system based on traditional rail circuit. The trains' interval is estimated and controlled by the safety margin plus the succeeding train's braking distance at the present speed. The block section occupied by the front train is protected from being advanced rashly. The starting point of braking is dynamic, the terminal point is fixed within the confines of a certain section. Train-tracking diagram is shown in Fig. 2.

Moving block ATC system incorporates medias, such as wireless communication, ground cross-induction loop line, and waveguide, to help transmit information to the mobile unit. The safe train interval is calculated according to the information of maximum permitted speed, the present docking station, and the operating line. The information is updated cyclically to ensure that the train receives instant message ceaselessly. By using the bi-directional data communication equipment between the train and the

Table 2 Operation status of each line

Lines	Passenger capacity (ten thousand person-time)	Design interval	Actual operation interval	Operation all day ranks (number)
Line 1	142	2 min	2 min 05 s	725
Line2	137	2 min	2 min	611
Line 5	98	2 min	2 min 30 s	562
Line 6	61	2 min	3 min	410
Line 8	24	3 min (transition system)	3 min 30 s	308
Line 9	40	2 min	4 min	382
Line 10	190	2 min	2 min 15 s	670
Line 13	86	3 min	2 min 40 s	525
Line 15	14	2 min	6 min 15 s	245
Ba tong line	34	4 min	2 min 50 s	434
Fang shan line	9	3 min	6 min	280
Chang ping line	15	2 min 30 s	5 min 30 s	263
Yizhuang line	18	2 min 30 s	5 min 50 s	246
Airport line	3	4 min	8 min 30 s	222
Line 4	128	2 min	2 min (peak period, one-way 1 min 43 s)	Ordinary days: 622: two-day weekend: 556
Da xing line	31	2 min	4 min	
Line 14	6	4 min (the first-period project, 12 km)	5 min (the first-period project)	Ordinary days: 362 two-day weekend: 372

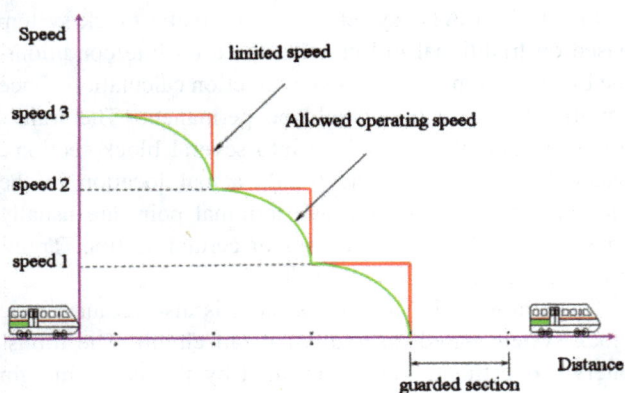

Fig. 1 Train-tracking diagram of fixed block system

ground, the ground signal equipment can get continuously the information of each train's current location. According to the information, the operating permissions of each train can be calculated, and dynamic updated information would be sent to the train. The train calculates the operating velocity curve according to the received information of operating permission and its own operating state, and then it can implement a complete-protection mode of train operation. The braking starting point and terminal point are dynamic. It is more advantageous for providing full play of rail capacity. Train-tracking diagram is shown in Fig. 3, monitoring the differences in guarded distance which is dynamic and shorter than that in Fig. 2.

The biggest difference among fixed block ATC system, quasi-moving block ATC system, and the moving block ATC system is the control over train's safety interval. Each line's signaling system's basic condition is compiled in Table 3.

At present, there are several types of the signaling systems which are operating online in Beijing.

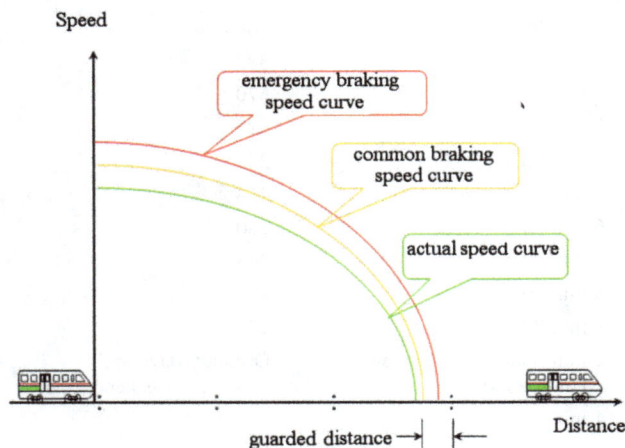

Fig. 2 Train-tracking diagram of quasi-moving block system

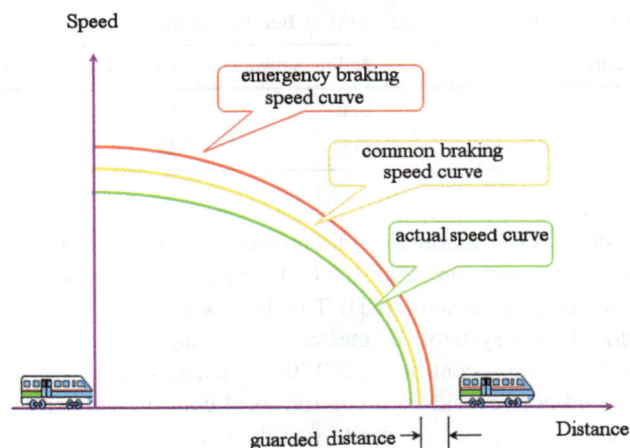

Fig. 3 Train-tracking diagram of moving block system

(1) Multi-systems coexist with each other. In light of the various technical levels of different constructing times and the different technical updating levels after being put into operation, there are multiple systems that coexist in Beijing's subway signaling system: Line 1, Line 13, and Ba tong Line belong to the fixed block ATC system; Line 5 belongs to quasi-moving block ATC system; and Line 2, Line 6, Line 8, Line 9, Line 10, Line 15, Yizhuang Line, Chang ping Line, Fang shan Line, Airport Line, Line 4, Da xing Line, and Line 14 all belong to moving block ATC system. Due to the large passenger traffic capacity and operating demand, the situation of coexistence of multiple systems would last for a relatively long time in Beijing's rail transit signaling system. As for the situation, the crucial problem faced by operating companies is that there is a wide range of trackside equipments and on-board equipments which differ from each other and they cannot be used universally, i.e., uniformly as different patterns of signaling systems coexist. Even if the systems share a same pattern, products from different manufacturers cannot be used universally. Therefore, for Beijing Subway, which operates many rail lines, the manufacturing of spare parts and updating instruments for each line would pose huge challenges. Furthermore, most of the equipments of fixed blocking system and quasi-blocking system are based on single-configuration—once signaling system's operation is influenced by faults, the whole normal operating process would be severely affected. This would bring huge pressure on the company in terms of daily repair and maintenance works of the signaling system.

(2) Coexistence of multiple suppliers. At present, there are more than ten suppliers who participate in the subway signaling system operations via bidding competition. Most of them are overseas-funded

Table 3 Signaling system's basic condition of each line

Lines	System mode	Backup mode	ATP supplier	ATO supplier	ATS supplier	Interlock supplier	Opening time of existing system/year
Line 1	Fixed block (speed step)	No	Westinghouse	No	CASCO	Electric relay interlocking	1993
Line 13	Fixed block(speed step)	No	Westinghouse (Dacheng)	No	CASCO	Academy of railway sciences	2002
Ba tong line	Fixed block (speed step)	No	Westinghouse (Dacheng)	No	CASCO	CRSC	2003
Line 5	Quasi-moving block (target-distance)	No	Westinghouse (Dacheng)	Westinghouse (Dacheng)	CASCO	CASCO	2007
Line 2	Moving block	Point ATP/ATO	Alston (CASCO)	Alston (CASCO)	CASCO	CASCO	2008
Line 6	Moving block	Point ATP/ATO	Alston(CASCO)	Alston(CASCO)	CASCO	CASCO	2012
Line 9	Moving block	Point ATP/ATO	Alston(CASCO)	Alston(CASCO)	CASCO	CASCO	2011 (south of the line opened), 2012 (all opened)
Fang shan line	Moving block	Point ATP/ATO	Alston (CASCO)	Alston(CASCO)	CASCO	CASCO	2010
Airport line	Moving block	No	Alston (CASCO)	Alston(CASCO)	CASCO	CASCO	2008
Line 10	Moving block	Point ATP/ATO	SIEMENS (railway signal communition corp)	SIEMENS SignalCommunitionCorp	SIEMENS (railway SignalCommunitionCorp)	SIEMENS (railway signal communition corp)	2008
Line 8	Point ATP/ATO	No	CRSC	CRSC	CRSC	CRSC	2011
Line 15	Moving block	Point ATP/ATO	Nippon signal (Jiaodamicrounion Tech. Co. Ltd.)	Jiaodamicrounion	Jiaodamicrounion Tech. Co. Ltd.	Jiaodamicrounion Tech. Co. Ltd.	2010
Yi zhuang line	Moving block	Point ATP/ATO	Traffic control technology	Traffic control technology	CASCO	CASCO	2010
Chang ping line	Moving block	Point ATP/ATO	Traffic control technology	Traffic control technology	CASCO	CASCO	2010
Line 4	Moving block	No	Thales Canada	Thales Canada	Thales Canada	Main track: Thales Canada; rail yard: Academy of railway sciences	2009

Table 3 continued

Lines	System mode	Backup mode	ATP supplier	ATO supplier	ATS supplier	Interlock supplier	Opening time of existing system/year
Da xing line	Moving block	No	Thales Canada	Thales Canada	Thales Canada	Main track: Thales Canada; rail yard: Academy of railway sciences	2010
Line 14	Moving block	Point ATP/ATO	Traffic control technology	Traffic control technology	CASCO	CASCO	2013

enterprises and joint-venture enterprises, while a few of them are independent domestic enterprises (less than 10 %). Objectively speaking, bidding competition for multiple suppliers is beneficial for reducing cost during the initial stage of construction; however, it is very disadvantageous for the cost control of entire life-cycle and is harmful for the operating management. According to our survey, overseas manufacturers usually employ strong research teams, have huge production capacity, and have overwhelming advantage in core technology. However, in the aspects of after-sales service, spare parts supply, and rapid emergency response, overseas manufacturers rely on local team's construction. Generally speaking, most overseas manufacturers perform well and render perfect service, but domestic manufacturers have prominent advantages in providing technical support, repair and maintenance, and rapid emergency response [4–6].

(3) The signaling system cannot open up all its functions when most lines start to operate. Being the capital of China, Beijing suffers from huge pressure of ground traffic as the combined urban and floating population is huge. Rail transit with large capacity and high efficiency is an important way of relieving the ground traffic pressure. Due to the short period of new line's construction and trial operation, subway lines' signaling system does not fulfill the condition of opening up full-functions. Here are the results: (1) Due to the shortage of time of debugging system's equipments, the functions are not implementable fully, which leads to unstable operation. (2) The system is updated too frequently after being put into operation, while some suppliers do not prepare well for updating.

(4) Signaling system's fault occurrence is inevitable. Signaling system's operating performance is not only influenced by equipments, but it is also related to running intervals, and the efficiencies of maintaining technician and assistants in dealing with faults. As there are various influencing factors, it is impractical and technically impossible for signaling system to realize "zero defect" and "zero risk".

3 The Condition of Equipment Repair and Maintenance

Despite following different managing mechanisms, both the above companies have strict institutions of equipment repair and maintenance. The signaling system equipments of the present 14 operating lines of Beijing Subway are maintained by its Communication signal branch company. The company follows a combined pattern of plant maintenance, status maintenance, and breakdown repair. The maintenance working schedule for the next year would be laid down by the company in the end of each year in order to complete the works of inspection, touring and maintenance. BJMTR has introduced the managing mechanism of MTR, which includes preventative maintenance and fault repair. Whether the equipment is within the quality guarantee period or not, the company would implement preventative maintenance regularly. At present, the preventative maintenance working plan directs the daily maintenance section which covers all the signal facilities completely. The employees follow those guidance rules to carry out the preventive maintenance work. When carrying out the fault repair work, if the facility is still within its quality guarantee period, the employees of BJMTR would point it out first, and notify the contractor to provide support if necessary. If the facility is beyond its quality guarantee period, the repairing work would be implemented by BJMTR employees.

4 The Condition of Spare Parts

In general, the newly built metro line would receive spare parts by following the stipulated percentage of construction contract, and apply them to use after the expiry of the system's quality guarantee period. According to the actual needs, the subway company would adjust the variety and the quantity of the spare parts, then follow the plan to buy more (exclusive capital would be allotted for those buying plans). The spare parts of key signaling system of BJMTR would be stored in concentrator station, and the related configuration would be completed beforehand, so that the replacing time could be minimized. The situation of the spare parts sites would be monitored daily, and spare part replenishment would be made at any time if necessary. According to various factors such as the number of facilities, the rates of fault incidences, and the supply/delivery periods, the signal department would manage to maintain the level of spare parts inventory reasonably, and review at regular intervals to obtain timely replenishment. Current problems are listed as follows:

(1) The total stock quantity of the spare parts received according to the stipulated percentage of the contract is relatively small, which leads to the severe shortage of spare parts of the operating company and affects the normal operation of the system.

(2) Due to the various electrical parts of the signaling system facilities and the rapid rate of replacement, some lines have encountered the situation of former model's production coming to a halt, such as Line 4.

(3) At the start of operation of certain newly built signaling systems, the suppliers do not comply with providing full range of complete instruments and software that can maintain the system, which leads to severe constraints in running the normal operation.

(4) After some lines of the signaling system come into operation, the rate of update is frequent, but the updating spare parts provided by the supplier fall short of actual requirement.

5 The Training of Employees

The operating companies all have strict training programs for signal operators. Beijing Subway has established a specialized base for Signaling system training courses, and conducts various forms of training, such as enlarged class, training given by the manufacturers, tutorial system, and theme training. Before the induction, professional signal maintainers should receive specialized training and get the certificate of qualification. During the working period, advanced training of operation skill would be provided according to plans, so that employees' skills would be improved. Competitions and examinations on operating procedures would be conducted at regular intervals. The operating level of employees would be judged according to the fixed programs.

The training program for the employees in BJMTR includes the following:

(1) For the new employee, in order to ensure the safety of production, induction training and safety education would be organized.

(2) During the probation period, the new employees would take fixed post's training, and only if they pass the training test they can work formally with certificates.

(3) The employees should receive training provided by suppliers.

(4) Professional maintenance department and company's training department review the professional knowledge levels of technicians, and organize related training.

(5) Generalizing the key experience of dealing with faults every month, and organizing exchange activities.

(6) Organizing exams at regular intervals, finding the problems, and providing specialized training. Besides, the signal repair department would conduct assessment of occupational skills, and analyze the weakness of every employee, and arrange for specialized training, so that the employees would have deeper understanding of the signaling system. The company would illustrate the working mechanisms of different functional patterns, and make its employees concentrate on the ideas of investigating faults in the signaling system. At the same time, the company would allow its employees to take part in the training provided by the suppliers.

The most crucial problem currently is the commitment of the personnel. In order to grasp the operation of each line craftily, the professional maintenance technicians need skill training for a long time as the multistandard signaling systems coexist with each other. However, many technicians would leave their jobs after specialized long-duration training. Strong guarantee of institutional support and payment is very important for forming a committed and loyal team of signaling system technicians.

The impending question is whether the demand of signal technicians can be met in the future or not. There is an analytic report of the present status of Beijing Subway: there are five senior engineers, 21 engineers, 61 associate engineers among the technician team, and with 0.22 person per kilometer, the team presents a pyramid-like structure, but the staff strength of the team is relatively small. There are five senior technicians, 25 technicians in the signal technicians team, and 0.07 person per kilometer, accounting for 2.99 % of the total staff strength. Compared with the actual percentage, 9.3 %, of Beijing's talent team building, the total percentage of signal technicians is relatively small. With the rapid growth of operating mileage, the percentage of new employees' growth would decline further, which cannot meet the demands of the present and future. If we estimate that Beijing's rail transit distance coverage would reach 900 km in 2020, the reasonable percentage of engineering and technical staff in the signal team should be 1 person per kilometer (exchanging posts is permissible), and the total staff number should be 900. The technical staff should make up 10 % of the total employee strength. According to the present estimate, Beijing Subway and BJMTR should employ 2900 and 460 technical staff each.

6 The Current Situation and Performance of ATC

Here is the current list of the main suppliers of Beijing's online signaling system: Alston (CASCO), SIEMENS (Railway Signal Communication Corp), The Nippon Signal Co. Ltd (Jiaodamicrounion Tech. Co. Ltd), Traffic control technology Co. Ltd. Bombardier provide the moving block ATC system; SIEMENS, Westinghouse (Dacheng), CASCO, and Bombardier provide the quasi-moving block ATC system; Westinghouse and China Railway Signal & Communication Corporation research and design institute (CRSC) provide the fixed block ATC system. Domestic companies which provide interlocking facility are the Academy of Railway Sciences, CRSC, and CASCO.

Table 4 shows the comparison and summary of the ATC system equipments, including all the present subway lines. On summarizing and comparing equipment conditions and main performances of the ATC systems of the present metro lines, we found that signaling systems provided by various suppliers do not have a uniform standard, either domestic or overseas. Although the systems share the same principle and function, they differ from each other in controlling, interface mode, and system's structure. Systems of different suppliers and models are not compatible with each other. Different maintenance facilities and spare parts cannot be applied universally.

According to the survey, however, domestic techniques of signaling system and equipment's management level are close to general international level. Specifically, it demonstrates that the *Two as average train kilometer between accidents delayed more than 5 min* ranks among the top of OMET members. The equipment fault rate is 0.088 piece/10,000 km (the international standard of which is 0.8 piece/10,000 km), which is much lower than the international standard.

7 Suggestions for Development

(1)   Implementing uniform formats of signaling system. As long as we follow the principles of safe operation, different signaling systems will all have alternative choices all over the world. [7, 8] In the future, safety and reliability are still the priorities in the choice of signaling systems in signal transit field where safety is highly important. Although the moving block signaling system is relatively advanced, we suggest that moving block signaling system be applied to the modification of the existing lines and the construction of new lines gradually.

(2)   Creating a uniform standard of equipments. Departments concerned with subway, such as investment, construction, design, operations, and management, should coordinate with each other, and take various factors into full account to make integrative plans. In the process of selecting models and updating software and hardware, with due consideration of the ability of managing, operating, repairing, and maintaining, the company can formulate the demand standard of system; set up large database; establish norms of user-requirement-oriented ATC system; construct uniform device interface of modularized management; carry out system's acceptance criteria; consider a uniform standard of maintenance and repair; and improve the collaborative guaranteeing mechanism of constructing, managing. and manufacturing.

(3)   Controlling signaling system suppliers' number, appropriately, strengthens the management and enables restriction in the number of system suppliers. In after-sales service, on the basis of constructing norms, the company should adopt standardized and universal products for the convenience of daily repair and maintenance. Besides, through signaling

Table 4 The ATC system equipments' comparison and summary of each line of Beijing Rail Transit

Lines	Line length/ km	Stations numbers	Numbers of ATC equipment interlocking control centers	Train detection system	The locomotive signaling system	TWC (train-ground communication) mode	On-board ATP redundancy mode	On-board ATC equipment	ATO system speed control mode	ATO system station locates the parking control mode	ATS subsystem
Line 1	31.58	23	23	Track circuit consist of send plate, after closing, weaving/code circuit, code generator, tuning unit, etc.	Speedometer, on-board computer, interface relays, speed motor, antenna, etc.	One-way communication from ground to train of track circuit	No head to tail redundancy	Speedometer, on-board computer, interface relays, speed motor, antenna, etc.	No ATO	No ATO	Consist of central server, local ATS extension, local workstation, OCC workstation, maintenance workstation by DCS network
Line 13	40.5	16	16								Consist of central server, local ATS extension, local workstation, OCC workstation, maintenance workstation by DCS network
Ba tong Line	17.2	13	13								Consist of central server, local ATS extension, local workstation, OCC workstation, maintenance workstation by DCS network
Line 5	27.6	23	10	Track circuit consist of TCOM computer, send plate, after closing, weaving/code circuit, code generator, tuning unit, etc.	Speedometer, on-board computer, interface relays, speed motor, doppler radar, beacon antenna, etc.	One-way communication from ground to train of track circuit, two-way communication of BIDI wireless	Head to tail redundancy	Speedometer, on-board computer, interface relays, speed motor, doppler radar, beacon antenna, etc.	Receive the ground speed code, control the speed according to the operating curve	Positioning according to the electronic map and the ground parking beacon. Accuracy of plus or minus within 300 mm	Consist of central server, local ATS extension, local workstation, OCC workstation, maintenance workstation by DCS network

Table 4 continued

Lines	Line length/ km	Stations numbers	Numbers of ATC equipment interlocking control centers	Train detection system	The locomotive signaling system	TWC (train-ground communication) mode	On-board ATP redundancy mode	On-board ATC equipment	ATO system speed control mode	ATO system station locates the parking control mode	ATS subsystem
Line 2	23	18	7	Axle counter consist of head, outdoor control box, axle host, etc.	\	Two-way communication from ground to train through waveguide wireless	Head to tail redundancy	On-board computer, human–machine interface, interface relays, code odometer, beacon antenna, etc.	Accurate positioning through the electronic map, speed motor and ground beacon, and control train SAcceleration, cruise, braking and parking under ATP protection curve	Positioning according to the electronic map and the ground parking beacon, and correct parking control curve timely. Accuracy of plus or minus within 300 mm	Consist of central server, local ATS extension, local workstation, OCC workstation, maintenance workstation by DCS network
Line 6	30.69	20	7			Two-way communication from ground to train through waveguide wireless	Head to tail redundancy	On-board computer, human–machine interface, interface relays, code odometer, beacon antenna, etc.			consist of central server, local ATS extension, local workstation, OCC workstation, maintenance workstation by DCS network
Line 9	16.5	13	6			Two-way communication from ground to train through waveguide wireless	Head to tail redundancy	On-board computer, human–machine interface, interface relays, code odometer, beacon antenna, etc.			Consist of central server, local ATS extension, local workstation, OCC workstation, maintenance workstation by DCS network
Fang shan Line	24.7	11	5			Two-way communication from ground to train through waveguide wireless	Head to tail redundancy	On-board computer, human–machine interface, interface relays, code odometer, beacon antenna, etc.			Consist of central server, local ATS extension, local workstation, OCC workstation, maintenance workstation by DCS network

Table 4 continued

Lines	Line length/ km	Stations numbers	Numbers of ATC equipment interlocking control centers	Train detection system	The locomotive signaling system	TWC (train-ground communication) mode	On-board ATP redundancy mode	On-board ATC equipment	ATO system speed control mode	ATO system station locates the parking control mode	ATS subsystem
Airport Line	28.1	4	4			Two-way communication from ground to train through waveguide wireless	Head to tail redundancy	On-board computer, human–machine interface, interface relays, code odometer, beacon antenna, etc.			Consist of central server, local ATS extension, local workstation, OCC workstation, maintenance workstation by DCS network
Line 10	57.1	45	14			Two-way communication of wireless free-field waves	Head to tail redundancy	On-board computer, human–machine interface, interface relays, speed motor, doppler radar, beacon antenna, etc.			Consist of central server, local ATS extension, local workstation, OCC workstation, maintenance workstation by DCS network
Line 8	20	13	5			Two-way communication of wireless free-field waves	Head to tail redundancy	On-board computer, human–machine interface, interface relays, speed motor, doppler radar, beacon antenna, etc.			Consist of central server, local ATS extension, local workstation, OCC workstation, maintenance workstation by DCS network
Line 15	30.5	13	4			Two-way communication of wireless free-field waves	Head to tail redundancy	On-board computer, human–machine interface, interface relays, speed motor, beacon antenna, GPS antenna, etc.			Consist of central server, local ATS extension, local workstation, OCC workstation, maintenance workstation by DCS network
Yizhuang Line	23.3	14	6			Two-way communication of wireless free-field waves	Head to tail redundancy	on-board computer, human–machine interface, interface relays, speed motor, doppler radar, beacon antenna, etc.			Consist of central server, local ATS extension, local workstation, OCC workstation, maintenance workstation by DCS network
Changping line	23	7	3			Two-way communication of wireless free-field waves	Head to tail redundancy	On-board computer, human–machine interface, interface relays, speed motor, doppler radar, beacon antenna, etc.			Consist of central server, local ATS extension, local workstation, OCC workstation, maintenance workstation by DCS network

Table 4 continued

Lines	Line length/ km	Stations numbers	Numbers of ATC equipment interlocking control centers	Train detection system	The locomotive signaling system	TWC (train-ground communication) mode	On-board ATP redundancy mode	On-board ATC equipment	ATO system speed control mode	ATO system station locates the parking control mode	ATS subsystem
Line 4	28	24	8	Axle counter equipment, beacon	VOBC	Train-ground communication through wireless access point	Three choose two in key parts	VOBC cabinet, TOD, speed sensors, proximity sensors, TI antenna, wireless antenna, the accelerometer			Consist of central server, local ATS extension, local workstation, OCC workstation, maintenance workstation by DCS network
Da xing line	22	11	4	Axle counter equipment, beacon	VOBC	Train-ground communication through wireless access point	Three choose two in key parts	VOBC cabinet, TOD, speed sensors, proximity sensors, TI antenna, wireless antenna, the accelerometer			Consist of central server, local ATS extension, local workstation, OCC workstation, maintenance workstation by DCS network
Line 14	12/47.3	6/37	3月14日	Axle counter (Keanda)		Wireless LAN(802.11B)	2 Sets of ATP at head and tail,three choose two mode	On-board ATP/ATO cabinets, the BTM BTM receiving unit, the BTM antenna, speed sensors, radar speed sensor, automotive wireless receiver, human–computer interface, etc.			Consist of central server, local ATS extension, local workstation, OCC workstation, maintenance workstation by DCS network

system's updating and reforming staff, the company should integrate the modes, control the supplier's number at the minimum, up to three if possible, so that the competition mechanism and the management of operating company would be applicable.

(4) Solving the updating problem of signaling system's spare parts and electrical parts. There are many questions raised on the qualities of after-sales-service, while spare parts' consumption reaches a large amount. Here are some suggestions:

(1) Increase the percentage to 5–8 % in the contract of new lines. Purchase contractors' exclusively offered spare parts which are important for the normal operation, reduce the percentage of universal market available spare parts.

(2) If more than 5 % of spare parts stock out of the total amount of equipments cannot meet the operational needs during the guarantee period of quality, it indicates that the manufacturing facility has existing flaw in the design of the equipment which leads to the high fault rate and therefore needs to implement thorough change in design. The contract of the new line should stipulate it explicitly.

(3) There are many electrical parts in signaling system equipments, and these parts have a high rate of replacement. In the new line contract, it should be stipulated that substituting products fitted for the system's demand should be provided by manufacturers during product's life-cycle period, and the substitution should be compatible with the former product.

(4) Operating company should make regular contact with the manufacturer, and ascertain the availability of spare parts for the next five years. As for the equipments which may encounter production halt, the company should initiate researching, testing, and purchasing work of the substituting products in advance.

(5) Enhance the research facilities of signaling system's fault warning, maintenance, and faultand-failure analysis. Add online real-time monitoring aids, especially the condition-monitoring method related to crucial signal equipments to strengthen system's fault prevention and warning function. To ensure working quality, the company should promote thematic study of repair and maintenance after the expiry of the manufacturer's quality-guarantee period. Initiate analytic research of signal fault's influence on transportation to improve the system's ability of identifying faults.

(6) Establishing a professional signaling system's maintenance team which is status wise stable, strength wise adequate, and technique wise experienced. Set up specialized personnel-training base. Enhance the team by institutional support and payment guarantees, such as institute setup, personnel ability, age structure, working age limit, technical grade to ensure the safe operation of subway.

(7) Promoting the engineering application of the signaling system which is researched and developed independently. At present, many CBTC (communications based train control) signaling systems produced by many domestic manufacturers are applied to many engineering projects in a mature manner. Domestication and engineering application should be propelled sustainably. In the future, according to related industrial policies, a certain domestication rate of signaling system should be required. In the bid-evaluation criteria, suppliers with multiple lines of CBTC operating performance should be given certain preference.

References

1. http://www.ccmetro.com (All Rights Reserved by Institute of Comprehensive Transportation of National Development and Reform Commission). Accessed Sept 2013
2. Tao T, Chunhai G, Kaicheng L, Fei Y (2005) Based on the communication of train operation control technology development strategy[J]. Urban Rapid Rail Transit 18(6):25–29
3. Chunhai G (2011) Study on core technologies for independent innovation in CBTC[J]. Urban Rapid Rail Transit 24(4):1–4
4. Yingming N, Youneng H, Guosheng Z, Yang Z (2011) Implementation of the CBTC demonstration project-Yizhuang line of Beijing Metro [J]. Urban Rapid Rail Transit 24(4):9–11
5. Mingbao F, Jun Y, Qinger L (2005) Strategic thinking of urban rail transit signaling system localization [J]. Urban Rapid Rail Transit 18(5):68–71
6. Fei Y, Chunhai G, Tao T (2011) Rail transportation signal project safety management and authentication mode[J]. Urban Rapid Rail Transit 24(4):12–16
7. Shijie J, Chunmei G (2007) Interconnection and resource sharing of the signaling system of urban rail transit[J]. Urban Rapid Rail Transit 20(2):92–95
8. Sinan S, Jinya L (2007) The reliability, availability, maintainability, and safety of rail transit signaling system[J]. Urban Rail Transit 11:66–69

5

Direct Observations of Pedestrian Unsafe Crossing at Urban Australian Level Crossings

Teodora Stefanova[1] · Jean-Marie Burkhardt[2] · Christian Wullems[1] ·
James Freeman[1] · Andry Rakotonirainy[1] · Patricia Delhomme[2]

Abstract The number of pedestrian victims at Australian and foreign level crossings has remained stable over the past decade and it continues to be a significant problem. To examine the factors contributing to pedestrians' unsafe crossing behaviours, direct observations were conducted at three black spot urban level crossings in Brisbane for a total of 45 h during morning and afternoon peak. In total, 129 pedestrians transgressed the active controls. More transgressions were observed at the crossings located in more populated suburbs in close proximity to large shopping centres and school zones, whereas the smallest number of transgressions were observed at the least populated locations. In addition to characteristics associated with the larger socio-economic area, the patterns of transgression could be associated with the properties of the existing safety equipment and the design of each level crossing (i.e. location of the platforms, number of rail tracks). Indeed, the largest number of crossed unoccupied but "at risk" rail tracks (where a train could have passed), was observed at the crossing with the least transgressions. Contrary to previous findings, younger adults were the most frequent transgressors. School children and elderly were most likely to transgress in groups. Potential directions for future research and more effective measures are discussed.

Keywords Direct observations · Level crossings · Pedestrian behaviour · Transgressions

1 Introduction

Level crossings (LCs) are generally classified according to the protection systems with which they are equipped. Active LCs are equipped with automatic controls (e.g. red flashing lights, boom gates), whereas passive LCs are signalled with passive signs (e.g. "STOP"). At passive LCs road users cross when there is no visible approaching train, whereas active LCs assist or enforce users' movement (i.e. crossing is prohibited in the presence of activated controls). In Australia and Queensland in particular, LCs in urban areas can be equipped with special form of protection for pedestrians. The pedestrian flow is directed through a pedestrian corridor surrounded by mazes. Additional pedestrian lights and gates positioned on each end of the pedestrian maze activate on the approach of a train, regulating pedestrian traffic independently of vehicular road traffic. In the Brisbane area, this measure is particularly important at sites where access to a train station is provided via the LC. In this case, the rail tracks are likely to be separated by a middle island and pedestrian traffic can therefore be regulated separately on each side of the middle island hosting a train station or a platform.

While such additional measures that specifically target the improvement of pedestrian safety at LCs have been taken in Queensland, the number of collisions involving pedestrians compared to those involving motorists has remained stable in the last decade. A similar trend has also been observed in other countries [1, 2]. In addition, more than half of the reported near-misses for 2011 in Queensland (54 %, $N = 253$) were between a train and a pedestrian, noting that such data are likely to be underreported (i.e. these reports are provided by rail staff

✉ Teodora Stefanova
t.stefanova@qut.edu.au

[1] Centre for Accident Research and Road Safety – Queensland, Queensland University of Technology, 130 Victoria Park Road, Kelvin Grove 4059, Australia

[2] IFSTTAR, AME, LPC, 78000 Versailles, France

Editor: Baoming Han

and therefore do not represent systematic counts) [3]. Collisions between rail vehicles and pedestrians are not only more likely to result in severe injuries and fatal consequences for victims (compared to other road crashes), but are also related to serious economic costs in the short and long term [4].

Each LC is unique, defined by the complex environment and surroundings comprising road and rail infrastructures and the actors involved in both systems [5]. Thus, safety constraints in this complex environment are subject to variability and are highly dependent on the dynamics of the larger system and the specificities of the crossing context. Building upon findings from a previous analysis of factors at play specific to LCs in Brisbane, the present paper presents the results from direct observations of pedestrian unsafe crossing behaviour at three actively protected black spot LCs [6]. Three sites with different, but common characteristics of the local Brisbane railway lines were selected to examine trends in pedestrian unsafe behaviour related to three main categories of factors.

After a brief review of the related literature, the study methodology is explained in detail and selected results are presented and discussed in context of previous findings and potential future research opportunities.

2 Related Work

A literature review on 23 papers related to pedestrian behaviour at LCs showed that, to date, a greater emphasis was directed towards studying the risky crossing behaviour of drivers as opposed to pedestrians' [6]. In most of the papers, the focus is on quantifying non-compliant behaviour according to legal norms—referred to as "*transgressions*", instead of looking at empirical evidence on the origins and the multiple factors contributing to unsafe pedestrian crossing behaviours. Seven of the studies included observation methods [1, 7–12]. Six of them were based on the analysis of video recordings of pedestrians crossing, and one (conducted in Australia) adopted a similar approach to ours, with observers coding the variables manually [7]. In the following paragraphs, main findings from observational and other studies are summarised in three large categories of factors that are likely to explain pedestrian unsafe crossing: (1) environmental and temporal characteristics of the crossing context; (2) pedestrian characteristics; and (3) social environment characteristics.

2.1 Factors Related to the Physical Characteristics of the Environment and the Dynamics of the Crossing Context

2.1.1 Presence of Active Controls—Pedestrian Gates

The presence of active pedestrian gates has been suggested as the most efficient type of controls by a number of

authors [7, 13]. Metaxatos and Sriraj [1] observed that the odds of transgression decreased with the larger numbers of pedestrian gates at the LC compared to LCs equipped with only one pair of pedestrian gates (i.e. on one side of the crossing) or without gates. However, automatic gates introduce three separate moments before the final stage of control's activation, which could be associated with a suboptimal safety performance. In some cases, the presence of pedestrian gates was suggested to increase the so called "beating the gate tendencies" or the perception of control over the risk as long as the gate is not fully closed. In line with this assumption, Edquist, Hughes [7] noted that 50 % of the observed transgressions (i.e. at LCs in Western Australia) occurred before the pedestrian gates had closed. Moreover, Metaxatos and Sriraj [1] observed more transgressions after the gates had started lowering and before they were in horizontal position than after. Transgressions in the riskiest moment (i.e. after the gates were fully lowered) were mainly observed after a train had already passed through the LC. Thus, the presence of pedestrian gates could be associated with an increase in risky crossing behaviours in the first moments of closure before the gates are fully closed (i.e. people assuming that "they can still make it safely on time"), but also after a train had passed through—often corresponding to the last moments of closure.

2.1.2 Position and Number of Trains During Crossing

Train position has been identified as a key factor influencing crossing decision [14]. One observational study demonstrated a significant effect of train position, such that the odds of transgression (versus safe crossing) were higher if crossing in front of an approaching train compared to behind an ongoing train [1]. Such behaviour could be explained by the lack of visibility of the approaching train or by a perception bias (i.e. a misjudgement of train speed or perception that the train is "far away"). Indeed, respondents in a survey conducted by Clancy, Dickinson [15] indicated that they had previously transgressed as they believed that they "had sufficient time to get across before the train reached the crossing" (p. 23). In relation to this, Clark, Perrone [16] have demonstrated that the estimation of the speed of large moving objects such as and specifically trains is likely to be erroneous. In their experimental simulation study, the same authors confirmed that consistent with Leibowitz' theory (1985), a visible approaching train is perceived to be moving slower that an approaching car and therefore could be a contributing factor towards pedestrians' low perception of risk.

While the risk of crossing in front of a second train has been largely demonstrated and discussed previously [10–13, 15], it might not be as important in the current crossing context at LCs in Brisbane (2014), given that often a single track is operated by a separate set of active controls (pedestrian gates and lights) which deactivate allowing crossing soon after a train had passed. Nevertheless, the separately operated pedestrian corridors on both sides of a middle island could engender a high risk of crossing in front of a "second train", considering that controls on the opposite side of the middle island could activate anytime. Moreover, at middle islands, the presence of a stopped "at station" train could hinder vigilance and the perception of the activation of the second pair of controls if pedestrians are transgressing in a hurry to catch the stopped train.

2.1.3 Platforms' Location

To our knowledge, only Edquist, Hughes [7] have, to date, correlated unsafe crossing with the platforms' location vis-à-vis the rail tracks. According to the authors, pedestrians are more likely to transgress if the rail tracks are between the station platforms than if they are separated by a middle platform forcing thus pedestrians to cross more than one track at the time, to access either of the platforms.

2.1.4 Temporal Characteristics of the Crossing Situation

Morning and afternoon peak hours are associated with an increased number of pedestrian transgressions [14]. Nevertheless, while Edquist, Hughes [7] observed more transgressions in afternoon peak hours, Metaxatos and Sriraj [1] demonstrated that transgressions in different times of the day correspond to pedestrian traffic volumes particularly high in the morning and more widely distributed in the evening peak hours.

2.2 Factors Related to Pedestrian's Characteristics and Motivations

Two types of unsafe crossing behaviours can be distinguished according to pedestrian's intention. The term "violation" is frequently used to distinguish deliberate crossing in the presence of active controls from unintentional rule breaches that are referred to as *"errors"*. In observational and other studies, young pedestrians are considered a high risk group of users who deliberately violate rules [9, 15]. Their crossing behaviours have been associated with sensation seeking tendencies (thrill-seeking) or perception of control, compared to elderly for example. Furthermore, male pedestrians are associated with higher risk-taking tendencies than females, however such a trend was only confirmed by one observational study in which male transgressors were identified slightly more often than females (59 %) [7, 14]. Finally, according to Clancy, Dickinson [15] as well as Metaxatos and Sriraj

[1], motivations to deliberately transgress are associated with the given journey context (e.g. being in a hurry, avoiding missing the next train, being on time at work/school). In contrast, errors are often associated with elderly pedestrians likely to experience hearing, motor or visual impairments [6, 8, 9, 14] or with distraction [1, 6].

2.3 Factors Related to the Social Context of Crossing and Interactions Between Multiple Factors

The presence of others has been shown to increase risk-taking likelihood in previous observational studies. Accounting for differences in the size of pedestrian flow in and out of peak hours, Metaxatos and Sriraj [1] and Khattak and Luo [8] found that the number of transgressions increase with an increasing platoon size. According to the observations of Edquist, Hughes [7], crossing in groups could be more common among school children encouraging each other to deliberately transgress. Similarly, Khattak and Luo [8] showed that group violations increased in the presence of young children. More generally, being in a hurry or trying to avoid missing the next train were associated with an increased number of transgressions in the presence of a stopped at station train [1, 14].

While previous observational studies provide some interesting insights on factors likely to impact unsafe crossing the behaviours of pedestrians, the current knowledge-base remains limited. Moreover, the generalisability of previous findings is questionable when comparing different countries, territories, or even urban areas with different environmental characteristics. Differences between the results from previous studies or their interpretation could be explained by the variability of the adopted research designs, procedures (e.g. the periods of data collection, utilisation of recording devices) or data analysis methods. For instance, the number of observation sites varied between one [9, 10, 17] and ten [1]. In addition, data collection was conducted between 1997 and 2011 and could last from several days (10) to several months (two and nine). The longest data collection period spanned three consecutive years [8]. Most of the previous observational studies were conducted in the USA where LCs have similar, but not identical, design compared to Australian LCs. At American LCs, pedestrian gates are similar to those for vehicles prohibiting pedestrian crossing while lowered, whereas pedestrian gates in Brisbane close horizontally blocking the access through the path. Arguably, the existing findings are unlikely to reflect the "current" and broad pedestrian crossing context at LCs. They are unlikely to relate to LCs, where specific measures targeting pedestrian safety have been taken, as is the case in Queensland.

Therefore, more in depth and context-centred research is needed.

2.4 Rational for the Adopted Research Method and Research Question

Compared to self-reported or crash data, direct observations allow for the detection of factors likely to impact decision-making without participants being necessarily aware of their influence (e.g. presence of others crossing unsafely). Providing more objective and descriptive information than any other methods, direct observations are fundamental for the investigation of pedestrian unsafe crossing, as a highly under-researched area.

This study is to our knowledge the most recently conducted in Australia, investigating multiple factors and their interactions that are likely to contribute to unsafe pedestrian crossing behaviours. Our main aim is to examine such factors and how they can be associated with different patterns of unsafe crossing, accounting for the specific crossing contexts of three typical LCs in Brisbane.

3 Method

3.1 Choice of Observation Sites

The first stage of site selection consisted in the review of the available indicators on unsafe crossing tendencies across LCs in Brisbane. According to the most recent data provided by the urban rail operator in Brisbane (Queensland Rail—QR), almost half of all reported near-misses with pedestrians for 2011 occurred at LCs on the same rail line, the Cleveland line (42 %). The second stage of site selection consisted in random direct observations at LCs black spot locations on the Cleveland line and other rail lines, during which information was collected on:

- Characteristics of the physical environment (e.g. number of rail tracks, location of the platforms and station, over bridge access and number of pedestrian corridors);
- Technical properties of the controls (e.g. progress of activation and duration of the active controls for pedestrians, presence of locking mechanisms on pedestrian gates);
- Characteristics of pedestrian-users (e.g. school children, dressed in business attire) and the most commonly adopted trajectories (i.e. in relation to pedestrian paths/shortcuts).

Finally, additional information was collected from rail professionals (e.g. train drivers, station masters and transit officers) and QR safety experts who contributed to our decision to select three intersections adjacent to suburban

train stations—all actively protected and part of the Cleveland rail line: Coorparoo, Cannon Hill, and Wynnum Central (Fig. 1). The selection of LCs that are part of the same rail line ensured that the observation sites had similar rail traffic characteristics and technical properties of the active controls (i.e. unlike the controls at other rail lines, the pedestrian gates on the Cleveland line do not lock when closed).

With a long history of reported accidents and the highest number of reported near-misses for 2011, the LC at Wynnum Central has been identified by QR as one of the worst black spots in Brisbane. By far, the largest percentage of near-misses reported on the Cleveland line occurred at Wynnum Central (41 %), compared to Coorparoo accounting for 8.5 % and Cannon Hill accounting for 5 %, noting that the number of reported near-misses should only be considered as an approximate indication of the risk rate, given the reporting reliability issues that have previously been raised [18]. The most recent fatal collision with a pedestrian in Queensland occurred at Cannon Hill LC in January 2014, raising significant safety concerns among rail authorities. Finally, QR provided information about an increasing number of pedestrian violations at Coorparoo in recent years—2013/2014.

All three LCs are equipped with pedestrian gate systems consisting of an entry pedestrian gate that closes when activated (but can be pushed open from outside) and an emergency pedestrian gate that remains closed at all times. The emergency gate can be pushed open from inside in the case that a pedestrian is caught inside the tracks during a *"closure"* defined here as: the period from the onset until the cessation of the controls. Pedestrian lights and audible alarms are installed in each pedestrian gate system (Fig. 2).

Fig. 1 The Cleveland rail line joining Cleveland—suburb of Redland city and with Brisbane the capital of the Australian state of Queensland. Part of the Queensland Rail City train network, the Cleveland line extends 37.3 km east-southeast from CBD (Brisbane Central Business District). In red are indicated the three selected LCs for observation sessions

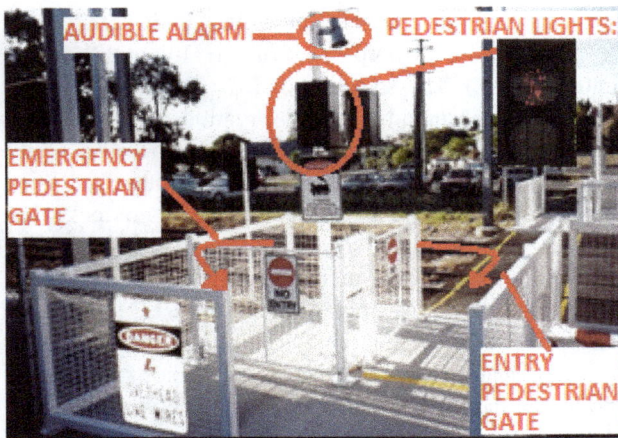

Fig. 2 Pedestrian gate system installed at the three LC observation sites

3.2 Architectural Characteristics of each LC and the Corresponding Larger Socio-Economic Areas

The suburbs of Wynnum Central and Coorparoo are more populated with 12,229 and 14,944 inhabitants (respectively) compared to Cannon Hill with a population of only 4507 inhabitants. All three LCs are in close proximity to schools and industrial zones. While Wynnum Central LC is positioned on a main road giving to a large shopping district, Cannon Hill and Coorparoo LCs are also in a close proximity to shopping centre zones.

3.2.1 Wynnum Central Level Crossing and the Adjacent Train Station

Wynnum Central LC has two rail tracks separated by a middle island giving access to the train station (Fig. 3). The middle (station) island comprises the two platforms typically giving access to passenger trains services in the direction to Cleveland—Outbound (i.e. Platform 1) and in the direction to Brisbane CBD-City (i.e. Platform 2). Two sets of pedestrian gate systems (i.e. one on the centre side and one on the residential side) activate simultaneously independently of the track or the direction of the approaching train. This implies that while an Outbound train (in direction to Cleveland) is stopped at station— pedestrian traffic is prohibited, whereas soon after a train in direction to the City had passed the LC (independently of whether the train is stationary or not), pedestrian traffic is allowed. A third set of pedestrian gate system regulates traffic on the opposite station road side.

The pedestrian corridor on the station road side is approximately 16 m long (8 m on both sides of the middle island) and the opposite station side pedestrian corridor is approximately 14 m long.

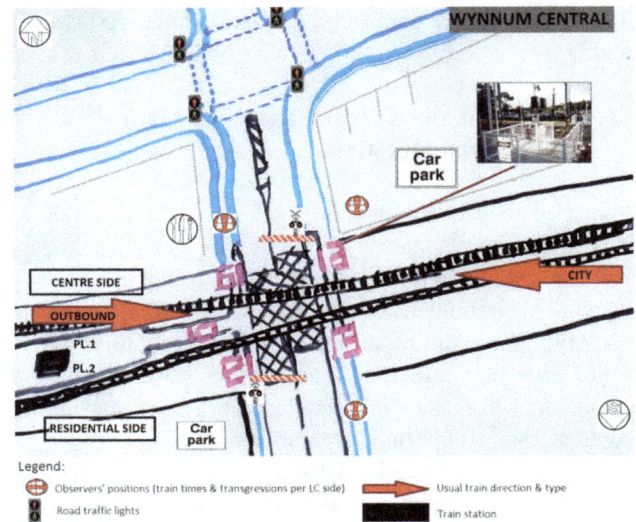

Fig. 3 Bird's eye graphic view of Wynnum Central LC based on a Google Earth photograph. Source: Google Earth (2009) corresponding to 151 m eye altitude

Two QR car parks are provided for users of the train station: one North and one South of the LC. A third car park, further West in the Centre side of the LC provides access to the station through an over bridge (not illustrated on Fig. 1).

3.2.2 Cannon Hill

Cannon Hill LC has three rail tracks separated by a middle island (Fig. 4). The station is external to the LC giving access to Platform 2 where typically passenger train services run in the direction to the City. Platform 1 located on the middle island typically gives access to Outbound trains.

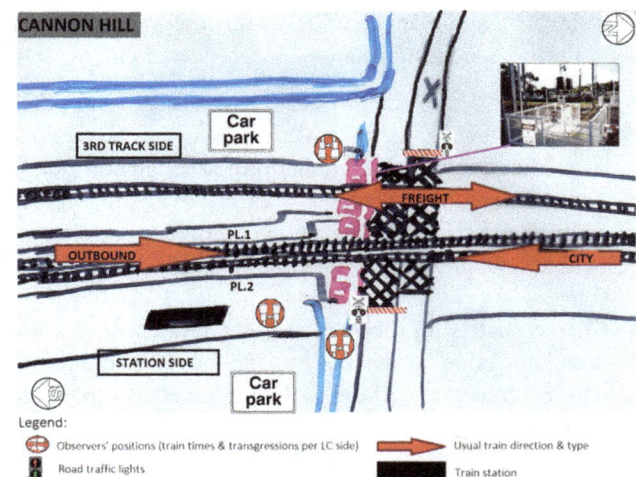

Fig. 4 Bird's eye graphic view of Cannon Hill LC based on a Google Earth photograph. Source: Google Earth (2009) corresponding to 151 m eye altitude

The third track serves only freight trains passing in both directions. Two sets of pedestrian gate systems activate separately prohibiting pedestrian traffic on either side of the middle island. Thus, pedestrian traffic is prohibited on the 3rd track side only during the rare passage of freight trains which do not follow a strict timetable. Similarly to Wynnum Central, when an Outbound train is stopped at the station pedestrian traffic is prohibited, whereas as soon as a City train has passed through the LC, pedestrian traffic is permitted. There is not a pedestrian corridor on the opposite station road side.

The pedestrian corridor is approximately 20 m long (7.50 m on the 3rd track side and 12.50 m on the station side of the middle island).

There are a number of primary schools on each side of the LC and a shopping centre is east from the LC (station side). Two QR car parks are provided for station users on both sides of the LC. An over bridge further south connects the two platforms and provides access to the middle island from the 3rd track side car park.

3.2.3 Coorparoo

Coorparoo LC (Fig. 5) has three rail tracks separated by a middle island giving access to the train station. The middle (station) island comprises the two platforms typically giving access to passenger Outbound (i.e. Platform 2) and City (i.e. platform 1) services. The third track serves only freight trains passing in both directions. Two sets of pedestrian gate systems activate separately prohibiting pedestrian traffic on either side of the middle island. Thus, every time an Outbound service is passing, pedestrian traffic through the freight track is also prohibited and inversely, every time a freight train is expected, the crossing of the Outbound rail track is prohibited. Similarly

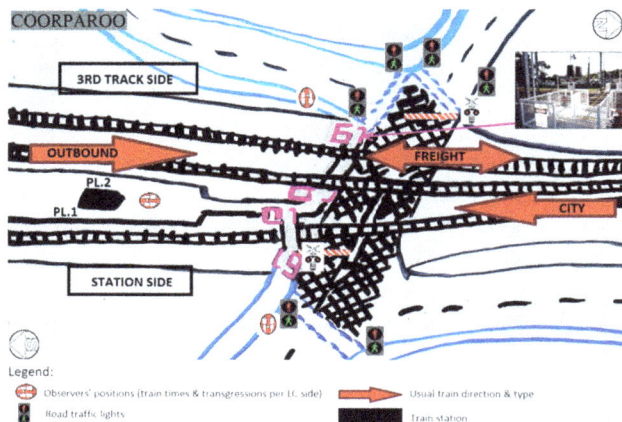

Fig. 5 Bird's eye graphic view of Coorparoo LC based on a Google Earth photograph. Source: Google Earth (2009) corresponding to 151 m eye altitude

to the other two LCs, pedestrian traffic is prohibited while there is a stopped Outbound train at station, and renewed—as soon as a City train has passed the crossing. There is not a pedestrian corridor on the opposite station road side.

The pedestrian corridor is approximately 26.5 m long (17.5 m on the 3rd track side and 9 m on the station side of the middle island). There are a number of schools mostly East from the LC (station side) and a shopping centre in the same direction.

3.3 Research Design and Participants

3.3.1 Choice of Time Frames for Morning and Afternoon Observation Sessions

To capture the busiest pedestrian traffic periods, observation sessions took place at morning and afternoon peak hours, respectively, from 7 am to 9.30 am and from 3 pm to 5.30 pm. They were conducted systematically every (working) Monday, Wednesday and Thursday in three consecutive weeks, thereby avoiding the collection data associated with specific social events likely to take place on weekends or public holidays. This organisation of the observation shifts allowed the conduct of one morning and one afternoon session at one of the three LCs on each of the 3 week days. All three LCs were visited during each week of observations following a random order.

Observations started in the first week after school holidays as students were among the targeted groups of potentially "at risk" pedestrians. The hours of the observation shifts were also planned in accordance with the crossing time frames of various socio-demographic classes (e.g. construction workers, office workers, school children and pensioners) and corresponded to the typical start/finish working (school) hours.

3.3.2 Observers

Five researchers from the Centre for Accident Research and Road Safety Queensland (CARRS-Q) were trained by the lead researcher for data collection and entry, during a week of pre-observation. To enhance familiarity, pre-observations took place at all LC sites and each observer was trained to code data related to two main observer's roles: (1) coding transgressions and (2) coding train times. Two *"Transgressions"* observers per session coded the personal and crossing characteristics of transgressors. They were positioned close to the pedestrian corridors on each side of the LC and coded: the gender and the approximate age of transgressors; the adopted crossing trajectory; the number of people crossing in groups; and the number of people waiting for the controls to deactivate (compliant crossing behaviour). One other *"Train times"* observer per session was in charge of coding the exact time (hh/mm/sec) when a train has reached the LC, stopped or left a station as well as the number and types of trains per closure and their respective direction and platform. Depending on the site, "Train Times" observers were positioned at a station (Figs. 4, 5) or at a nearby car park (Fig. 3). The variables related to **Closure characteristics** (e.g. the exact hour of each control's activation) were taken either by observers coding train times (Fig. 3) or by observers coding transgressions—where the controls on the two sides of the middle island activate separately (Figs. 4, 5).

3.4 Material

Observation sheets and chronometers (on android mobile devices) were used for data collection. Variables related to each closure were coded on a separate sheet independently of whether a transgression took place or not. A closure identification number was coded on each observation sheet, facilitating the synchronisation of data between observers during data entry. All observers were equipped with a set of observation sheets in the form of a notebook.

3.4.1 Transgression Sheets

Transgression sheets (Appendix 1) had two main parts. In the first part, a rough plan of each LC's platforms and pedestrian corridors served to trace the *trajectory* of transgressions. The same method was used to code *the number of people at each angle of the LC who did not transgress* (compliant crossing group) at the end of each closure. It is important to note that where pedestrians waiting at the angle exceeded ten, the counts should be considered approximate due to poor visibility.

In the second part of the sheet were coded demographic and other characteristics of the pedestrians who transgressed: *gender*—male versus female; *approximate age*—baby/toddler (0–4 years old) versus school children (5–15 years old) versus young adult/teenager (16–30 years old) versus older adult (30–70 years old) versus elderly (70+ years old); *exact time of transgression*—exact hour when the pedestrian stepped on the LC platform (hh/mm/sec); *status of controls' activation at the moment of transgression*—Moment 1 (pedestrian lights flashing) versus Moment 2 (pedestrian gates closing) versus Moment 3 (pedestrian gates fully closed). It is worth noting that the time difference between the three moments is typically 8 s, meaning that 16 s after the activation of the pedestrian lights, the pedestrian gates are fully closed. In addition, observers were trained to identify a minimum set of variables related to the description of the transgressors: *crossing pace*—walking versus speeding/running; *social influences*—crossing alone *vs.* in group, *journey purpose*—

on the way to catch a train (yes vs. no, where possible to identify).

3.4.2 Train and Closure Times Sheets

Train time sheets (Appendix 2) were used to code the following variables: *order of train passing at the LC* (the order of arrival at the LC or at the station); *number of platform*; *direction*—City versus Cleveland; *type of train*—stopping at station versus express, independently of whether it was an empty service, a train that does not serve the station or else, a freight train (i.e. typically long trains passing on the 3rd track at Cannon Hill and Coorparoo); *hour of train passing*—three times were taken for stopping trains (arrives at LC vs. stops at station vs. leaves station) and one for express trains—the hour it arrived at the LC.

Closure times were coded by multiple observers at each LC and included the following variables: *start closure*—hour of the activation of the pedestrian flashing lights corresponding to Moment I (hh/mm/sec); *gate closing*—hour when the pedestrian gate starts closing corresponding to Moment II (hh/min/sec); *gate closed*—hour when the pedestrian gate is fully closed corresponding to Moment III (hh/min/sec); *end closure*—hour when the pedestrian lights deactivates (hh/min/sec). To avoid mistakes in data entry, these variables were entered on the observation sheet only after the end of each closure, given that the times remained recorded on the chronometer screen.

3.5 Procedure

Having obtained permission from QR to conduct this study on their property, all visits of LC sites were preceded by safety instruction sessions for observers. Observers were in contact with rail staff at all times. Pre-observations were conducted for one week prior to the actual observations. During this period, the first researcher familiarised the four assistant observers with the objectives of the study, the coding process and the specificities related to each LC site. The actual observation sessions were conducted by three of the five researchers each. The larger number of observers allowed the shuffling of shifts and thus to avoid fatigue related issues. Each observation session was preceded by a synchronisation of all chronometers. No breaks were taken during observations. It is likely that the presence of observers was noticed by pedestrians even though the most discrete positions were selected considering safety procedures (e.g. remain in a significant distance from roadside) and the visibility of the targeted variables. After the end of the sessions, all observers were debriefed by the first researcher. Questions around data were discussed and resolved. All observers together started data entry shortly after the end of each session using a laptop and pre-established Microsoft Excel sheets. Data entry took approximately 1 h and 30 min. This study was approved by the university ethics committee.

3.6 Collected Data and Statistical Analysis

The data were collected during three consecutive weeks between 28 April 2014 and 15 May 2014, representing a total of 45 h of observations across all sites. In total 438 closures were observed, each lasting from 12 s at Wynnum Central, where crossing through the two passenger services tracks is prohibited simultaneously, to 3 min and 51 s at Coorparoo, where the two passenger services tracks close autonomously. There was not a significant difference between the average duration of closures at all three LCs ($M = 75.06$ s, $SD = 35.62$ s), $F(2, 435) = 1.23$, ns. It should be noted that during the last afternoon observation session at Wynnum Central, a cancellation of all train services following an incident resulted in a smaller number of closures and a higher volume of passengers leaving the train station (after having disembarked a City train). Nevertheless, the number of closures at each site was relatively constant over the three days of observation, χ^2 (4, $N = 438) = 2.17$, ns. One "false closure" was observed at Cannon Hill during which a train did not pass. Instead, both sides of the LC were closed for maintenance during 21 s, noting that no transgression took place.

Most of the closures were for the passage of a single-train (84 %), two trains passed in 15 % of the closures, and only on three occasions did three trains pass during the same closure (Table 1). Because of this small number of three train closures, they were considered together with two train closures for the remainder of the analysis. Regarding the types of train passing during closure, most of the closures included at least one stopping train, accounting for 76 % of the single-train closures and for 93 % of the multiple trains closures (Table 1). Closures involving only express trains represented 21 % of all 437 closures with passing trains. The distribution of number and types of trains passing during closures did not differ according to the three LCs (Fisher, ns.).

For the analysis of the collected data, a series of Chi-square tests (χ^2) were performed to test the significance effect between two discrete variables. Fisher's exact test was used for contingency tables that contain small expected values (<5) in more than 20 % of the cells (i.e. only p value is reported). Cramers' V^2 statistic was used to report the strength of association between discrete variables typically applied to $2 \times n$ tables, which is conventionally considered to be low if < 0.04, medium if between 0.04 and 0.16, and high if > 0.16 [19]. Relative Deviations (RDs) were used to inform on the strength of association between the modalities of the two discrete variables. Relative deviations are calculated on the basis of the comparison between the observed and expected frequencies in each cell. By convention, there is a high positive or negative association when the absolute RD value is >0.20. Only associations >0.10 are described in the results section. Finally, analysis of variance tests (ANOVA) and correlations were used to test the effects on continuous variables. Post-hoc tests using Bonferroni correction were used to examine the relationships between the modalities of continuous variables (only p value is reported where the means are presented in tables).

4 Results

4.1 Frequency and Proportions of Observed Transgressions at the Three LC Sites

As per Table 2, the largest number of transgressions was observed at Wynnum Central and Coorparoo accounting for, respectively, 46.5 and 41.9 % of all 129 observed transgressions across the three LC sites. In contrast, Cannon Hill was characterised with a low number of transgressions representing only 11.6 % of all transgressions. Twenty percent of all closures included at least one person in transgression. The proportion of closures with at least one transgression varied significantly between sites, χ^2 (2, $N = 438) = 28.03$, $p < 0.000$, with the largest ratio of closures with transgressions observed at Wynnum Central

Table 1 Type of trains observed during closures	One train closures		Two train closures		Three train closures		Total	
	N	%	N	%	N	%	N	%
Express train	87	92.5	7	7.4	0	0	94	100
Stopping train	283	94.6	16	5.3	0	0	299	100
Both	0	0.0	41	93.1	3	6.8	44	100
Total	370	84.6	64	14.6	3	0.6	437	100

Legend The false closure" has been omitted in the table as not implying a train passage

Table 2 Counts and percentages of closures with at least one pedestrian in transgression per LC site

| | Closures | Closures with transgression | | Transgressions | | Transgressions per closure | |
| | | | | | | Among all closures | Among closures with transgression |
	N	N	%	N	%	M (SD)	M (SD)
Wynnum central	117	40	34	60	46.5	0.51 (0.87)	1.97 (1.21)
Cannon hill	149	13	9	15	11.6	0.10 (0.36)	1.40 (0.83)
Coorparoo	172	35	20	54	41.9	0.32 (0.73)	2 (0.95)
Total	438	88	20	129	100		

and the least—at Cannon Hill, the strength of association between the variables being moderate, $V^2 = 0.06$.

Looking into the number of pedestrians in transgression during the same closure, a maximum of five were observed at Wynnum Central and four at Coorparoo, both on a single occasion. Most commonly, between one and three transgressors were observed per closure with no significant difference in the distribution across the three LCs, Table 2 (Fisher, ns.).

4.2 Transgressions Associated with the Physical Characteristics of the Environment and the Specific Crossing Context

4.2.1 Transgressions According to the Status of the Controls (Moment of Crossing)

Comparing transgressions according to the three moments of controls' activation, more than half were observed during the first seconds after the activation of the pedestrian lights (Moment 1); almost one quarter were observed in the riskiest moment while the gates were closed (Moment 3); and the smallest amount occurred in Moment 2 while the gates were in the process of closing (Table 3). The distribution of transgressions according to the moments of crossing differed significantly among the LC sites (Fisher, $p < 0.01$), with an intermediate strength of association between the variables, $V^2 = 0.08$. The analysis of the RDs revealed that Cannon Hill was particularly associated with crossing in Moment 1, Coorparoo with crossing in Moment 2 and Wynnum Central with crossing in Moment 3.

4.2.2 Transgressions According to Train's Position

According to the position of the train during a transgression, only one pedestrian was observed crossing in front of a stopped at station train and a small number of transgressions were observed behind a passing (express) train. These crossing situations were merged as "other train position" modality for further analysis. Globally, the large

Table 3 Counts and percentages of transgressions according to the status of active controls

| | Moment 1 ped. Lights | | Moment 2 gates closing | | Moment 3 gates closed | |
	N	%	N	%	N	%
Wynnum Central	32	53.3	6	10	22	36.6
Cannon Hill	13	86.6	2	13.3	0	0
Coorparoo	30	55.5	15	27.7	9	16.6
Total	75	58.1	23	17.8	31	24

Legend Moment 1—from activation of pedestrian lights until activation of pedestrian gate; Moment 2—from activation of pedestrian gate until full closure (period of closing); Moment 3—from the full closure of pedestrian gate until deactivation of pedestrian lights

majority of transgressions (85 %) occurred in front of an approaching train (Table 4). However, there was a significant difference in the number of transgressions according to train's position between the three sites (Fisher, $p < 0.05$), with an intermediate strength of association between the two variables, $V^2 = 0.05$. The estimation of the RDs showed that among the three sites, Wynnum Central was the one preferentially associated with transgressions behind a stopped train and in "other positions", all of these situations characterised by the presence of a visible train.

4.2.3 Transgressions According to Crossing Trajectory and LC Angle

Looking into the adopted trajectories during transgressions (Table 5), the largest proportion of pedestrians were observed on their way towards a middle island (71.3 %), whereas crossing out of a middle island (15.5 %) and just crossing the road (13.2 %) were less frequently observed trajectories during transgressions. A Fisher's exact test showed a significant difference in the adopted trajectories between the three LCs (Fisher, $p < 0.001$). The association between the modalities of the variables was moderate ($V^2 = 0.08$), suggesting that Cannon Hill, contrary to the other two LCs, was associated with the two less common

Table 4 Counts and percentages of transgressions according to train position across the three LCs

	In front of an approaching train		Behind a stopped train		Other train position	
	N	%	N	%	N	%
Wynnum central	44	73.3	12	20	4	6.6
Cannon hill	15	100	0	0	0	0
Coorparoo	51	94.4	3	5.5	0	0
Total	110	85.3	15	11.6	4	3.1

Legend the category "Other train position" included (1) transgressing behind an express passing train, and (2) transgressing in front of a stopped at station train

Table 5 Counts and percentages of transgressions according to the adopted crossing trajectory

	To middle island (train station) N	Out of middle island (train station) %	Just crossing the road N	To middle island (train station) N	Out of middle island (train station) %	Just crossing the road N
Wynnum Central	46	76.6	6	10.0	8	13.3
Cannon Hill	4	26.6	8	53.3	3	20.0
Coorparoo	42	77.8	6	11.1	6	11.1
Total	92	71.3	20	15.5	17	13.2

trajectories (i.e. out of a middle island or just crossing the road).

The patterns of the adopted trajectory could be associated with the specific design of each LC. To examine further these patterns, Fig. 6 illustrates graphically the distribution of the transgressions among the three LCs, according to the three trajectories and the crossing angle. While at Wynnum Central and Coorparoo the majority of transgressions occurred on the way to a middle island (i.e. corresponding to the emplacements of a train station), pedestrians at Wynnum Central adopted visibly more variable trajectories, particularly when crossing from the Centre side of the LC (i.e. diagonal through the road and crossing in the middle of the road). In contrast, at Coorparoo more transgressions were observed from the Station side of the LC and none on diagonal which could be explained by the absence of pedestrian path on the opposite station side. However, the only transgression on a diagonal out of a station line was observed at the same LC, which

could be associated with an impatience to wait at the adjacent road traffic lights. The majority of transgressions at Cannon Hill out of the middle island or just crossing the road seemed to be associated with accessing the train station positioned externally to the rail tracks or the large car park adjacent to the Station side of the LC.

4.2.4 Transgressions According to the Number of Crossed Tracks

A significant difference was found between the number of rail tracks crossed while transgressing between the three LCs, such that transgressions at Wynnum Central implied the least number of crossed tracks ($M = 1.15$, $SD = 0.36$), followed by Coorparoo ($M = 1.54$, $SD = 0.66$) and the largest number of crossed rail tracks per transgression was observed at Cannon Hill ($M = 2.13$, $SD = 0.51$), $F(2, 126) = 23.06$, $p < 0.000$, $\eta^2 = 0.26$, the difference comparing all three sites being significant at $p < 0.000$.

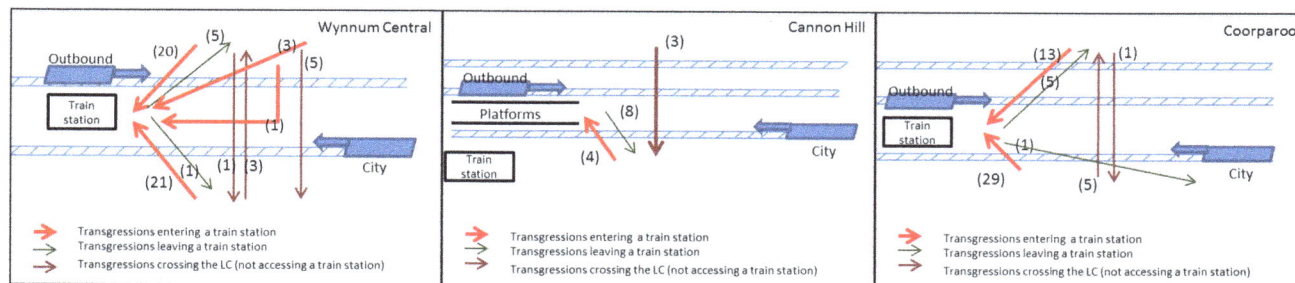

Fig. 6 Patterns of transgressions according to LC angle at each LC

To investigate further the risk-taking tendencies accounting for the number of crossed tracks, an additional variable was computed corresponding to the number of crossed *"Unoccupied tracks"*. This variable corresponded to the counts of crossed tracks where a train could have passed during the closure given that crossing through the same track after a train had already passed is not associated with a real risk of being hit by a train. As shown in Table 6, more than half of the pedestrians across all three LCs crossed at least one unoccupied track (48 + 4.6 %). Here again, a significant difference was found in the number of crossed unoccupied tracks according to the LC (Fisher, $p < 0.01$), with an intermediate association between the variables ($V^2 = 0.05$). The estimation of the RDs revealed different risk-taking patterns across the three sites. Consistent with the total number of crossed tracks during transgressions, Wynnum Central was moderately associated with crossing one unoccupied track, whereas Cannon Hill was at the same time moderately associated with the crossing of one and strongly associated with the crossing of two unoccupied tracks. In contrast, Coorparoo was associated at the same time with crossing none and two unoccupied tracks.

4.2.5 Transgressions According to Time of the Day

More than two thirds of the closures with at least one transgression took place in morning peak hours (69.7 %), χ^2 (1, $N = 438$) = 9.67, $p < 0.01$ (Table not provided). Similarly, two thirds of all transgressions were observed in morning peak hours (Table 7). Although systematically more transgressions were observed in the morning than in the afternoon, there was a significant difference between the three sites according to the time of day, χ^2 (2, $N = 129$) = 7.04, $p < 0.05$, with an intermediate strength of association between the variables $V^2 = 0.05$. The estimation of the RDs showed that unlike the two other sites, Wynnum Central is more associated with transgressions in the afternoon, (Table 7).

4.2.6 Transgressions According to Exposure

In total, 2446 pedestrians were counted crossing compliantly during all observed closures (i.e. closures with and without transgressions). The number of pedestrians crossing compliantly per closure varied between 0 and 77 ($M = 5.58$, $SD = 8.13$). As indicated in Table 8, the largest number of pedestrians crossing compliantly per closure was observed at Wynnum Central $F(2, 435) = 23.17$, $p < 0.000$, $\eta^2 = 0.10$. Also, more compliant crossings were observed during the afternoon closures, $F(1, 436) = 4.09, p < 0.05, \eta^2 = 0.02$. The interaction between the two variables (Sites * Time of the day) was also significant, $F(2, 432) = 10.14$, $p < 0.000$, $\eta^2 = 0.05$, suggesting that the largest number of pedestrians crossing compliantly was counted at Wynnum Central compared to the other two LCs ($p < 0.000$). This result could be related to the exceptional cancellation of the train services. In contrast, there was a similar number of people in the morning peak hours at the most and least populated LCs (i.e. respectively Coorparoo and Cannon Hill).

The 129 observed transgressors represented around 5 % of all people crossing during the closures. Accounting for compliant crossing, at Wynnum Central was observed the highest percentage of transgressors in the afternoon peak hours and at Coorparoo—the highest percentage of transgressions in the morning peak hours (Table 8).

4.3 Transgressions Associated with Pedestrians' Characteristics and Motivations

4.3.1 Transgressions According to Demographics

All 129 transgressors were distributed among five approximate age groups. Two babies (toddlers) were merged for further analysis with the young adults group as they were accompanied by adults of this age group. Male transgressors were slightly more numerous than females, and young adults were the most numerous among all age groups, χ^2 (3, $N = 129$) = 2.59, ns., (Table 9). Similarly, there was not a significant difference in the number of transgressors according to age (Fisher, ns.) or gender p[χ^2 (2, $N = 129$) = 1.41, ns.] between the three LCs (Table not presented).

Table 6 Counts and percentages of crossed unoccupied tracks during transgressions at the three LCs

	None		1 Track		2 Tracks	
	N	%	N	%	N	%
Wynnum central	27	45	33	55	0	0
Cannon hill	4	26.6	8	53.3	3	20
Coorparoo	30	55.5	21	38.8	3	5.5
Total	61	47.2	62	48	6	4.6

Table 7 Counts and percentages of transgressions according to time of the day (morning vs. afternoon peak hours)

	AM (7-9.30)		PM (3-5.30)	
	N	%	N	%
Wynnum central	36	27.9	24	18.6
Cannon hill	12	9.3	3	2.3
Coorparoo	44	*34.1*	10	*7.8*
Total	92	71.3	37	28.7

Table 8 Counts of pedestrians crossing compliantly and proportion of transgressions per LC

| | Compliant crossing/Closures | | | | | | Transgressions/Compliant crossing | | | | |
| | AM (7-9.30) | | PM (3-5.30) | | Total (AM + PM) | | AM (7-9.30) | | PM (3-5.30) | | Total (AM + PM) |
	N	M	N	M	N	M	N	%	N	%	%
Wynnum Central	425/68	6.25	663/49	13.53	1088	9.29	36/425	8.47	24/663	3.49	3.61
Cannon hill	330/80	4.12	277/69	4.01	607	4.07	12/330	3.63	3/277	1.32	2.47
Coorparoo	420/93	4.52	331/79	4.19	751	4.37	44/420	10.47	10/331	3.02	7.19
Total	1142/241	4.88	1304/197	6.45	2446	5.58	92/1175	7.82	37/1271	2.91	5.27

Table 9 Counts and percentages of transgressors according to gender and approximate age groups

| Age groups | Male | | Female | | Total | |
	N	%	N	%	N	%
School children	16	21.9	9	16	25	19.3
Young adults	33	45.2	26	6.4	59	45.7
Older adults	21	28.7	15	26.7	36	27.9
Elderly	3	4.1	6	10.7	9	6.9
Total	73	56.6	56	43.4	129	100

Legend The approximate age of transgressors was coded according to five pre-determined age groups as follows: baby/toddler (0–4 years old); school children (5–15 years old); young adult/teenager (16–30 years old), older adult (30–70 years old); elderly (70 + years old)

4.3.2 Journey Context and Crossing Pace

Among all 129 transgressors, 91 were seemingly going to catch a train with most of them ($N = 86$) accessing the train station through a middle island (at Wynnum Central and Coorparoo). The remaining five accessed the station at Cannon Hill either on their way out of a middle platform ($N = 2$), either crossing all LC tracks to access the station on the opposite road side ($N = 3$). Only 66 of all pedestrians going to catch a train appeared to hurry while crossing, while the remaining more than a quarter crossed at a walking pace.

4.4 Transgressions Associated with Pedestrians' Social Context

Globally, pedestrians crossing alone (not in groups) accounted for more than three quarters of all transgressions (Table 10). However, there was a significant difference between the three LCs in the number of transgressions while alone, in a group of two, and in a group of more than two pedestrians (Fisher, $p < 0.05$). The association between the variables was weak ($V^2 = 0.03$), with the estimated RDs indicating more likelihood to transgress

alone at Cannon Hill, and in groups of two and more pedestrians—at Coorparoo.

4.5 Transgressions Accounting for the Interactions Between Factors

4.5.1 Time of the Day, Moment of Transgression and High Risk Groups of Pedestrians

The distribution of transgressions in the three moments of controls' activation differed significantly according to the time of the day, $\chi^2 (2, N = 129) = 9.98, p < 0.01$, with an intermediate association between the variables ($V^2 = 0.07$). The estimation of the RDs, revealed that transgressions in morning peak hours were likely to be observed in Moment 1 of the controls' activation, whereas afternoon transgressions were associated with Moment 3. Pedestrians of different age groups also showed significantly different crossing patterns according to the moment of transgression (Fisher, $p < 0.05$), the association between the variables being also intermediate ($V^2 = 0.05$). The RDs associated school children with transgressing before and until the pedestrian gates are closed (Moment 1 and 2), whereas older adults and elderly were associated with transgressing in Moment 2, and younger adults with transgressing in the riskiest Moment 3. On the contrary, there was not a significant difference in the moments of transgression between male and female pedestrians, $\chi^2 (2, N = 129) = 1.08$, ns.

In contrast, the two genders showed different patterns of transgression according to the time of the day, $\chi^2 (1, N = 129) = 5.66, p < 0.05$, the association between the variables being weak ($V^2 = 0.04$). According to the RDs, female pedestrians were more likely to be observed transgressing in the morning, whereas male pedestrians - in the afternoon. Concretely, the odds of observing a male pedestrian transgressing in the afternoon peak hours were 0.37 times higher than observing a female. Pedestrians of different age groups also appeared to be likely to transgress in different times of the day, $\chi^2 (3, N = 129) = 8.31, p < 0.05$. The strength of association between the two

Table 10 Counts and percentages of transgressions alone and in group of pedestrians

	Alone		In group 2 pedestrians		In group 3–4 pedestrians	
	N	%	N	%	N	%
Wynnum Central	40	39.6	7	6.9	2	1.9
Cannon Hill	13	12.8	1	0.9	0	0
Coorparoo	27	26.7	7	6.9	4	3.9
Total	80	79.2	15	14.8	6	5.9

variables being intermediate, the estimation of the RDs showed that young adults/teenagers were associated with transgressing in afternoon peak hours, whereas older adults and elderly were associated with transgressions in the morning peak hours ($V^2 = 0.06$).

4.5.2 Train Position, Trajectory and Number of Crossed Tracks

The number of transgressions was significantly different according to the train's position in interaction with: moment of crossing, time of the day and crossing trajectory. The moment of crossing was strongly associated with train's position, Fisher, $p < 0.000$, $V^2 = 0.27$. The estimation of the RDs showed that transgressions in front of an approaching train were particularly associated with crossing before the gates are closed (i.e. Moment 1 and 2), whereas transgressions in the presence of a visible train (i.e. behind a stopped train and other positions) were strongly associated with Moment 3. In contrast, a weak association between train position and time of the day suggested that transgressions in the presence of a visible train (behind a passing train and other positions) were likely to be observed in afternoon peak hours, Fisher, $p < 0.05$, $V^2 = 0.04$. The adopted trajectory was also weakly associated with train position, χ^2 (4, $N = 129$) $= 9.04$, $p < 0.05$, $V^2 = 0.04$. The estimation of the RDs showed that crossing behind a stopped train was associated with going towards a middle island, whereas other train positions (i.e. crossing behind a passing express or in front of a stopped train) were associated with going out of a middle island.

The adopted transgression trajectories differed significantly according to time of the day, χ^2 (2, $N = 129$) $= 6.82$, $p < 0.05$, $V^2 = 0.05$. According to the estimated RDs, the intermediate relationship between the variables suggested that leaving a middle island and just crossing the road were associated with transgressions in the afternoon peak hours. There was not a significant difference in the adopted trajectories according to the moment of transgression.

The total number of crossed rail tracks during transgression was significantly different according to the Moment of transgression, Fisher, $p < 0.05$, the association

between the variables being moderate, $V^2 = 0.04$. The RDs revealed that the crossing of more than one rail track (i.e. two or three) was more likely to be observed during the closure of the pedestrian gates (Moment 2), whereas crossing in Moment 3 was associated with crossing one rail track. However, the number of crossed unoccupied tracks was similar independently of the moment of transgression (Fisher, ns.), noting that 11.6 % of the pedestrians crossed one unoccupied track after a first train had passed, taking the potential risk of crossing in front of a second train.

4.5.3 Crossing Alone and in Group, Demographics and Time of the Day

There was a significant relationship between transgressions alone or in group and the age of pedestrians [χ^2 (3, $N = 129$) $= 23.20$, $p < 0.000$, $V^2 = 0.17$], the adopted crossing trajectory [χ^2 (2, $N = 129$) $= 6.70$, $p < 0.05$, $V^2 = 0.05$] and the time of the day [χ^2 (1, $N = 129$) $= 4.11$, $p < 0.05$, $V^2 = 0.03$]. The strong association with the age groups of participants indicated that school children and elderly were more likely to transgress alone, whereas older adults were more likely to transgress in groups. According to the RDs, the intermediate relationship with the adopted trajectory revealed that transgressing alone was associated with going out of a middle island or just crossing, whereas group transgressions were associated with going towards a middle island. Finally, according to the RDs the weak association with time of the day indicated that group transgressions were more likely to be observed in the morning, whereas pedestrians transgressing alone were associated with afternoon hours.

5 Discussion

Pedestrians' unsafe crossing at LCs has been identified as a highly under-researched area lacking notably in the understanding of the key factors influencing decision-making of this particular population. This paper presented the results from direct observations conducted at three key black spot LCs in Brisbane, providing novel and contextual relevant evidence on the role of multiple factors contributing to risky crossing behaviours. Despite the short duration of the observations, a relatively large number of transgressions ($N = 129$) was observed corresponding to more than 5 % of all pedestrians present at the LCs at the end of the observed closures, noting that information on the moment of their arrival at the LCs was not collected.

The following sections contrast the simple effects of risk factors on unsafe crossing at the three LCs (generalised case) with the effects of the same risk factors on unsafe crossing according to the specific characteristics of the crossing context at each of the three LCs.

5.1 The Simple Effects of Risk Factors at LCs in Brisbane

The observed transgressions seemed to differ according to the moment of crossing, the time of the day and the adopted trajectory, which were directly or indirectly associated with different demographic profiles of pedestrians. The links between moment of transgression and other risk factors are described in the following paragraphs, representing thus findings informing on three potential key at risk transgression patterns adopted by pedestrians, users of the Cleveland rail line.

5.1.1 Transgressions in the First Moments After the Controls' Activation

In line with previous findings, the largest proportion of transgressions occurred before the gates are closed and even active. Such transgressions were particularly associated with crossing in front of an approaching train (unlikely to be visible), and with transgressions in morning peak hours. In fact, contrary to what has been demonstrated by Edquist, Hughes [7] and Metaxatos and Sriraj [1], the largest proportion of transgressions at all three LCs occurred at morning peak hours and this was even after accounting for the number of pedestrians crossing compliantly during all closures. The observed transgressions in morning peak hours were associated with female pedestrians and school children. Unlike previous findings, school children were linked with crossing alone. The summary of all these simple effects could explain transgressions motivated by a fear of missing the next train and of being late for school. Transgressions before the gates have started moving and in front of an approaching train were consistently associated with crossing towards a middle island. This was globally the predominantly adopted trajectory during all transgressions potentially related to the motivation of catching the next train as was visible in 70 % of the cases, noting that the journey purpose was not identifiable for all transgressions.

5.1.2 Transgressions During the Closure of the Pedestrian Gates

A larger number of pedestrians were observed transgressing once the gates were fully closed compared to while they were closing. Such findings are in contradiction with "beating the gates" tendencies and the obtained results by Metaxatos and Sriraj [1]. Nevertheless, the results from these observations associated older adults and elderly with crossing during gate closure and with afternoon peak hours. Older adults were associated with crossing in groups and elderly associated with crossing alone. The combination of these results could explain an increased perception of control (e.g. "I could make it on time") before the gate is fully closed, rather than sensation seeking tendencies.

5.1.3 Transgressions After the Closure of the Pedestrian Gates

Crossing after the gates are closed was also associated with afternoon peak hours and with the presence of a visible train (stopped or passing through). Crossing in this last moment of the activation of controls was common to young adults/teenagers, who themselves were also associated with crossing in the afternoon peak hours. Transgressions of young adults/teenagers in afternoon peak hours corresponded to crossing out of a middle island or just crossing. All these results taken together could be associated with impatience to wait for the controls to deactivate after disembarking from a train in the afternoon peak hours, potentially taking the risk of crossing in front of a second train. Examining the risk of crossing in front of a second train according to the number of crossed unoccupied rail tracks, no significant difference was found according to the moment of transgression, meaning that independently of the moment of transgression, pedestrians were equally likely to cross one or more unoccupied potentially "at risk" of second train tracks. Similarly, transgressing after the gates are closed was strongly associated with the crossing of one rail track, most likely after a train has passed the LC, which could explain a certain awareness of the risk of second train. Still, in total, a large number of pedestrians crossed one unoccupied track after the gates were closed (11.6 %) embracing the risk of crossing in front of a second train. Taken together, these results suggest that crossing once the gates are fully closed is highly influenced by the train's visibility and is indeed a serious potential threat for crossing in front of a second train. Being associated with younger adults, such risk-taking behaviours could be explained by the perception of control or familiarity with the LC design and rail traffic. It could also be explained by sensation seeking tendencies or "recreational" risk-taking in the late afternoon peak hours. It would be worth looking further into the patterns of transgressions according to whether one or both sides of the middle island were closed during transgression in front of a second train. Such evidence would contribute to a better understanding of the pros and cons of having a separate regulation of pedestrian traffic at both sides of a middle island, especially in the case that there is a train station on the middle island.

5.2 The Effects of Interacting Factors Associated with the Crossing Context at Three Typical Black Spot LCs Within Brisbane Area

Different transgression patterns across the three LCs were identified depending on the characteristics of the larger area, the LC and station environment, as well as according to rail traffic characteristics. The largest number of pedestrians crossing compliantly and transgressions were counted at Wynnum Central, the second most populated suburb, giving access to the train station through a middle island where crossing is prohibited along the pedestrian corridors for each train passage independently of its direction. In contrast, the largest proportion of transgressions accounting for the total number of people crossing during closures was observed at Coorparoo, the most populated suburb where the total number of people crossing compliantly and transgressing was lower compared to Wynnum Central. Having a similar design comprising the train station at a middle island, the main difference between Coorparoo and Wynnum Central is the presence of a third track and the separately operated pedestrian corridors on the two sides of the middle island at Coorparoo. On the contrary, Cannon Hill was the least populated suburb with the lowest number of pedestrians observed to cross compliantly and transgressing at this LC not giving access to the train station but to a middle platform separating the three tracks. Contrary to previous findings, the location of the platforms outside of the rail tracks was not associated with a larger number of transgressions. However, a more in depth analysis of the results revealed that at Cannon Hill pedestrians were observed to take the most risk by crossing the largest number of rail tracks where a train could have passed (unoccupied tracks).

5.2.1 Transgression Patterns Related to the Crossing Context at Wynnum Central

Transgressions at Wynnum Central were associated with crossing in the last moment of controls' activation, with the presence of a visible—stopped or passing train (i.e. the majority observed behind a stopped at station train) and with afternoon peak hours. At this site, pedestrians crossing right after the train has passed the LC (City train) could still catch it from the station. Such transgressions behind a

stopped train and after the gates are closed were also associated with younger adults. Moreover, in the afternoons more transgressions were observed in groups. Wynnum Central stood out as the LC where most variability was observed in the adopted trajectories towards the station. A large number of transgressors came from the large shopping Centre side and crossed on diagonal or even through the centre of the LC. Crossing on diagonal could be explained by the motivation to avoid waiting to cross at a nearby intersection with four pairs of pedestrian traffic lights connecting the different sides of the road (Fig. 3). The large number of transgressions from either side of the crossing could be associated with catching the City Train service (i.e. if crossing behind a stopped train). For those transgressing from the Centre side of the LC, being in a hurry to catch the City train implies crossing through the Outbound track. If pedestrians are familiar with such crossing situation, they can easily assume that even if there is an approaching second train (coming from the City) it will stop at station before reaching the LC platform. However, in reality an express train could be approaching anytime at full speed. Consequently, it can be argued that the simultaneous regulation of pedestrian traffic on both sides of a middle island could lead people to underestimate the risk of second train arrival even it is visible. Such risk could potentially be avoided if pedestrians were to use the existing over bridge that provides access to the platforms from the car park. However, pedestrians might be unlikely to cross the overbridge given its distant location from the main road, which is a main adopted trajectory if coming from the shopping centre. Consequently, a more adequate location of the over bridge or a separate regulation of both tracks at this LC could potentially minimise the risk of transgressions and especially—in front of a second train.

5.2.2 Transgression Patterns Related to the Crossing Context at Cannon Hill

Cannon Hill was associated with transgressions before the gates have started closing (Moment 1). Transgressions in the first moments were predominantly observed in morning peak hours, in front of an approaching train and by school children. School children were likely to be seen crossing alone and so were in general transgressors observed in the morning peak hours. Contrary, to previous findings the location of the train station externally to the rail tracks was associated with a lower number of transgressions compared to the other two LCs. However, looking into the adopted trajectories a strong pattern of transgressions was identified, corresponding to the crossing of multiple tracks to access the station (City train service platform), including crossing the road and going out of a middle island. In fact, among the three LCs, only Cannon Hill was associated with just crossing the road or exiting the middle platform, trajectories corresponding to the emplacement of the train station at this particular site externally to the rail tracks. While the existing over bridge linking the middle island platform and the train station could have contributed to decrease transgressions, the separation of the third track from the passenger services tracks could potentially be associated with an increased level of risk during transgressions as pedestrians crossing through the passenger services corridor are obliged to cross both tracks at the same time.

5.2.3 Transgression Patterns Related to the Crossing Context at Coorparoo

Finally, Coorparoo was associated with crossing before the pedestrian gates are fully closed (Moment 2). Transgressions in this moment were predominantly observed in afternoon peak hours and by older adults and elderly. Moreover, Coorparoo was mostly associated with crossing in groups of two and more pedestrians, noting that group transgressions were also associated with older adults and with crossing towards a middle island. Coorparoo was also the only LC associated at the same time with crossing none (not at risk) and two unoccupied tracks. Thus in addition to the trajectory corresponding to crossing from either side to catch a City train, pedestrians at Coorparoo also transgressed on their way out of the middle island after disembarking a train. Thus, this LC seems to be associated with two different transgression patterns: one describing transgressing in groups to catch a train, and one crossing towards a car park in the afternoon hours (Fig. 5). Indeed, the diagonal transgression towards the LC's angle without a pedestrian path could explain the motivation to avoid waiting at pedestrians' road traffic lights on the way to a nearby smaller car park (not illustrated on Fig. 5). Compared to Wynnum Central where the station is also on the middle island, at Coorparoo more transgressions occurred on the City train rail track than on the Outbound rail track. Therefore, the introduction of an external platform similarly to Cannon Hill could help improving the safety of City train passengers.

6 Limitations and Future Perspectives

A number of limitations can be addressed to the collected data and the adopted observations method. The presence of observers could unduly influence participants' behaviour. Indeed, it is possible that pedestrians have refrained from transgression in the presence of observers. Moreover, given that there is a legal sanction for crossing at red signal, such bias should not be underestimated. In terms of the adopted

procedure, data could not be considered as representative to the larger Queensland area, as the observations were conducted at only three LC sites and the data collection period was limited. Nevertheless, the results give an approximate indication on the number and proportion of transgressions at each LC site, given that observation sessions lasted for five hours per day. Also, the method facilitated gathering a detailed body of data, including description of potential risk prone crossing situations at LCs part of the riskiest Brisbane railway line, although not exhaustive. For instance, no indication was collected on the patterns of behaviour out of the two peak time zones. In addition, an estimation of the size of pedestrian flow, not only during the closures, would enhance the understanding of the proportion of transgressors among pedestrians crossing compliantly. Furthermore, a more in depth analysis of the characteristics of the respective populations at the three LC sites would enhance the understanding of high risk groups of pedestrians. Video data could provide complementary information on the proportion of transgressors compared to compliant pedestrians from each demographic group, and is therefore a potential path for future research. Moreover, the interactions between multiple factors could be further tested in simulation studies with the possibility to recreate various realistic crossing situations. Such studies are likely to provide a more in depth explanation of the precursors of behaviour and would therefore enhance the development of more effective safety measures (be it through safety campaigns aiming at the reduction of motivational factors, be it through updates of the environment improving pedestrian traffic conditions). Moreover, simulation studies would allow to pre-test the effects of already identified risk factors on a wider range of crossing situations (e.g. passive LCs, different active or passive controls).

7 Conclusion

The interactions between different factors were examined, contributing to the better understanding of the larger pedestrian crossing context likely to be influenced at the same time by the environmental properties of the LC, by personal motivations and characteristics of pedestrians themselves or else, by the presence of other individuals. As opposed to a large part of previous studies' emphasising on a single factor's contribution to unsafe crossing, this analysis of the interactions between factors illustrates potential highly "at risk" crossing situations, taking into consideration similarities and differences across typical for the area LC designs and socio-economic contexts. Arguably, the discussed interactions between risk-contributing factors suggest that independently of the LC site and its design, transgressions correspond to the fastest and most convenient path of accessing the platforms in order to catch a train. However, the analysis of the specific crossing context also reveals that such transgressions can be associated with a different level of risk-taking. In addition, transgressors at the three observation sites adopted different crossing trajectories likely to be associated not only with the design of each LC in terms of the location of the platforms and rail tracks, but also with characteristics of the larger area, notably in relation to the provided access points to the station's platforms. Thus, arguably, the role of characteristics of the larger area, such as the presence of car parks, road traffic lights, over bridges and main roads are often underestimated as potential risk-contributing factors to pedestrian crossing. Therefore, to improve safety, each LC environment should be optimised according to the characteristics of the area and the population.

Acknowledgments The authors are grateful to the CRC for Rail Innovation (established and supported under the Australian Government's Cooperative Research Centres program) for the funding of this research. Project No. R2.120: Understanding Pedestrian Behaviour at Level Crossings. The authors would like to thank Queensland Rail for their permission to conduct this study at the three selected level crossing sites. The authors would also like to thank all train stations' rail staff for their cooperation and diligence. The authors are grateful to Julian Pearce, Laurel English, Michael Leo and Wanda Griffin for their outstanding performance as assistant observers.

Appendices

Appendix 1

Transgressions observation sheet

Appendix 2

Train times observation sheet

Coorparoo – Train Times Date: 13 05 **Shift:** AM/ PM **Observer:** ____

CLOSURE N: 131

N	PL	Direction	TYPE	LC	STATION - STOP	STATION - GO
1	1	CITY	Exp Stop	08 / 12 / 21	08 / 12 / 14	08 / 12 / 09
	1	CITY	Exp Stop	_ / _ / _	_ / _ / _	_ / _ / _

N	PL			STATION-STOP	STATION - GO	LC
	2	CLEV	Exp Stop	_ / _ / _	_ / _ / _	_ / _ / _
	2	CLEV	Exp Stop	_ / _ / _	_ / _ / _	_ / _ / _

N	PL			LC		
	3	CITY/CLEV	Exp	_ / _ / _		
	3	CITY/CLEV	Exp	_ / _ / _		

CLOSURE N: 132

N	PL	Direction	TYPE	LC	STATION - STOP	STATION - GO
	1	CITY	Exp Stop	_ / _ / _	_ / _ / _	_ / _ / _
	1	CITY	Exp Stop	_ / _ / _	_ / _ / _	_ / _ / _

N	PL			STATION-STOP	STATION - GO	LC
1	2	CLEV	Exp Stop	08 / 19 / 00	08 / 20 / 11	08 / 20 / 15
	2	CLEV	Exp Stop	_ / _ / _	_ / _ / _	_ / _ / _

N	PL			LC		
	3	CITY/CLEV	Exp	_ / _ / _		
	3	CITY/CLEV	Exp	_ / _ / _		

133

References

1. Metaxatos P, Sriraj PS (2013) Pedestrian/bicyclist warning devices and signs at highway-rail and pathway-rail grade crossings in civil enginneering studies. Illinois Center for Transportation Series, Illinois
2. ATSB (2012) Australian rail safety occurrence data 1 July 2002–30 June 2012 in Australian rail safety occurrence data. Australian Transport Safety Bureau, Canberra, p 48
3. Queensland Rail (2012) Level crossings near miss table (2011). Queensland Rail website, Brisbane
4. Iorio L, De Marco S, Cosciotti E (2012) Life momentum at level crossing: human factor, road-rail safety policies, available technologies. A cross-sectorial challenge playing a rewarding role for the upgrade of safer mobility options, in Global Level Crossing and Trespass Symposium 2012. London
5. Edquist J et al (2009) A literature review of human factors safety issues at Australian level crossings. MUARC, Melbourne
6. Stefanova T et al (2015) System based approach to investigate pedestrian behaviour at level crossings. Acc Anal Prev 81:167–186
7. Edquist J, Hughes B, Rudin-Brown CM (2011) Pedestrian non compliance at level crossing gates. CURTIN-Monash accident research centre, Perth
8. Khattak A, Luo Z (2011) Pedestrian and bicyclist violations at highway-rail grade crossings. Transportation research record. J Transport Res Board 2250(1):76–82
9. McPherson C, Daff M (2005) Pedestrian behaviour and the design of accessible rail crossings. 28th Australian Transport Research Forum, Sydney
10. Parker A (2002) Second train coming warning sign demonstration projects. In: Administration FT (ed) Transportation Research Board of the National Academies. p 39

11. Sposato S, Bien-Aime P, Chaudhary M (2006) Public education and enforcement research study in safety of highway-railroad grade crossings series 2006. U.S Department of transportation, Federal Railroad Administration, p 110

12. Stewart S, Brownlee R, Stewart D (2004) Second train warning at grade crossingsIn: Canada T (ed) Transportation Development Centre, p 103

13. Basacik D, Cynk S, Flint T (2012) Spotting the signs: situation awareness at level crossings, in Global Level Crossing and Trespass Symposium, London

14. Clancy J, Kerr W, Scott M (2006) Study of pedestrian behaviour at public railway crossings. Lloyd's Register Rail Limited

15. Clancy J, Dickinson S, Scott M (2007) Study of pedestrian behaviour at public railway crossings. Lloyd's Register Rail Limited

16. Clark HE, Perrone JA, Isler RB (2013) An illusory size–speed bias and railway crossing collisions. Acc Anal Prev 55:226–231

17. Khattak A (2009) Comparison of driver behavior at highway-railroad crossings in two cities. Transportation research record. J Transport Res Board 2122(1):72–77

18. Wullems C, Toft Y, Dell G (2013) Improving level crossing safety through enhanced data recording and reporting: the CRC for rail innovation's baseline rail level crossing video project. Proceedings of the institution of mechanical engineers, Part F. J Rail Rapid Transit 227(5):554–559

19. Kotrlik JW, Williams HA, Jabor MK (2011) Reporting and interpreting effect size in quantitative agricultural education research. J Agric Educ 52(1):132–142

6

Attitudes of Metro Drivers Towards Design of Immediate Physical Environment and System Layout

Aleksandrs Rjabovs[1] · Roberto Palacin[1]

Abstract In this study, the authors examined attitudes of the Tyne & Wear (T&W) Metro drivers towards system design-related factors and their influence on the propagation of driver-related incidents. The system design features assessed include the position of running signals, visibility of different signal types, and platform location in relation to the travelling direction. The methodology based on data gathering through a self-administered questionnaire distributed among the drivers has been used. These data have been evaluated using multivariate analysis techniques against historic data on incidents to uncover potential relationships between drivers' perceptions and incident occurrence. The results show that the participants do not tend to consider system design factors as influential towards incident propagation. However, the analysis shows correlation between the driver responses and historical incident data such as corroboration of the increased incident propagation risks during the engineering works and the possessions.

Keywords Urban railways · Metro · Human–system interface · Human factors · Safety · PSF

✉ Aleksandrs Rjabovs
a.rjabovs@ncl.ac.uk

[1] NewRail – Centre for Railway Research, School of Mechanical and Systems Engineering, Newcastle University, Newcastle upon Tyne NE17RU, UK

Editor: Baoming Han

1 Introduction

Despite significant progress in the automation of different processes, railways still rely heavily on the performance of front-line staff, especially drivers. Being safety critical systems, the railways are assessed on their safety-related performance, and as such, the amount of incidents and accidents has serious consequences. Most incidents can be linked to the front-line staff as they still carry approximately 80 % of the risk in the industry [1]. It is accepted that numerous performance shaping factors (PSFs), which also include human factors (HF), can affect train drivers and influence incident propagation. The operation of the railways involves a variety of PSFs, including very important system design-related factors [2]. However, there appears to be no holistic understanding of the influence of system design-related PSFs on train drivers' performance. Even though the field of PSFs and HFs has been growing significantly recently, existing research still appears to be fragmented and studying a single railway physical environment aspect at a time.

Metro systems have been even less successful in attracting human factors & ergonomics (HF&E) specialists so far. It is important to treat metro systems separately from the mainline railways as they have certain differences affecting incident propagation and consequences. Considerably smaller variability of rolling stock, routes, track layout, and infrastructure in a closed system enhances drivers' route knowledge. The high capacity nature of metro systems leads to shorter headways and distances between stations thus increasing amount of signals and station stops encountered by drivers as well as risks of incidents associated with those. However, the use of automatic train protection (ATP) along with highly efficient brakes creates a risk profile that is different from that of the mainline railways.

This paper explores the Tyne & Wear Metro (T&W Metro) drivers' perception of causal factors behind some driver-related incidents. The paper provides an overview of drivers' attitudes towards different design-related PSFs which have been identified as potential latent failures in previous research.

The questionnaire surveys are an established approach to source attitude data from train drivers. Questionnaire studies have been carried out to investigate train drivers' motivations [3], organisational factors in incident reporting [4], influences of praise the drivers receive [5], physiological reactions to accidents [6], and effects of job stress on train drivers [7]. Self-administered questionnaires were used by Yum, Roh [8] to explore symptoms of post-traumatic stress disorder among Korean train drivers. Further questionnaire study by Jeon, Kim [9] showed that post-traumatic stress, in conjunction with sleep deprivation, is a major factor in human errors among train drivers. Effects of driver's chronotype on performance and quality of life had been assessed through a series of questionnaires by De Araújo Fernandes Jr, Stetner Antonietti [10]. Design of immediate physical environment, specifically cab environment, and train drivers' attitudes towards it had been assessed by Stevenson, Coleman [11]. They had used numerous mixed methods (quantitative and qualitative questions combined) questionnaires which asked train drivers to assess ergonomics and design changes of the new Tangara train in Australia. More examples of questionnaire used for evaluation of physical design can be found in automotive industry. For instance, attitudes on car design requirements have been studied among ageing demographics of drivers [12].

Section 2 briefly introduces the Tyne & Wear Metro system; Sect. 3 describes the methodology used including the questionnaire-based data gathering; and Sect. 4 introduce the results discussing them leading to some conclusions in Sect. 5.

2 Tyne & Wear Metro System

The Tyne & Wear Metro is located in the Tyne & Wear conurbation connecting Newcastle, Sunderland, Gateshead, South and North Tyneside. It is first opened in 1980 and mostly adapted existing heavy rail infrastructure. Today, the system spans 77.5 km and has 60 stations. The fleet consists of 90 twin-section articulated Metrocars, which currently run in pairs when in service. The rolling stock is currently undergoing its ¾ life refurbishment. The system has two routes. The South Gosforth to Pelaw section of the network is considered the "core of the system" as both routes pass through it.

The majority of the stations in the T&W Metro system are located overground. There are only eight underground stations in the network. Most stations have two platforms with a length suitable for two-car train sets. Some of the underground stations, the "legacy" stations adapted from the older heavy rail system, and some other stations have longer platforms. There are twelve line and service terminus stations in the Metro. Line terminuses have either a single platform or a layout allowing trains to arrive at any of the two available platforms. The service terminuses are used for short services and have turn-back facilities at a station or in sidings. The majority of stations on the network have a standard design with a running signal and dispatch equipment at the platform's edge. An example of such a standard design is shown in Fig. 1. However, some of the adapted metro stations have retained the previous heavy rail design, e.g. Tynemouth station has a canopy roof over the station and significantly wider platforms. Despite the standardised approach, many design aspects change from station to station, e.g. the point where passengers enter to the platform, the position and presence of a running signal, and the type of dispatch equipment.

The Tyne & Wear Metro operates on its own infrastructure as well as some sections using shared track with Network Rail. Thus there is a variety of signalling used, all fixed block. Most of the system has simple two-aspect signalling with occasional fixed distants and three-aspect signals. However, the Pelaw to South Hylton route uses Network Rail infrastructure and subsequently utilises standard mainline four-aspect signalling with yellow and double yellow signals. As of May 2014, signals using LED technology were only installed on the section shared with Network Rail as well as at the depot. The Metro drivers do

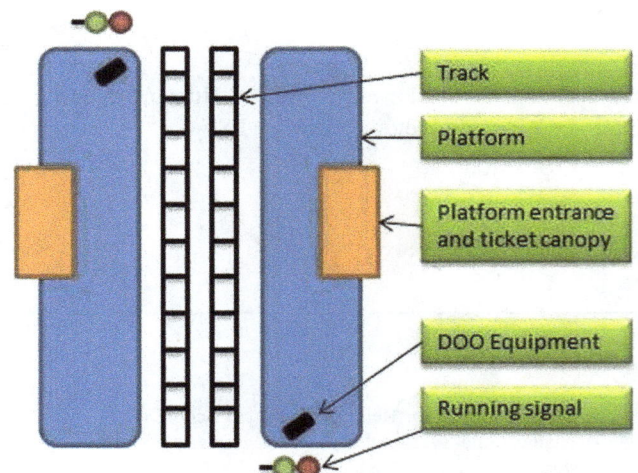

Fig. 1 An example of the standard station layout in the T&W Metro

not have the benefit of automatic warning system (AWS) or train protection and warning system (TPWS) available to the mainline train drivers on the same route. The automatic train protection (ATP) system controls overspeeding and signal passed at danger (SPaD). However, the speed control infrastructure is at certain locations only. The ATP system used is the Indusi system which is a version of German mainline railway warning and supervision system Induktives Sicherungssystem.

More information on the T&W Metro can be found in [13–15]. Fenner [16] describes the features of ATP used in the T&W Metro.

3 Methodology

The data for the study have been gathered using a custom-made questionnaire distributed among the T&W Metro drivers. Questionnaire surveys are an established practice to source attitude data. It has been extensively used in the railway industry in the past. The results were analysed against historical data to uncover relations of statistical significance. Descriptive statistics and the multivariate analysis have been used to explore underlying structures of the data collected.

3.1 Historic Incident Data

The questionnaire survey described in this paper followed up an analysis of the past incident statistics in the T&W Metro. The historic incident data included 1282 incident reports from the T&W Metro for a period between April 2011 and 2013. The focus of the past incident data analysis was on the location of driver-related incidents and potential system design-related factors affecting incident propagation. The driver-related incidents include signal passed at danger (SPaD), overspeeding, station overrun, failure to call, passenger entrapments, wrong-side door activation, and wrong-route incidents. Findings from the past incident data were used to create the questionnaire in order to source further data and reinforce some results. In summary, the incident data analysis revealed that in the T&W Metro, which is highly standardised in terms of the design of the physical environment (signals, stations and station infrastructure, track layout), an elevated rate of driver-related incidents occurs at locations deviating from a standard design. Some of the results are included in the discussion to provide more insightful analysis of the drivers' responses and whether their perceptions are similar to what incident statistics suggests.

3.2 Data Gathering

In the beginning of the questionnaire, the respondents were asked to assess a list of statements about their perception of different aspects of the system. They had to select an appropriate answer from a 7-point Likert scale (from strongly agree to strongly disagree). The scale used in the study is shown in Table 1. Due to the ordinal nature of the Likert scale, the descriptive statistics selected for the statements are mode and median. The statements along with the descriptive statistics are shown in Table 2.

Secondly, additional questions were asked in order to understand the perceived risks regarding passenger entrapment and wrong-side door incidents. The metro drivers had to mark the list of pre-selected PSFs based on their importance as a potential cause for each incident type. The marking scale was 1–10, with 1 being not important and 10 being very important. Due to the interval nature of the marks, the descriptive statistics selected for these questions were the mean and standard deviation. The mean marks for PSFs associated with wrong-side door incidents and passenger entrapment are shown in Tables 3 and 4, respectively. The respondents also had an opportunity to add other factors that they feel to be important.

Finally, this particular metro system allows for a direct comparison between different signal types as the drivers have to operate on both mainline and metro-only infrastructure. Hence, the drivers were asked to give a mark to some of the most frequent signal types in the system. Table 5 contains descriptive statistics for this comparison. Similar to the previous questions, there was a non-compulsory follow-up open question for the respondents to explain their choice of marks.

The drivers also were asked to tick the driver-related incidents they have been involved in the past 3 years. This information was later used to analyse whether previous involvement in certain incidents changes drivers' perception of effects arising from the physical environment. The Mann–Whitney U test has been used to compare the samples. It is important to note that samples for all the incident types, apart from wrong-side door activations, were significantly different in size. However, the U test does not require equal sample sizes [17].

The questionnaire study has been self-administered, but participants had means of contacting the authors if they had any issues. In total, 26 metro drivers participated in this study (17.3 % of all the T&W Metro drivers). Out of the 26 respondents, 23 were males and 1 female. Two respondents did not state their gender. 42.3 % of the participants have been metro drivers less than 3 years. The overwhelming majority of the respondents were aged between 26 and 35.

Table 1 The 7-point Likert scale used in the study

Strongly agree	Agree	Just agree	Not sure	Just disagree	Disagree	Strongly disagree

Table 2 The drivers were asked to assess statements on 7-point Likert scale

Statement	Mode	Median
My route knowledge of the Metro is good	Strongly agree	Strongly agree
My confidence reduces while driving during possessions or engineering works	Just agree	Just agree
The training provided for operations in degraded mode is adequate	Just agree	Just agree
The moment I enter or leave a tunnel, I feel more alert	Just disagree	Disagree
Running signals between the stations are easy to interact with	Agree	Agree
Running signals at the stations are easy to interact with	Agree	Agree
I am less alert if the outside physical environment is monotonous	Not sure	Just disagree
I prefer varied outside environment, such as a mix of vegetation and buildings	Just agree	Agree/ just agree
The recent change in door closing procedure from 2 to 1 button sequence is easier to operate	Just agree	Just agree
A 1-button sequence might increase the occurrence of passenger entrapment	Agree	Strongly agree
I like mirrors as station dispatch equipment	Just agree	Just agree
I like monitors as station dispatch equipment	Just agree	Just agree
When coming to a scheduled stop I pay attention to a running signal at the platform end	Agree	Strongly agree
Running signals located far from the platform end can make selection of a stopping position difficult	Disagree	Disagree
It is difficult to choose which side doors to open when station signals and the platform are on opposite sides	Just disagree	Disagree
Signalling at ground level can be confusing after driving a train in passenger service	Just disagree	Just disagree
The change of platform side does not affect my ability to select correct side to open the doors	Just agree	Agree/just agree
The stations differ a lot in terms of driver visibility of passengers on a platform	Strongly agree	Strongly agree
If the time between two stations is more than 2.5 min, this improves my alertness	Not sure	Not sure
I feel more alert when the time between stations is less than 1 min	Not sure	Not sure
I prefer steep change in speed limits rather than gradual change	Not sure	Not sure
My familiarity with operational protocols at sidings is very good	Agree	Agree
I have good familiarity with the layout of the depot	Agree	Agree
I find it harder to keep within higher speed limit than lower speed limit	Just disagree/ disagree	Disagree

Table 3 The drivers were asked to mark causal factors for wrong-side door incidents in terms of importance

Potential causal factor	Mark/10	SD
Attention lapse	8.19	2.433
Lack of reminders for drivers at stations	3.73	2.523
Layout of the door control	6.35	3.019
Distractions	7.88	2.142
Inadequate training	2.96	2.891

4 Discussion

4.1 Metro Drivers' Safety Performance During Engineering Works

Route knowledge is one of the most important skills of a train driver as a considerable part of their movement authority is often hidden from a driver's view [18]. Taking into account the more closed nature of metro systems in comparison to mainline railways, it is safe to assume that

Table 4 The drivers were asked to mark causal factors for passenger entrapment incidents in terms of importance

Potential causal factor	Mark/10	SD
Night time	5.85	3.283
Snow	2.96	2.375
Rain	4.27	2.647
Mist	5.96	2.793
Direct sunlight	8.04	2.490
Vegetation overgrowth	3.77	2.971
Location of station infrastructure, e.g. CCTV cameras	7.69	2.462
Overcrowding at a platform or in a train	8.73	1.733
Winter clothing on passengers	4.12	2.889
Shopping bags and suitcases	5.88	2.718
Layout of the stations	6.35	2.652
Design of passenger approaches	7.19	2.871
Mobility aid equipment, e.g. walking sticks, crutches	6.73	2.677
Station dispatch instructions/procedures used in the Metro	4.92	2.560
Low height passengers, e.g. children	5.42	2.656

Table 5 The drivers were asked to assess how easy to interact and interpret different signal types

Signal type	Mark/10	SD
Running signal on Network Rail infrastructure	9.42	1.332
Running signal on Tyne & Wear Metro infrastructure	8.54	1.529
Repeater	8.77	1.583
Advance warning signal	8.35	1.696
Flashing aspects	9.04	1.685
Ground position lights	7.62	2.041
Junction indicators	8.46	1.772

the route knowledge of the metro drivers is at a very high level. This is also supported by a relatively small amount of category A SPaDs in T&W Metro with a considerably higher frequency of the running signals than in the mainline railways. The drivers also believe that they know the system very well. Moreover, involvement in category A SPaD incidents does not make the drivers reconsider their route knowledge ($U = 21.5$, $p = 0.234$). On the other hand, the incident data analysis suggests that the increased familiarity with the day-to-day running causes a decrease in drivers' safety-related performance during non-routine operations. For example, engineering works and possessions affect the routine operation protocols in the system. The drivers' assessment corroborates a reduction in confidence in such operational conditions.

4.2 Operations at the Depot and Sidings

Even though driving into sidings/the depot is a frequent task for drivers working shifts around rush hours, it does not account for a significant percentage of their shift. Hence, when the large percentage of all category A SPaDs and the low percentage of the shift spent in these locations are considered, this suggests that there are issues in the

design of the locations and operational procedures. The signalling is the main difference at the depot and the sidings compared to the rest of the network with the majority of the signals being ground position lights. Even though the majority of the drivers disagree with added complexity of those locations and ground position lights, they gave this type of signals the lowest mark. Moreover, participants' involvement in category A SPaDs does not influence the attitudes of the drivers towards the ground position lights ($U = 28$, $p = 0.663$). The respondents mention that the ground position lights are hard to see at times due to low brightness and a poor choice of location. It is possible that the drivers do not see any hazard in this type of signalling in the context of day-to-day driving. However, when compared with different types of signals, they perceive the ground position lights as the hardest to interact with. It is also possible that the ground position lights are the hardest to interact with in the beginning of a shift before driving in passenger service.

4.3 Running Signals

Further investigation into the category A SPaD accidents in the T&W Metro showed a large proportion of these being start against a signal SPaD (SASSPaD) accidents. SAS-SPaD occurs when a train leaves a station against a red signal at the station's running signal. However, the drivers did not see the difference between running signals at and between the stations. Taking into account a very low number of SPaDs compared to the overall amount of signals encountered by the metro drivers in an operational year, it is possible that many drivers even do not notice any difference. However, the drivers previously involved in category A SPaDs are more inclined to evaluate both running signals at the stations ($U = 9.5$, $p = 0.032$) and between the stations ($U = 11.5$, $p = 0.051$) less positively. Moreover, the SASSPaDs are potentially caused by non-signalling-related factors as stations have more dynamic environment than tracks between those.

Analysis of the historic category A SPaD statistics also revealed that only 4 % of such incidents happen between Pelaw and South Hylton (mainline railway owned by Network Rail). This part of the network accounts for almost 25 % of the length of the system and 21 % of the signals in the system. This line differs from the rest of the network in several aspects. Those include bigger distances between stations and straighter alignment of the track. However, the biggest difference is use of mainline 4-aspect signals. These mainline signals use LED lights, whereas conventional light bulbs are used in the rest of the network. Exploring mean marks for the running signals, it is possible to claim that drivers prefer the running signals on Network Rail infrastructure. The drivers also expressed concerns that the signalling on the metro infrastructure does not give as much advance warning and is more prone to overgrowth. The incident data analysis returned no correlation between approach distance and any incident type propagation. The drivers also have not reported any effect on their alertness due to the distances between stations.

4.4 Effects of the Physical Environment on the Drivers' Performance

The results show that the drivers disagree slightly with negative effects of the monotonous physical environment but prefer a varied landscape around a train. Research from the automotive industry indicates that a long exposure to the monotonous physical environments negatively affects arousal levels of car drivers [19]. The metro drivers are exposed to such a physical environment frequently, for example, when driving through tunnels and walls of vegetation. In fact, locations associated with prolonged driving in a tunnel were among the worst performing in terms of driver-related incidents. It is possible that the respondents struggled to assess their alertness level in retrospect, considering that it decreases with time [20]. However, a change of physical environment is known to boost drivers'

arousal levels for a limited period of time [21]. Hence, alertness and safety-related performance were expected to rise in locations associated with the most extreme change of the physical environment in the system—tunnel exit or entrance. Conversely, the metro drivers disagree that their alertness rises at such moments. Furthermore, statistics from the historic incident data analysis support this statement. The locations associated with tunnel exits/entrances in the T&W Metro displayed an increase in incident levels.

One way in which a decrease in alertness manifests itself is via an increased rate of drivers' lapses. However, drivers' perception of importance of the attention lapse causal factor (Table 3) has no influence on their views on the effects of the physical environment on their state of alertness. None of the five related statement displayed any statistical significance in a Kruskall–Wallis H test with low (1–3), medium (4–7), and high (8–10) marks for the attention lapse factor as a categorical variable.

4.5 Station Overruns

Even though station overruns are usually associated with low rail adhesion (LRA), there were several incidents in the studied period which were outside the LRA season. An in-depth investigation of the locations revealed unusual positions of a running signal. Instead of being at a platform's end, the running signal at such locations was located further down the line. The respondents indicated that while attention is paid to the running signal at the platform while coming to a scheduled stop, their perception is that the two (the running signal and the stopping position) are not related. The station overruns have been historically treated as LRA-related incidents in T&W Metro which could have led to the participant not being able to consider other factors. However, the statements were worded in a way that does not make the association with a certain driver-related incident type obvious. Most respondents have never been in a station overrun incident outside the LRA season. On the other hand, the most important factor in choosing the stopping point for a driver is the ability to interact with driver only operations (DOO) dispatch equipment.

4.6 Passenger Entrapment Incidents

The T&W Metro uses both monitors and mirrors as DOO equipment. The responses indicate that these are equally liked and no preference has been highlighted. However, poor-quality monitors and inappropriate positioning of cameras have been mentioned as a causal factor often, but not as often as passengers' behaviour. Many passenger entrapment incidents in the T&W Metro are door misuse by passengers. Three of the four worst-performing stations, in terms of passenger entrapment, in the T&W Metro were the stations with very high patronage levels and monitors as the dispatch equipment. However, all of these stations have issues with the design of passenger approaches. The approaches to a platform at these stations are located outside of driver's view. Along with poor positioning of the cameras, it causes situations when the passengers, who run for a train, emerge suddenly into a driver's field of view. The respondents strongly agree that there is a significant discrepancy in terms of visibility of the passenger approaches throughout the system. Hence, the PSFs associated with cameras and the station design received very high marks as shown in Table 4.

Analysis of the time when passenger entrapment incidents happen in the Metro revealed a midday peak (12–3 pm). From this information and the description of the incidents, it was assumed that a higher percentage of elderly passengers creates additional entrapment risks. The drivers tend to agree with this giving the related causal factor fifth highest average mark. The high mark for direct sunlight does not align with the midday peak statistics as the sun is in zenith during that time. Hence, there is little risk of direct sunlight affecting the DOO equipment. However, morning and evening rush hours also have significantly elevated levels of passenger entrapments. This is when the sunlight causal factor can be the most important.

T&W Metro has recently changed the door closing procedure from two-button to one-button operation. In the past, the drivers had one more button to press after a door closing tone has sounded. This means that they were able to react if a passenger tried to misuse the doors by running for a train after the warning tone. The respondents admit that the new procedure left them with less control over a situation and increased the risk of passenger entrapment. This could be the reason behind a doubling of the number of passenger entrapment incidents between the two operational years studied.

4.7 Speed Control

The historic incident data analysis has revealed that the most problematic locations, in terms of overspeeding incidents, are predominantly stations with a lower than usual speed limit. The metro drivers agree that it is harder to keep to low speed limits than to the higher speed limits. It is not a case of perception of 10 km per hour (km/h) speed as very slow after driving at 80 km/h moments ago. The ATP system utilised by T&W Metro has a tolerance level of ± 2 km/h. The operator encourages the drivers to travel 5 km/h under the speed limit in locations with speed measurement magnets. Hence, at the 10 km/h locations, the drivers have to stay 50 % under the speed limit, whereas at the 80 km/h, the same decrease in speed accounts for less than 10 %. Moreover, when the problematic locations were compared, the worst incident rates were demonstrated by a

station with the fastest drop in speed limits. However, the drivers could not answer what rate of change of speed limits they prefer. It is possible that the statement was worded incorrectly with adjectives "steep" and "gradual" to define rates of changes.

4.8 Wrong-Side Door Activations

Similar to the overall driver-related incident statistics, wrong-side door activations have been localised to the stations with designs deviating from the standardised design. It could be an unusual position of a signal, Victorian built environment at the legacy stations or other deviation. Furthermore, stations with a different platform side, compared to a previous station, demonstrated increased levels of such incidents. The drivers disagree that either change of the platform side or an unusual positioning of a signal affects their door-opening duties. They scored the attention lapse and distraction high as the most important potential causes for this type of incident. Difference in the layout of the door control in different cars has been marked high too. However, when the respondents were asked what stations they believe to carry the highest risk and why, they mostly mentioned stations associated with the platform side change for the same reason.

Throughout the questionnaire, the respondents assessed their own performance very high stating that they do not get affected by different design-related factors. It is known that the respondents tend to be mildly positive with Likert-type questions [22]. Perhaps the drivers struggled to project the statements on themselves or were not fully convinced that the survey is fully anonymous. Moreover, lack of experience in being involved in particular incident type, and subsequent lack of retrospective analysis of an incident, affected drivers' answers. Mann–Whitney U test ($U = 35.5$, $p = 0.017$) shows that the drivers previously involved in the wrong-side door incidents tend to agree more that it is harder to choose the correct side doors to open when the platform side changes.

5 Conclusions

The data gathered in the survey provide a valuable insight into how drivers perceive design-related risks in the system. It is possible to claim that, in general, drivers do not perceive design-related factors to be those which notably affect them. They predominantly rated their performance to be independent from the effects of various features of the station design, track layout, and the signalling. However, when asked to compare various PSFs outside of a situational context, they are able to discriminate between those. Thus the respondents assign various risk levels to different

PSFs related to the same incident type. It is possible that the drivers struggle to associate themselves with the situations described in the statements.

Findings from the historic incident data analysis have been supported by the drivers' responses. Those include the effects of distance and change of physical environment on alertness levels, problems with ground position light signals and the passenger approaches at the stations, increased risks during engineering works and possessions. Moreover, the drivers agree on the increased risk of passenger entrapment due to new procedures, potential effects from the change of platform side, and the increased difficulty of keeping to a low speed limit.

The drivers, similar to the incident data analysis, do not agree on negative effects of long distances between the stations or positive effects from entering/exiting a tunnel. The drivers disagree that a monotonous outside physical environment affects their safety-related performance negatively but still prefer a varied physical environment.

The participants previously involved in some of the driver-related incidents seem to perceive the physical environment features associated with the incident type less positively. They are able to assess a situation presented based on their experience. Most likely they had to analyse an incident in retrospective as a part of the compulsory debrief by a safety manager.

6 Further Research

Differences have been found between effect magnitude suggested by the historic data and driver attitudes for some of the features of system design. Namely, additional attention needs to be focused on complexity of the sidings and the depot, factors involved in SASSPaD incidents, and effects of a running signal on a stopping position. Further steps are required to investigate whether the drivers do not perceive the physical environment as something that introduces performance shaping factors or the hypotheses drawn from past the incident data analysis are incorrect. Non-intrusive psychometric methods like eye-tracking or posture sensors can be useful for this type of investigation.

References

1. De Egea BG, Holgado PC, Suárez CG (2013) Humanscan®: a software solution towards the management of human reliability in the rail industry, in 4th International Conference on Rail Human Factors. CRC Press, London, pp 718–724
2. Gourlay C, Cole C, Rakotonirainy A (2013) Special Issue on work of the cooperative research centre for rail innovation, Australia. In: Proceedings of the Institution of Mechanical Engineers, Part F: Journal of Rail and Rapid Transit, vol 227(5), pp 405–406
3. Fujino H et al (2013) Study of train drivers' work motivation and its relationship to organisational factors in a Japanese railway company
4. Clarke S (1998) Organizational factors affecting the incident reporting of train drivers. Work Stress 12(1):6–16
5. Horishita T, Yamaura K, Kanayama M (2013) Study of effective praise in train driver's workplace, in 4th international conference on rail human factors, CRC Press, London, pp 614–620
6. Vatshelle Å, Moen BE (1997) Serious on-the-track accidents experienced by train drivers: psychological reactions and long-term health effects. J Psychosom Res 42(1):43–52
7. Chang HL, Yang CY, Wu JG (2005) Job stress and its influence on train drivers. J China Railw Soc 27(2):21–27
8. Yum BS et al (2006) Symptoms of PTSD according to individual and work environment characteristics of Korean railroad drivers with experience of person-under-train accidents. J Psychosom Res 61(5):691–697
9. Jeon HJ et al (2014) Sleep quality, posttraumatic stress, depression, and human errors in train drivers: a population-based nationwide study in South Korea. Sleep 37(12):1969–1975
10. De Araújo Fernandes S Jr et al (2013) The impact of shift work on Brazilian train drivers with different chronotypes: a comparative analysis through objective and subjective criteria. Med Princ Pract 22(4):390–396
11. Stevenson MG et al (2000) Assessment, re-design and evaluation of changes to the driver's cab in a suburban electric train. Appl Ergon 31(5):499–506
12. Herriotts P (2005) Identification of vehicle design requirements for older drivers. Appl Ergon 36(3):255–262
13. Powell JP, González-Gil A, Palacin R (2014) Experimental assessment of the energy consumption of urban rail vehicles during stabling hours: influence of ambient temperature. Appl Therm Eng 66(1–2):541–547
14. Howard DF (1976) Tyne and wear metro: a modern rapid transit system. Proc Inst Mech Eng Proc 190(18):121–136
15. Mackay KR (1999) Sunderland metro: challenge and opportunity. In: Proceedings of the ICE-Municipal Engineer, vol 133, pp 53–63
16. Fenner D (2002) Train protection. IEE Rev 48(5):29
17. Mann HB, Whitney DR (1947) On a test of whether one of two random variables is stochastically larger than the other. Ann Math Stat 18:50–60
18. Naweed A (2013) Simulator integration in the rail industry: the Robocop problem. In: Proceedings of the Institution of Mechanical Engineers, Part F: J Rail Rapid Transit, vol 227(5), pp 407–418
19. Thiffault P, Bergeron J (2003) Monotony of road environment and driver fatigue: a simulator study. Accid Anal Prev 35(3):381–391
20. Yang HK et al (2012) A study concerning analysis of arousal state of locomotive engineering during operating train. Trans Korean Inst Electr Eng 61(6):891–898
21. Keun Sang P, Ohkubo T (1994) A case study on the relationship between neuro-sensory work and work load. Comput Ind Eng 27(1–4):393–396
22. Gillham B (2000) Developing a questionnaire. Continuum, London

7

Driving Simulator Evaluation of the Failure of an Audio In-vehicle Warning for Railway Level Crossings

Grégoire S. Larue[1,2] · **Christian Wullems**[1,3]

Abstract It is impracticable to upgrade the 18,900 Australian passive crossings as such crossings are often located in remote areas, where power is lacking and with low road and rail traffic. The rail industry is interested in developing innovative in-vehicle technology interventions to warn motorists of approaching trains directly in their vehicles. The objective of this study was therefore to evaluate the benefits of the introduction of such technology. We evaluated the changes in driver performance once the technology is enabled and functioning correctly, as well as the effects of an unsafe failure of the technology? We conducted a driving simulator study where participants ($N = 15$) were familiarised with an in-vehicle audio warning for an extended period. After being familiarised with the system, the technology started failing, and we tested the reaction of drivers with a train approaching. This study has shown that with the traditional passive crossings with RX2 signage, the majority of drivers complied (70 %) and looked for trains on both sides of the rail track. With the introduction of the in-vehicle audio message, drivers did not approach crossings faster, did not reduce their safety margins and did not reduce their gaze towards the rail tracks. However, participants' compliance at the stop sign decreased by 16.5 % with the technology installed in the vehicle. The effect of the failure of the in-vehicle audio warning technology showed that most participants did not experience difficulties in detecting the approaching train even though they did not receive any warning message. This showed that participants were still actively looking for trains with the system in their vehicle. However, two participants did not stop and one decided to beat the train when they did not receive the audio message, suggesting potential human factors issues to be considered with such technology.

Keywords Railway crossing · Safety · Intelligent transport systems · Compliance · Driving simulation

1 Introduction

There are currently 23,500 level crossings in Australia [11], broadly divided into one of two categories: active level crossings with flashing lights; and passive level crossings controlling traffic solely with stop and give way signs. Crashes at level crossings continue to result in significant human and financial cost to society. According to the ATSB, there were 601 road vehicle collisions at Railway Level crossing (RLX) between July 2002 and June 2012 [1]. Many of these collisions between road vehicles and trains occur at level crossings with passive controls.

Closure of crossings, grade separation and installation of active protection at level crossings with passive controls are undoubtedly the most effective approaches to reducing the risk of collisions at railway level crossings. However, the feasibility of such approaches is questionable given economic and logistical implications [4]: it is impracticable to upgrade the 18,900 passive crossings due to various challenges such as the high number of private and

✉ Grégoire S. Larue
   g.larue@qut.edu.au

1 Centre for Accident Research and Road Safety - Queensland, Queensland University of Technology, Brisbane 4000, Australia

2 Australasian Centre for Rail Innovation, Canberra 2600, Australia

3 Cooperative Research Centre for Rail Innovation, Brisbane 4000, Australia

Editor: Marin Marinov

occupational crossings (13,000), the remote location nature of such crossings, the lack of power available on site and the difficulty to reach viable cost-benefit ratios for crossings with low road and rail traffic.

Several analyses have demonstrated that errors or violations on the part of the road user represent the largest contributor to RLX crashes [1, 12, 19], indicating the need for innovative interventions targeted at drivers to complement current railway interventions. The rail industry is therefore interested in new approaches to reduce the number of crashes at passive level crossings. One of them is to use Vehicle to Infrastructure (V2I) and Vehicle to Vehicle (V2V) communications to warn drivers about the approach of a train directly in their vehicles. Significant research has been conducted to study driver behaviour and responses to various traditional road-based interventions at different types of RLX. This resulted in some evidence to suggest the effectiveness of traditional approaches to level crossing safety, however, much more research is required to properly evaluate emerging technologies. Emerging technologies might be optimally used as a comparatively low-cost approach to increasing safety at passively protected crossings.

While in-vehicle technologies helping drivers perform the driving task become more pervasive, there is a lack of evaluation of the effects such interventions would have on driver behaviour if applied to railway crossings, particularly in the event of an unsafe failure of the technology [19].

When evaluating an intervention at railway crossings such as an in-vehicle Intelligent Transport System, it is important to ensure that such systems are efficient in attracting drivers' attention and ensure compliance [16, 19]. In Australia, passive crossings with 'Stop' signs require the driver to completely stop their vehicle at the appropriate road marking and to look for the presence of a train in both directions. If a train is entering the railway crossing or an approaching train is visible from the crossing, the driver is expected to give way to the train. After the train has cleared the crossing, and no other trains are approaching, the driver may traverse the crossing. Also, such systems must be developed to avoid over-reliance and subsequent reductions in the performance of protective crossing behaviours, such as stopping compliance rate, monitoring approach speeds and scanning the crossing environment [2]. Indeed, familiarity with a level crossing has been shown to reduce the perception of risk, and encourage drivers to engage in risk-taking behaviours. This is particularly the case for passively protected crossings, where regular road users encounter trains infrequently and tend to know the train timetables [3], leading to reduced number of visual checks (a factor which was attributed to one of the fatalities analysed in [18]).

Various studies have been conducted in order to evaluate the effect of various railway level crossings interventions on driver behaviour, such as traffic lights [9], rumble strips and in-vehicle systems [6, 17]. Traffic signals at railway level crossings do not appear to offer any safety benefits over and above flashing red lights, and rumble strips seem to be effective in reducing approach speed but not compliance at passive crossings. On the other hand, in-vehicle interventions tend to result in driver behaviour similar to active crossings, which result in higher compliance when trains are approaching railway crossings. Among in-vehicle interventions, an audio message has been shown to be more effective than visual information on a GPS-like device [6]. With an audio message, drivers tended to use the in-vehicle warning as assistive information, while using visual feedback resulted in drivers using the in-vehicle system as the main control of the crossing, disregarding the STOP sign present at the crossing. Research has also shown that drivers exposed to in-vehicle warning system for railway crossings held a strong intention to use such technology if available, particularly in the context of passive crossings. Such positive intention was largely explained by the fact that such technology was perceived to be useful, easy to use and socially acceptable [7].

Such studies provide a good rationale for the development of in-vehicle interventions, but they are lacking an evaluation of the potential familiarisation of road users with the technology, which could then result in driver complacency. It is the aim of this study, which familiarises participants with a new audio in-vehicle technology for an extended period of time, and then trials an unsafe failure of the technology.

The objective of this paper is to provide a human factors study evaluating operational performance of an audio in-vehicle message warning participants about approaching trains at railway level crossings using a driving simulator. In particular, this study will evaluate driver behaviour in case the technology fails to provide warning of the approaching train. A simulation-based study was conducted to address these questions through the development of specific scenarios in a controlled simulation environment. This approach allowed evaluation of the following research questions with statistical power:

- What is the current performance of drivers at railway level crossings with passive controls (RX2)?
- What changes are evident in driver performance once the technology is enabled and functioning correctly?
- What is the road user performance when the audio ITS does not provide the warning message about the approach of a train (failure mode of the technology) while approaching a level crossing.

This was done through the observation of changes of driver behaviour in terms of compliance, gaze patterns, approach speed and safety distances. Questionnaires were also used to obtain feedback from drivers after they trialled the technology.

2 Experiment

2.1 Trialled In-vehicle Intervention

The audio in-vehicle warning system implemented in this study used the speakers of the simulator positioned inside the vehicle. When the driver was within 200 metres of the crossing and a train was within 20 s of the crossing, which is the minimum warning time required in the Australian standard (AS1742.7) for single track crossing [15]. At activation, the system provides the following audio stimuli:

- The sound of a warning bell similar to the bell used at Australian railway level crossings; followed by
- A spoken message: ''Warning. Train at crossing. Prepare to stop.''.

This sequence was repeated up to three times until the speed of the driver was below 20 kilometres per hour.

2.2 Experimental Design

This study was designed to assess the changes to road user behaviour when the warning fails and the RX2 (passive level crossing control with stop sign and cross-bucks) is the primary control at a level crossing. The familiarisation phases were designed to enable participants to initially become accustomed to the simulator and to provide them with an environment in which they are exposed to RX2 crossings before and after the in-vehicle warning system was operational. Participants were only told that the vehicle was fitted with an in-vehicle system providing audio messages when they were required to stop at a crossing due to the presence of a train. Data were not analysed during these phases.

A within-subjects design was used in order to evaluate the effects of familiarisation on driver decisions with the technology. The study was composed of two sessions, which were composed of 12 and 13 repetitions of a similar scenario respectively, varying in terms of presence of the technology and presence of a train. The crossing with train approaching was varied between subjects. Details of these scenarios are provided in Table 1 for the first session and Table 2 for the second session. The first half of the first session was a baseline of driver behaviour at traditional passive crossings with a Stop sign. The second half was used to introduce participants with the operational

in-vehicle technology. The second session was also using the audio in-vehicle warning. This session focused on creating a habit where the technology worked properly, until it failed to provide the warning message as a train was approaching (trial 2e) in a way that the driver could not be aware of the failure until they saw the train (unsafe failure of the system).The technology started failing in the middle of the second session, but the participant could not have noticed the change, as no train was approaching crossings. During the last trial, the train was programmed to arrive at the crossing 12 s after the driver reached a full stop at the level crossing.

This type of study had to be implemented in a driving simulator, as the effects of unsafe failures of the technology were investigated with participants. On-road evaluation would not be possible due to ethical and safety issues. The rural level crossing located on Lane road, Lanefield, Australia was replicated in the driving simulator environment (see Fig. 1).

A power analysis was used to determine the number of participants required for this study. The required sample size N to obtain a significant result at level $\alpha = 0.05$ for a power β when comparing the mean of a variable of interest at railway crossings with and without the intervention is given by [5], assuming similar variance

$$N = \frac{2(u + v)^2}{d^2}$$

where $u + v = 2.8$ for $\beta = 0.8$ and $u + v = 3.24$ for $\beta = 0.9$ and $d = (\mu_1 - \mu_0)/\sigma_0$ is the standardised mean difference.

For detecting high size effects ($d > 1$), 16 participants are required at power $\beta = 0.8$ and 21 participants are required for a power $\beta = 0.9$.

2.3 Participants

Twenty subjects volunteered to participate in this study. Five subjects (4 females and 1 male) were not able to complete the study due to motion sickness in the driving simulator, resulting in a sample size of 15 participants (11 males and 4 females) aged between 27 and 53 (mean age = 37.5 years, SD = 7.0). Participants were recruited from the Queensland University of Technology (students and staff members) as well as from the general public. Participants had an open Australian driving licence. All subjects provided written consent for this study, which was approved by the Queensland University of Technology ethics committee. Participants were paid AUD $100 for completing the two driving sessions; and AUD $20 in case they were not able to complete the study.

Table 1 Design of the first session	Session 1: RX2 and ITS
	1a. Familiarisation (3 RX2 crossings and one train at crossing)
	1b. Five trials with 3 RX2 crossings and no train approaching
	1c. Familiarisation with 3 RX2 crossings with audio ITS and one train at crossing
	1d. Five trials with 3 RX2 crossings with audio ITS and no train approaching

Table 2 Design of the second session

Session 2: RX2 and ITS (partial coverage/failure)
2a. One trial with 3 RX2 crossings with audio ITS and one train at crossing
2b. Five trials with 3 RX2 crossings with audio ITS and no train approaching
2c. One trial with 3 RX2 crossings with audio ITS and one train at crossing
2d. Five trials with 3 RX2 crossings with the audio ITS failing and no train approaching
2e. One trial with 2 RX2 with the audio ITS failing, no train approaching. A third RX2 with the audio ITS failing and a train approaching

Fig. 1 Typical Australian passive crossing as implemented in the simulator

2.4 Ethical Clearance

Ethical clearance for this research was obtained from the QUT Human Research Ethics Committee (reference number 1300000298).

2.5 Procedure

Each participant took part in two sessions taking approximately 2 h each. During each session, the participant was performing a repetitive succession of simulated driving tasks consisting of driving an itinerary from start to end. The itinerary was a typical rural Australian road, with 3 passive railway level crossings with stop signs and two road intersections (one with a stop sign, one with right-of-way).

Upon arrival, participants provided written consent to participate in the study, and were then provided with a short familiarisation drive in the simulator allowing them to become accustomed to accelerating, stopping and driving though intersections, passive railway crossings and curves.

In the middle of the first session, the audio system was implemented in the driving simulator, and the system was presented to the participants. The message was played to the participant, and they were then given a familiarisation drive with the ITS switched on to enable them to feel confident whilst driving with the system activated. Participants subsequently drove with the system active until the end of both sessions.

The second session took place on another day, or a couple of hours after the participant finished the first session. At the start of the second session, the participant was reminded that the car was equipped with a system giving audio messages when trains are approaching a level crossing.

While participants were driving, a researcher was manually recording head movements as drivers were driving toward railway crossings. The researcher had a table where he could tick the following categories: both sides checked, only one side (left or right) checked and no sides checked. From these observations, compliance was defined as checking both sides of the crossing before going through the crossing.

Surveys were administered post-trials for the human factors evaluation study. The questionnaires assessed driver demographics, general driving experience and exposure to passive crossings in Australia, and experience of the technology in accordance with the Technology Acceptance Model. They were administered to participants directly as a paper version at the end of the second session. Questionnaires took 10–15 min to complete.

2.6 Materials

2.6.1 Driving Simulator

The study was conducted on CARRS-Q's Advanced Driving Simulator (see Fig. 2). The simulator included a complete Holden Commodore vehicle with automatic transmission, working controls and instruments. The simulator is based on SCANeR^{TM} studio software (Oktal) with eight computers, three projectors and a six degree of freedom motion platform. When seated in the simulator vehicle, the driver is immersed in a virtual environment which includes a 180 degree front field view composed of three screens, simulated rear view mirror images on LCD screens, surround sound for engine and environment noise, real car cabin and simulated vehicle motion. The road and surrounding environment were designed to represent, as closely as possible, a rural railway crossing in accordance with Australian Standards at railway crossings (see Fig. 1).

2.6.2 Questionnaires

A general questionnaire was first given to participants to cover general demographics such as age, gender and driving experience, with particular focus on railway level crossings (with a distinction between active and passive crossings). Experience of incidents at railway crossings in

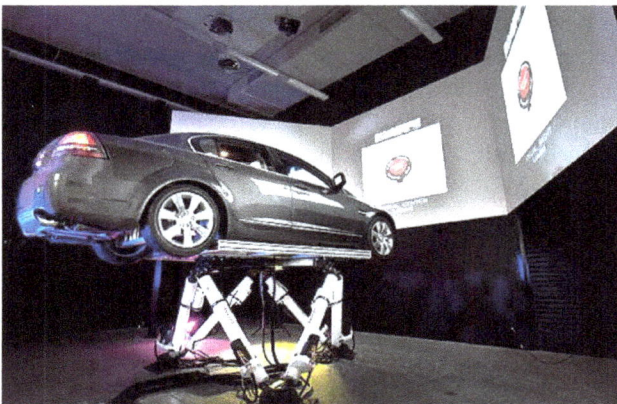

Fig. 2 CARRS-Q's advanced driving simulator

the last 5 years was also investigated. Questions gauging such experience included

- You thought there was no train approaching the crossing, but when you could not stop anymore, you realised a train was approaching.
- The signal at the crossing was off, but when you could not stop anymore, the signal started flashing.
- You got stuck on the crossing because of the traffic in front of you.
- You had a collision with a boom gate.
- Other. Please describe.

As this study focuses on malfunctions of technology at level crossings, participants were also asked about their perception of likelihood of signal failure at crossings; their expected behaviour if the flashing lights were always on and no train was coming or if the lights were always off, even if a train was approaching. Participants were also asked about their perception of the issues at level crossings, and how predictable are trains arriving at crossings in their area. Knowing the times when trains are present at crossings is a known contributing factor to crashes at crossings, which can also be correlated with their perception of the usefulness of the system.

Questions were then presented about the technology trialled. It focused on the Technology Acceptance Model, which targeted the perceived usefulness and the perceived ease of use of the technology. Perceptions of advantages and disadvantages of the system were also investigated. The failure of the system was also of interest in this questionnaire, with questions such as: How often would this system have to fail for you to consider not useful anymore?

We also asked to what extent the system would change their behaviour, as participants trialled the system and should have a good idea of the potential effects of such system on their driving behaviour. Changes in their perception of system usefulness after they saw it failing were also investigated.

2.6.3 Software

Data analyses were conducted with the statistical package R, version 3.1.1.

3 Data Analysis

Data analysis was performed on data collected from the advanced driving simulator as well as surveys conducted after the simulator study.

The simulator study comprised a baseline data collection with traditional RX2 signage at the railway crossings, followed by the introduction of a new audio in-vehicle technology designed for railway level crossings. After a familiarisation with the technology, the failure of the technology was trialled during the last run of the second driving session. The driving simulation provided a controlled environment that allowed participants to experience the system working correctly, familiarise themselves with the system and then facilitated assessment of their behaviour once the system was failing.

The dependent variables under investigation in this data analysis were the following:

- The variation of the approach speed when no train was approaching the crossing. By comparing to the baseline, the perception of the level crossing as being actively or passively protected can be inferred. For the driving simulator, speed profile can be measured at any distance to the crossing, and speed 20 and 40 m away from the crossing were investigated, as they have been shown by various studies in the literature to be distances where speed varies for passive crossings. Speed 150 m to the crossing was also investigated in order to ensure that no differences were present before approaching the crossing;
- Compliance with the crossing signage;
- Gaze behaviour when participants approach a railway crossing;
- Safety distances: reaction time distance, minimum possible braking distance and safety margin, as defined in [8] and
- Subjects' perceptions of the systems from the surveys.

The analysis evaluated how the following independent variables (type of signal at the RLX) had effects on the dependent variables presented above:

- RX2 with no train;
- RX2 with the in-vehicle audio warning with no train and
- RX2 with the in-vehicle audio warning failing while a train is approaching the crossing

Statistical analyses were conducted to evaluate the effects of the failure of the technology. Generalised Linear Mixed Modelling was used in order to take into account the repeated measures study design.

4 Results

4.1 Driving Simulation

4.1.1 Observed Driving Behaviours

The following behaviours have been observed in the driving simulator when the audio in-vehicle message failed to be played by the vehicle:

- Stopped without any issues (12 participants);
- Decided to beat the train (1 participant);
- Did not change their behaviour and did not stop at the crossing, without any indication as to whether they realised a train was approaching with the in-vehicle system failing (2 participants).

With the participants completing the study, we obtained 80 % of participants stopping without any problems when the system failed to provide the warning message; and 7 % saw the train and decided to beat it. It was not possible to ascertain the behaviour of 13 % of participants as they were not complying before the occurrence of the failure.

4.1.2 Effects on Approach Speed

Participants had a similar approach 150 m to the crossing. The in-vehicle technology is used only to increase awareness of drivers while approaching crossings, and participants were expected to stop the same way as for the RX2 signage. Speed closer to the crossing were very similar with the ITS system implemented, whether it was working or failing with a train approaching (see Table 3).

The average speed 150 m to the crossing was 72.5 kph during the baseline with RX2 signage. No differences were observed with the in-vehicle warning provided. It has also to be noted that speed increased with the number of trials in a parabolic manner with a factor .02 ($p < .001$). This means that on average, participants speed increased by 2.9 kph by the time they reached the last trial run.

The same trend was observed 40 and 20 m to the crossing. At these locations, speed was on average 33.8 and 19.3 kph, respectively. Such value held with or without the technology. Speed also increased with the number of runs in a parabolic manner with factor .02 ($p < .001$).

4.1.3 Effects on Stopping Compliance

The stopping probability observed in the driving simulator for the different sign conditions is reported in Table 4. On average probability of stopping with the RX2 signage was 69.8 %. Statistical analyses for repeated measures were conducted and showed that the probability of stopping changed with the introduction of the in-vehicle audio warning. It decreased to 53.4 % ($p < .001$) showing that participants started to rely on the technology for their decision-making in this mode of operation. While participant compliance reduced with the technology, the proportion of participants stopping at the crossing with the train approaching and the in-vehicle system not working was very high (80 %) and suggest that participants were still actively analysing the situation at the crossing with this technology installed (at least for the time involved in this trial; this could

Table 3 Approach speeds (kph, standard deviations in brackets)

Level crossing	150 m	40 m	20 m
RX2, no train	71.3 (7.4)	33.9 (13.2)	18.4 (16.6)
RX2, ITS, no train	72.9 (9.7)	34.6 (12.9)	18.9 (15.7)
RX2, failing ITS, train approaching	71.2 (9.8)	32.5 (16.8)	21.2 (19.5)

Table 4 Stopping probability (standard deviation in brackets)

Level crossing	Stopping probability
RX2, no train	69.8 % (.46)
RX2, ITS, no train	53.4 % (.48)
RX2, failing ITS, train approaching	80.0 % (.38)

change with familiarity with the system over longer periods of time). We did not observe any participants failing to detect the train with the system failing.

4.1.4 Effects on Head Movements Toward Rail Tracks

Rates of gaze compliance in the different conditions are reported in Table 5. Statistical analyses (taking into account repeated measures from the same participants) showed that participants checked on average both sides of the rail track 96.9 % of the time, independently of the factors under investigation in this study. No statistically significant differences were observed with the technology in place.

4.1.5 Effects on Safety Distance

Reaction time distance, minimum braking distance and safety margins in the different conditions are reported in Table 6. Statistical analyses (taking into account repeated measures from the same participants) showed that all of the factors under investigation in this study had no statistical effect on these variables (Fig. 3).

4.2 Feedback from Drivers

4.2.1 Positive Feedback

Participants were asked about the advantages they saw in using such an in-vehicle system, and their feedback is presented in the following subsections.

Table 5 Compliance in gaze behaviour

Level crossing	Compliance
RX2, no train	100 % (–)
RX2, ITS, no train	96.5 % (.18)
RX2, failing ITS, train approaching	93.3 % (.26)

4.2.1.1 Additional Support as a Complementary System
The advantage most reported by participants (6 participants) is that this system can be used as a backup system, which would support drivers approaching and traversing crossings. One male participant (46 years old) reported that this system acts as a *backup system for events where driver inattention or drowsiness has the potential to result in a collision with a train*. A female (29 years old), reported that this system is a *safety net for driver distraction*, while another one (53 years old) said that *it is reinforcement; the two together would probably help. However, failure rate would have to be nil*.

4.2.1.2 Better Awareness at the Crossing and Better Visibility of the Crossing
Five participants felt that this technology would increase their awareness when approaching a crossing. A male, 36, reported that this equipment is *certainly useful in providing awareness amongst drivers who are paying less than adequate attention, or drivers who are fatigued*. Another one (27 years old) wrote that *it provides an additional cue and hence improves safety*. A female participant (34 years old) added that this system made her *aware of train crossing and made her more careful when crossing train track*.

Participants also reported potential positive effects in case visibility is reduced while approaching the crossing, such as in night time conditions: *if it works, it's a helpful aid particularly at night time when vision may be impaired if passive crossing*.

4.2.1.3 Mitigation of Driver Inattention
This system is particularly perceived as having benefits for inattentive drivers, whether due to sleepiness or distractions. It is *useful for drivers who may be tired or are momentarily inattentive. It could save their life if they miss the stop sign* (male, 43). A female participant (42) reported that this *additional audio warning is harder to ignore even if your eyes are heavy*. It was perceived as very useful if integrated within the car: *This may help some people, especially if there are other distractions in the car. This should be built into the radio in the car so it temporarily mutes the music to make it very obvious* (male, 35).

4.2.1.4 Ease of Use
It was perceived as an easy system to use, which could be useful in rural areas rather than metropolitan areas where railway crossings are already

Table 6 Safety distances (standard deviation in brackets)

Level crossing	Reaction time distance	Minimum braking distance	Safety margin
RX2, no train	88 (54)	28 (7)	130 (36)
RX2, ITS, no train	92 (53)	29 (7)	124 (35)
RX2, failing ITS, train approaching	84 (29)	31 (14)	132 (39)

Fig. 3 Effects of the technology on safety distances

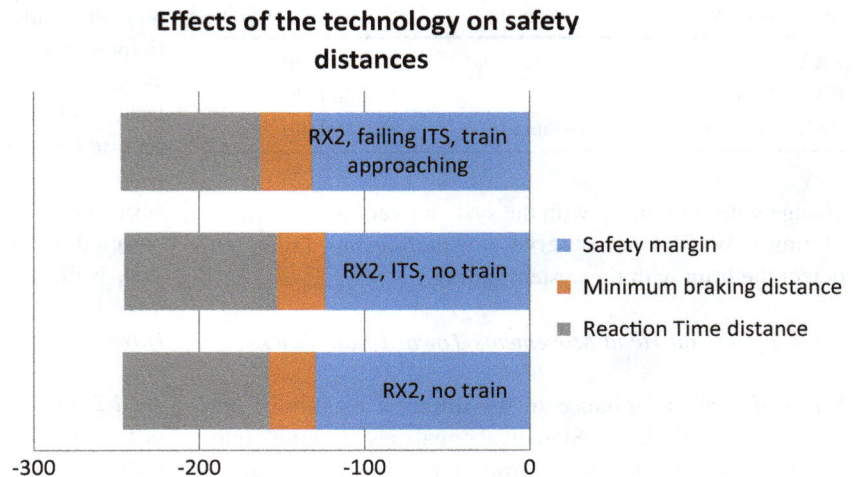

Effects of the technology on safety distances

equipped with flashing lights and boom gates. It was perceived useful particularly if it was *integrated into other technological aids already being used, such as GPS* (male, 36).

4.2.2 Negative Feedback

After experiencing the failure of the technology, participants also reported a number of potential issues with such systems.

4.2.2.1 Facilitating Complacency The highest concern was related to complacency, with 10 participants saying that such system could lead to complacency: *many people may depend on the technology and fail to stop or may not stop on time when a train is actually coming and the technology failed to activate.* Participants felts that drivers may *rely too much on the system and not use their eyes and ears for situational awareness*, which highlights the potential risk of reduction of active checks while approaching a crossing, once used to having a reliable warning message in the vehicle. One female participant reported that *drivers could become reliant on it and less attentive to visual warnings such as signs, road markings,* etc. It could result in *a tendency to perhaps not stop at a crossing, despite signs, if the alarm does not sound* (male, 36). Implementing such technology could *encourage people to rely on it which could be dangerous if it fails* (male,

42), which could even make *crashes happen more often* (male, 32).

4.2.2.2 Cost Participants reported that this system should only be considered as a backup system that should only *complement traditional solutions in place such as lights or stop sign* (male, 43). However, the risk that drivers use this system as the primary control when approaching the crossing suggests that such technologies would be *a good initiative if it works 99.99 % (at least)*[1] of the time and if it fails in a safe manner, as for flashing lights installed at level crossings. This raises concerns about the cost such technology would have on the driver, as the *cost to install such system in a car could be expensive*, and could require *maintenance and repairs by the owner*.

4.2.2.3 Potential Failures Experiencing the failure of the technology in the last trial resulted a decrease in the perception of the usefulness of the technology for 47 % of participants, which shows the need for a reliable system which would fail in a safe manner.

4.2.2.4 Annoyance This device could also annoy drivers if it was activated quite often: *it will become so annoying if the warning comes up too often* (male, 36).

---

[1] This is only opinion from the participant. Safety reliability analysis is required to determine how reliable equipment needs to be.

5 Discussion

This driving simulator study has shown that with the traditional passive crossings with RX2 signage, the majority of drivers complied (70 %) by stopping at the crossing and all participants looked for trains on both sides of the rail track. A high number of participants (30 %) did not comply at passive crossing, and this is consistent with observations made at the Lane road, Lanefield crossing, which the scenarios in the driving simulator were based on. This can largely be explained by familiarity with the crossing and the high visibility that is characteristic of this crossing (sighting distance above 1.5 km in each direction).

With the introduction of the audio in-vehicle warning for railway level crossings, no differences were observed for approach speeds, gaze behaviour and safety margins. Drivers did not approach crossings faster and did not reduce their gaze towards the rail tracks. However, participant compliance at the stop sign decreased with the technology installed in the vehicle: stopping compliance reduced to 53.5 %, which is a 16.5 % reduction in stopping compliance.

The effect of the failure of in-vehicle audio warning was also assessed with the simulator study. It showed that most participants did not experience difficulties in detecting the approaching train even though they did not receive any warning message. This showed that participants were still actively looking for trains with the system in their vehicle. Most participants stopped, but two participants did not (it was their usual behaviour at the crossing without trains after familiarisation) and one decided to beat the train. Such behaviour suggests that these participants might have been re-enforced in their decision to go through the crossing by the lack of message from the system. Two of these three participants have interpreted that it was safe to cross, as they later reported that they did not notice that the technology had failed. However, if the crossing was actively protected, it would have been closed (flashing lights). Further, drivers are required by law to not proceed through level crossings when a train is visible. This highlights the need for such technology to be failsafe.

The characteristics of the level crossing in terms of sighting distance made it easy for participants to see the train. Further research should be conducted in conditions where the train could be seen only when closer to the crossing (by reducing the sighting distance to the minimum distance required in the Australian standard). Further, this study induced a certain level of familiarity with driving through crossings with the technology, using two repetitive sessions. This can be confirmed by the reduction in stopping compliance, which suggests that drivers started to rely on the technology. However, the study was of a limited time, and effects on a longer term could not be estimated. The reduction of compliance at the crossing with the introduction of the technology as well as the feedback obtained from participants at the end of their trial suggest that with time drivers may start to largely rely on the technology if they perceive it reliable. This could potentially reduce the ability of drivers to detect approaching trains when the system fails. The decrement of compliance with the introduction of the technology should be carefully taken into consideration if such system was to be implemented at crossings. Reliability and integrity of the technology should be analysed and used to model risks involved with the implementation of such technology. In particular, overall effects on safety at each individual crossing should be evaluated in order to ensure that risk does not increase.

This study also suggests that in the long term, there could be negative effects on driver performance of introducing such technology. Hence the system needs to be able to manifest its failure modes to the driver in a clear manner. This could be mitigated by providing drivers with interfaces that display contextual information about the reliability of the systems, as it has been shown that complacency becomes less likely when information about the reliability of the instructions is provided [13]. Various methods have shown promising results for reducing complacency, such as displaying dynamically the system's confidence in its recommendations [10] or making the automation failure more salient [14]. Other human factor issues, such as the effects the availability of the system at some crossings and not others, while outside the scope of this study, should be considered if such device was to be implemented.

6 Conclusion

This driving simulator study has shown that the current performance of driver behaviour at traditional passive crossing with a stop sign is low after being familiarised with a crossing with low train traffic and with high sighting distances. This is similar to the behaviour that can be seen on roads with this type of level crossings. The introduction of the audio in-vehicle warning resulted in a reduction of stopping compliance at the stop sign, suggesting that drivers tend to use such system not only as an assistive device but more as a primary control. The warning device has no significant effect on gaze behaviour, approach speeds or safety distances. Participants were still actively checking the rail environment in order to detect trains, and drivers were all able to detect the approaching train when the in-vehicle system was failing unsafely. However, two participants did not stop and one decided to beat the train when they did not receive the audio message. Two of these participants did not realise the technology was failing, and they decided to proceed through the crossing, as the technology suggested it was safe to do so. This study has highlighted potential complacency issues with the introduction of new in-vehicle technologies for railway level crossings, both with the driving performance observed after familiarisation with the crossing, and self-reports from participants at the end of the driving task. This shows the need for taking into consideration the failure modes of such in-vehicle systems during the engineering design phases and also design how to provide this information to the road users.

Conflict of interest   The authors declare that they have no conflict of interest.

References

1. Australian Transport Safety Bureau (2012) Australian rail safety occurrence Data 1 July 2002 to 30 June 2012. In: Australian Transport Safety Bureau AG (ed) Australian Transport Safety Bureau. Statistical Publication Investigation number: RR-2012-010
2. Caird JK, Creaser JI, Edwards CJ, Dewar RE (2002) A human factors analysis of highway-railway grade crossing accidents in Canada. Transp Canada, Canada
3. Caird JK, Smiley A, Fern L, Robinson J (2011) Driving simulation design and evaluation of highway–railway grade and transit crossings. In: Fisher DL, Rizzo M, Caird JK, Lee JD (eds) Handbook of driving simulation for engineering, medicine, and psychology. CRC Press, Hoboken
4. Cairney P (2003) Prospects for improving the conspicuity of trains at passive railway Crossings. *Road safety research report.* Australian Transport Safety Board, Vermont South
5. Kirkwood BR, Sterne JAC (2003) Essential medical statistics. Blackwell Science, Malden
6. Larue GS, Kim I, Buckley L, Rakotonirainy A, Haworth NL, Ferreira L (2014) Evaluation of emerging Intelligent Transport Systems to improve safety on level crossings—an overview. In: 2014 Global level crossing symposium, 4–8 August 2014, Urbana
7. Larue GS, Rakotonirainy A, Haworth NL, Darvell M (2015) Assessing driver acceptance of Intelligent Transport Systems in the context of railway level crossings. Transp Res F 30:1–13
8. Lay MG (2009) Handbook of road technology. Spon, New York
9. Lenné MG, Rudin-Brown CM, Navarro J, Edquist J, Trotter M, Tomasevic N (2011) Driver behaviour at rail level crossings: responses to flashing lights, traffic signals and stop signs in simulated rural driving. Appl Ergon 42:548–554
10. Mcguirl JM, Sarter NB (2006) Supporting trust calibration and the effective use of decision aids by presenting dynamic system confidence information. Hum Factors 48:656–665
11. Railway Industry Safety and Standards Board (2009) Level crossing stocktake
12. Railway Safety Regulators' Panel (2008) Review of national level crossing statistics
13. Rovira E, Cross A, Leitch E, Bonaceto C (2014) Displaying contextual information reduces the costs of imperfect decision automation in rapid retasking of ISR assets. Hum Factors 39(6):1581–1584
14. Seppelt BD, Lee JD (2007) Making adaptive cruise control (ACC) limits visible. Int J Hum Comput Stud 65:192–205
15. Standards Australia (2009) Manual of uniform traffic control devices, Part 7: railway crossings, 3rd edn. Standards Australia, Sydney
16. Tey L-S, Ferreira L, Wallace A (2011) Measuring driver responses at railway level crossings. Accid Anal Prev 43:2134–2141
17. Tey L-S, Kim I, Ferreira L (2012) Evaluating safety at railway level crossings using micro-simulation modelling. In: Transportation research board conference, 2012 Washington DC
18. Wigglesworth EC (1979) The epidemiology of road-rail crossing accidents in Victoria, Australia. J Saf Res 11:162–171
19. Wullems C, Wayth R, Galea V, Nelson-Furnell P (2014) In-vehicle railway level crossing warning systems: can intelligent transport systems deliver? In: Conference on railway excellence, Adelaide

8

The Circumvention of Barriers to Urban Rail Energy Efficiency

Paul Batty[1] · Roberto Palacin[1]

Abstract As energy prices rise, urban rail energy efficiency becomes even more important. Many technological, operational and policy-based energy efficiency measures are well known and can have a notable positive effect on the urban rail systems. However, these measures can remain unimplemented. This lack of action can often be attributed to a variety of conflicting stakeholder opinions and a lack of knowledge transfer. This paper firstly focusses on the energy efficiency requirements of various stakeholders, before discussing about how such conflicts can be circumvented to ensure the success of future energy efficiency projects.

Keywords Mobility · Urban rail systems · Energy efficiency · Requirements · Barriers · Solutions

1 Introduction

Improving energy efficiency is an important goal for all transport systems worldwide, particularly urban transport systems, given the forecasted rise in urbanisation and car ownership [1, 2]. Urban areas currently contain approximately 50 % of the global population, which is expected to rise to 70 % by 2050, leading to a tripling of the travel-km in urban areas [3]. Therefore, it is vital that urban transport systems adapt their entire operations to address these challenges and prevent negative economic, social and environmental consequences.

The importance of urban rail systems—specifically their superior capacity and energy efficiency—should be exploited further in order to achieve these goals. However, despite their high level of energy efficiency, urban rail systems nevertheless consume huge amounts of energy. For example, the London underground consumes over 1.2 TWh of energy annually, which currently costs almost £100 million, and is expected to rise to £140 million by 2020 [4].

Research by the European Commission highlights that the greatest potential for energy savings lies in buildings, whilst the second greatest lies in transport [5]. Given that the urban rail systems consist of a mixture of both (the traction:non-traction split for the two largest urban rail systems in the UK is approximately 75:25), there is great scope to exploit this savings potential and further enhance their efficiency levels [6]. However, it is difficult to implement many energy efficiency improvements in urban rail systems which can be attributed to numerous, often interrelating, factors.

Firstly, this paper provides a brief background on the energy efficiency requirements for urban rail systems in Sect. 2. This is followed by a summary of the main problems associated with improving urban rail energy efficiency in Sect. 3. Subsequently, in Sect. 4, the potential solutions to these challenges are discussed and analysed.

✉ Roberto Palacin

Paul Batty
paul.batty2@ncl.ac.uk

[1] School of Mechanical and Systems Engineering, NewRail – Centre for Railway Research, Newcastle University, Stephenson Building, Newcastle Upon Tyne NE1 7RU, UK

Editor: Baoming Han

2 Background to Energy Efficiency Requirements

Urban rail energy efficiency improvements can largely be split into two separate categories: energy consumption reduction and reduction of energy consumption per unit

Table 1 The main keywords used during the literature search

Topic	Keywords used
Technical energy efficiency solutions	Regenerative braking; energy recovery; energy storage; retrofitting; air conditioning, escalators, lighting; innovative technology*; flywheel
Operational energy efficiency solutions	Peak travel; power peaks; timetable optimisation, service frequency
System characteristics	AC power; DC power; legacy system; rail*; urban rail; metro; tram; light rail
Political issues	Political strength; political will, electoral cycle, funding, legislation, partnerships; tender*; franchis*

* The use of asterisks at the end of keywords means that different suffixes are included in the search

output. Urban rail energy consumption can be reduced through both traction- and non-traction-related measures, although constraining circumstances mean it is one of the numerous trade-offs, also including capacity, safety, journey time, reliability and comfort; the prioritisation of which varies between the systems. Further information on a comprehensive set of 22 energy consumption-related key performance indicators that enable a multilevel analysis of the actual energy performance of the system, an assessment of the potential energy saving strategies and the monitoring of the results of implemented measures is detailed in [7]. Energy efficiency can subsequently be considered as the energy consumption per unit of output (e.g. kWh per passenger-km), and can therefore be improved by increasing the passenger density on the existing rolling stock for a given level of energy consumption through, for example, modal shift. Modal shift is the movement of travellers from private cars to public transport and can greatly increase the energy efficiency of the overall transport sector, while reducing congestion and emissions levels within cities.

Energy efficiency requirements come from numerous stakeholders and can be summarised into three main categories: economic requirements, environmental requirements and political requirements. The economic case for greater energy efficiency is the predominant requirement; less energy used leads to cost savings, increases business competiveness and protects against rising energy prices, which, for example, in the UK, are forecasted to rise by up to 104 % by 2030 [8]. This is attributable to a number of factors, including additional charges on electricity to encourage large-scale renewable energy generation; these currently cost Transport for London an additional £16 million annually [4]. Reducing energy consumption can lower the power peaks within an urban rail system, which can lead to cost savings [9]. From an environmental perspective, energy efficiency allows the reduction of energy consumption, and hence CO_2 and other associated emissions. Finally, from a political perspective, greater energy efficiency can help satisfy numerous political energy

efficiency and emission requirements at local, national and international levels, while helping in increasing energy security and reducing dependency on fossil fuels [10].

3 Methodology

This section presents the methodology of the investigation carried out to develop a comprehensive assessment of the barriers to greater urban rail energy efficiency from the available body of research. Urban rail systems are highly complex; energy efficiency improvements can be effected in a plethora of different ways and as such, it was important to find the state-of-the-art energy efficiency solutions and to analyse the barriers to their implementation. This was achieved through a variety of means.

An academic literature search—which constitutes the main reference source for this paper—was primarily conducted using international, online databases such as Scopus (http://www.scopus.com) and the Newcastle University Library search tool, which is linked to the major electronic resources worldwide. The main keywords used in this literature search are shown in Table 1.

Furthermore, relevant unpublished information from dedicated conferences, seminars and workshops was examined. In addition, as the topic is not only of academic interest, the literature search also included international databases of research and industrial projects, such as the transport research portal (http://www.transport-research-portal.net) and Spark (http://www.sparkrail.org). Documents by organisations such as the European Commission, the United Nations and the International Energy Agency were also considered, in addition to press releases and reports from manufacturers and operators.

In general, the literature search was focused on the last 15 years, although older resources were also consulted where their relevance could be determined. In total, over 150 documents and websites were reviewed for the purpose of this paper.

4 Problems

4.1 Uniqueness of Systems

Problematically, urban rail systems are intrinsically unique, with solutions suiting one system often being inappropriate for others. Nevertheless, a range of solutions—both technological and operational—exist that can allow urban rail systems to improve their energy efficiency. However, while these technologies exist, they may remain unimplemented for many reasons, including a lack of capital preventing their purchase/lack of subsidies to enable retrofitting, and a lack of awareness/full understanding of the technology. Most urban rail systems use direct current (DC) traction, via a catenary or third rail, with examples of AC traction being a relatively new phenomenon (i.e. S-Bahn Systems in Germany, Delhi Metro). DC systems are commonly used for small, dense rail networks with many trains as transformers and rectifiers are lineside rather than on-board, enabling lightweight rolling stock compared to AC systems, where such equipment is located onboard. However, AC systems exhibit lower losses in the power supply system (due to higher voltages) and are able to recover regenerative braking energy much simpler than in DC systems. As such, different strategies need to be applied. Further technologies to aid energy efficiency improvements in urban rail systems include energy-efficient driving, reducing power supply losses, lightweighting rolling stock, improving the energy efficiency of HVAC/lighting and those that lower the energy consumed whilst stabled. For a more comprehensive look at such measures, see [11, 12].

4.2 Awareness and Understanding of Technologies

It is also important that appropriate guidance is given to those in decision-making positions to ensure that new technologies are implemented appropriately; this was notably not the case in Ottawa, Canada, whereby the publicly owned urban transport service (OC Transpo) purchased 177 hybrid buses in the belief that they would produce the stated achievable fuel savings when operated in any manner [13]. Many of these buses were then operated on long expressways, with minimal braking and accelerations where the benefits of the hybrid engine were exploited very little and fuel consumption was notably greater than comparable journeys in diesel-powered buses [13]. A retrofit programme to then convert all these to diesel-powered engines was then implemented, expecting to cost over £7 million, proving extremely wasteful and time consuming, simply because insufficient guidance was given to those in charge of procurement. It can also be more difficult to make decision-makers aware of the current situation; in, for example, certain legacy urban rail systems, it has been found

that there is a lack of accurate knowledge on the energy flows within the system itself, which can obstruct efforts to implement energy efficiency measures [6]. However, in certain cases, the modelled energy savings have actually been inferior to the energy savings achieved in real measurements of the system; it was reported that in Bielefeld, Germany, energy savings from a newly installed flywheel delivered annual savings of 360 MWh—63 % greater than the expected 220 MWh [14].

4.3 Lack of Exploitation of Energy Savings from Existing Vehicles/Infrastructure

It appears that, in certain cases, there is insufficient impetus from governments to facilitate energy efficiency improvements; both through the aforementioned lack of funding mechanisms, and the lack of legislation requiring improvements in energy efficiency, particularly in the case of existing assets. Necessitating such energy efficiency improvements is very important for urban rail systems; given the 30–60-year working life of rolling stock, many vehicles currently in use in light rail, tram and metro systems will still be in service for decades [15]. However, current energy efficiency directives fail to place sufficient emphasis on increasing the energy efficiency in the existing infrastructure, rolling stock and equipment. It is widely acknowledged that the greatest energy savings can come from the existing buildings and transport yet, until 2012, the only legislation to instigate such actions was voluntary [16]. More recent legislation has failed to properly address this, which appears to be due to inappropriate implementation, and the weakening of requirements throughout the legislative process. This can be attributed to a lack of both political will and a sufficient understanding of the urban rail systems [16, 17]. However, there are significant costs associated with the retrofitting energy efficiency technologies to the existing rolling stock; where the new technology is not replacing the existing technology at the end of its life, often the energy/cost savings are insufficient to allow the new technology to deliver benefits during the remaining life of the rolling stock. Additional funding mechanisms may be required, given the scale of the energy savings delivered.

4.4 Effects on Service Quality

Certain energy efficiency measures can detract from the quality of service, which can dissuade customers to continue using the system [18]. Such measures include switching off escalators during off-peak times, forcing passengers to walk up them, lowering the use of air conditioning during summer months, resulting in uncomfortable climatic conditions, reducing the frequency of service, increasing average waiting times and operating smaller trains off-peak, potentially leading to cramped conditions. Furthermore, construction, repair and upgrading work carried out on urban rail systems can also negatively affect the passengers, despite the outcome leading to a higher quality system (e.g. the time required to replace old, inefficient escalators). Friman [19] describes how passenger opinions of the PT system quality declined in the face of such improvements, due to the level of disruption caused during their implementation.

4.5 Political Issues

It is also vital to develop methods to circumvent the current prioritisation of short-term economic issues over long-term environmental and economic sustainability. This involves addressing the numerous political distractions, including the highly disruptive effects of political campaign cycles and the need to curry favour with voters to aid re-election, at the expense of the implementation of potentially controversial projects. The political campaign cycles can be numerous; for example, in London, there are general, mayoral and borough elections, each taking place every 4–5 years, which is significantly shorter than the timescales for planning/funding that rail systems often work on. A notable example of the problems caused by the political cycle is the London congestion charge, which was viewed by many as an extra tax that brought no benefits. The scheme (and especially the important 2006 Kensington and Chelsea extension) became a political issue during the 2008 mayoral elections, which led to the conservative opposition candidate, Boris Johnson, to state he would remove the charge in the Kensington and Chelsea area if was voted into power, which he achieved [20].

5 Discussion of Solutions

5.1 Shifting Peak Travel to Facilitate Modal Shift

When attempting to utilise urban rail systems to encourage modal shift, it should be ensured that sufficient capacity exists, or can be made to exist. During peak times in, for example, London this may not be possible without additional capacity improvements/mobility management solutions. Singapore has demonstrated success in using soft measures to move travellers out of peak time operations; the Singapore Land Transport Authority introduced a 2-year trial scheme in 2013 to provide free travel for those passengers who end their journey at one of the 18 central metro stations before 07.45 on weekdays, with discounted travel for those exiting between 07:45 and 08:00 [21]. This benefits the passengers through cost and time savings, and a more pleasant ride, and resulted in a permanent moving

of approximately 7 % of the peak-time ridership (between 08:00 and 09:00) to the pre-peak (07:00–8:00). It was found that over 66 % of those who stated that they did not switch had set times when they must be at work [21].

However, notable levels of modal shift are very difficult to be achieved on a long-term basis, due to the numerous societal and political challenges. These are discussed further by Batty et al. [22], who analyse how best to attract people to public transport and dissuade car usage using 'push' and 'pull' mechanisms. This highlights the necessity for significant improvements in the public transport system as a whole, in terms of quality, capacity and level of integration, to help remove the perception that public transport is unclean, unreliable and of low comfort [23–25].

5.2 Innovative solutions

Problematically, urban rail systems are often very unique in their design, and so prescribing solutions suitable for all or most systems is rarely effective. This necessitates decision-makers to work with all the relevant stakeholders to develop innovative solutions specific to the system in question, in all areas of the system from finance mechanisms to technological measures. Research also highlights that the public react positively toward energy-saving measures that they perceive to be clever or innovative; for example, floor tiles containing piezoelectric mats that produce energy when stepped upon have been installed in busy locations worldwide, with a positive reaction from the public [26]. This suggests that the way in which energy efficiency and energy-saving schemes are marketed can help determine their level of social acceptance.

Innovative funding solutions should also be investigated, which can provide the required capital to fund the implementation of energy efficiency measures. For instance, Southeastern Pennsylvania Transportation Authority (SEPTA) of Philadelphia sold 250,000 3-day passes to the international deal website Groupon for $1.8 million, with the aim of encouraging more people to try its PT services and become more open to engaging in future modal shift [27]. The £1 billion funding for the Northern Line extension to the London underground was also considered to be an innovative method of funding; with the entire project being funded with the intention of no detriment to the British taxpayer. This involved the Greater London Authority borrowing the £1 billion, with a repayment guarantee provided by the UK Government to minimise borrowing costs [28]. The loan repayments are then to be made through contributions from local developers (to be collected by the local authorities) and through the growth in business rates revenue within the enterprise zone in which the extension is to be built. Over time these funding sources are expected to cover the complete repayment of the loan [28]. However, the transferability of the model to dissimilar cities is less clear, as it is only the soaring demand for property and high land values that are considered to have made the scheme feasible in London [29].

While innovative technologies exist that could lead to a notable impact in energy consumption reduction, a lack of certainty regarding their operation and ease of implementation can dissuade operators from implementing them. This demonstrates the needs for governments to encourage the development of collaborations between original equipment manufacturers and end-users to allow for demonstration projects of technologies to facilitate their market uptake by developing a business case for them. For example [14] highlight how energy-saving technologies can be made financially viable using the economies of scale principle, with several operators working together; committed investment from each operator, and a standard design for an energy storage system (ESS) would allow the cost to each operator to be significantly reduced. Similarly, the use of flywheels for on-board ESSs is becoming a more widely investigated topic, but a prototype, developed in Rotterdam for their tram system, caused significant damage when it became detached from the tram during testing in the workshop. However, new, safer composite flywheels are in development, and the DDFlyTrain flywheel program is of particular note, which involves a UK Government-funded collaborative effort between a flywheel developer (Ricardo), a hydraulic transmission developer (Artemis Intelligent Power) and Bombardier.

Other types of innovative partnerships between stakeholders have also led to fruition; in 2010, the SEPTA—the public transport operator for Philadelphia—aimed to further increase the energy efficiency of their network operations by further utilising the regenerative braking capabilities of their rolling stock in their metro system. The plan involved linking the third rail system to a wayside ESS, thereby allowing excess regenerated energy to be stored and used at a later time rather than wasted in rheostats when no other rail vehicles are in the same electrical section [30]. However, while this plan would successfully save large amounts of energy, it could not be made economically viable in its proposed form. Therefore, after dialogue between numerous stakeholders, a collaborative, profitable plan was developed to use the ESS for multiple purposes, such as voltage stabilisation and peak shaving on the external electrical grid [31]. The success of this plan is centred around the money generated from participating in the local electricity market, which is 3–4 times greater than the value of the energy savings themselves, and equates to approximately $200,000 per annum [32].

The initial investment came from each involved party: Envitech Energy (Power controls and power conversion systems), Viridity Energy (Smart Grid Technology) and Saft Batteries (Lithium-Ion battery), each of them invested its own funds in the project to provide the necessary capital to install the proposed infrastructure [30]. Funding was also provided through the Transit Investment for Greenhouse Gas and Energy Reduction programme. Indeed, this scheme proved so successful that a second, hybrid ESS will be installed on the same line, consisting of both a supercapacitor and a battery [33].

However, the transferability of such projects should be considered thoroughly before implementation; preliminary testing of the ESS in Philadelphia demonstrated that the revenue from the frequency generation market was strongly influenced by the external climatic conditions in the region, with the revenue generation in the coldest month (January— average daily temperature 0 °C) being six times that of the warmer months [32]. This is in stark contrast to other, more conventional, schemes, where energy recovery was highest in the summer months, due to the higher auxiliary energy consumption in the winter months [14].

5.3 Solutions the Public Do Not Notice

The travelling public are much more likely to accept energy efficiency measures that do not negatively affect their travel experience and as such, maximising the level of energy recuperated from the regenerative braking of urban rail rolling stock appears to be a promising solution. Combining regenerative braking, storage and reuse with basic timetable optimisation can lower energy consumption by up to 45 %, depending on the individual characteristics of the system in question [11]. However, it is important that due consideration should be given to the side-effects of the chosen ESSs, particularly larger, wayside ESSs; for example, the noise produced by flywheel ESSs during normal operation can be as high as 96 dB and as such their effects on the surrounding environment should be taken into account [14].

Other solutions include energy-efficient driving, aided through the use of driver advisory systems or automatic train operation, which can reduce or remove the negative effects of inefficient driving styles. However, drivers do not always utilise the guidance available to them appropriately, a challenge which was circumvented in the Helsinki tram system by training all staff in public transport, energy and environmental issues, so as to help motivate drivers and other staff to take practical actions for the environment, a scheme which proved to be successful.

5.4 Political Strength

Political strength and support is critical to the successful implementation of numerous energy efficiency measures. Politicians are responsible for defining laws, engaging stakeholders and developing funding mechanisms. For example, politicians should further facilitate dialogue between relevant stakeholders (i.e. industry, operators, research institutes) to share expertise and help develop successful urban rail energy efficiency solutions; it is prohibitively costly to address the problems the transport sector faces individually; addressing problems as part of a connected approach across modes is vital for success. This should also include the development of funding mechanisms to support the whole process, helping in bridging the gap between research and testing of energy efficiency-related technologies in universities, and the final technologies developed and sold by private companies.

Longer-term energy efficiency projects, especially those requiring high levels of investment, may need a cross-party consensus to ensure their continued implementation in the face of a change in government. In this sense, there is a great need to be able to ensure the incumbent politicians continue to spend public money, potentially in the face of public criticism, when the benefits may only manifest in a number of years, when future MPs may take the credit.

The numerous advantages of driverless trains over conventionally driven trains include the ability to program their operation to maximise their energy efficiency, to operate regardless of time of day without the need for expensive personnel, and their equal (or even superior) safety record [34, 35]. However, the process of introducing driverless trains is commonly stagnated or stopped by drivers' unions. The notable current exception is the Paris Metro System, where Ligne 1 was converted to fully driverless operation after 10 years of preparation and consultation with unions. It is recommended that politicians take a firm stand on such matters to ensure that, where necessary, driverless operations can be implemented.

To circumvent the issues regarding the political cycle, and to provide MPs decision-makers with confidence, it could be legislated that all recommendations from panels of experts (e.g. the UK committee for climate change) should be compulsory to implement, ensuring necessary energy efficiency legislation is implemented. Additionally, a greater usage of passenger advisory bodies to predetermine the effectiveness and acceptability of policies would also ensure that a more accurate understanding of the opinions of passengers is developed. However, this may not account for those people initially against a certain measure or programme, but who could, over time, adapt and accept it.

5.5 Competitive Tendering

Although sometimes considered a contentious topic, placing the operations of urban rail systems out to tender can lead to significant benefits. Tendering can instigate private sector investment and encourage innovative working practices, which can relate to energy efficiency. Private investors hold

public transport in high regard, due to its demonstrable strengths, such as its stable revenue and cash flow, the clear potential for growth and its status as a provider of essential services. However, the potential negative consequences of tendering should also be considered, such as concession/franchise failure, contract rigidity scuppering innovation and the costs of the tendering process for bidding companies. Nevertheless, tendering the operation of the system can increase the competitiveness of the bids and, if not undertaken previously, can provide the local authority with a better understanding of the costs required to run the system at the increased level of efficiency [36]. However, research undertaken by ERRAC and the UITP found that 17 % of the urban rail systems in Europe are operated without a public service contract, and of the remaining 83 %, only 17 % of those contracts were awarded after having been put out to tender, the rest being directly awarded [37]. Therefore, the possibility to tender operations of urban rail systems should be explored to a greater extent in the future, although compulsory tendering of operations has been postponed by MEPs during the weakening of the revision to the Public Service Obligation Regulation 1370/2007 [38]. It should also be noted that impetus can be given to the concessionaire/franchisee to improve the energy efficiency of the system; if they are not paying for the energy consumed, there may exist little incentive to develop energy efficiency measures. However, this is only practical where a sufficient understanding of energy flows in the system is already known.

6 Conclusion

The need for greater energy efficiency has been gaining in prominence for many decades, spurred on by rising energy prices and advances in technology. While urban rail is perhaps the superior mass transit system, the energy consumption in many systems is still able to be reduced significantly. This paper has aimed to summarise the current challenges in developing greater urban rail energy efficiency, and has discussed a range of solutions that appear to be applicable to other urban rail systems.

References

1. UN (2007) Realizing the potential of energy efficiency: targets, policies, and measures for G8 countries. Expert group on energy efficiency. United Nations Foundation, Washington

2. UN (ed) (2008) World urbanization prospects: the 2007 revision. United Nations, New York

3. van Audenhove FJ, Baron R (2014) The future of Urban mobility. Logist Transp Focus 16(4):24–31

4. Transport for London (2014) TfL energy purchasing 2017 to 2020. https://www.tfl.gov.uk/cdn/static/cms/documents/fpc-20141014-part-1-item-17-tfl-energy-purchasing.pdf

5. EC (2011) Energy efficiency plan 2011. European Commission, Brussels

6. Powell JP, González-Gil A, Palacin R, Batty P (2015) Determining system-wide energy use in an established metro network. In: The Stephenson conference: research for railways, London, 21–23 Apr 2015

7. González-Gil A, Palacin R, Batty P (2015) Optimal energy management of urban rail systems: key performance indicators. Energy Convers Manag 90:282–291

8. DECC (2013) Estimated impacts of energy and climate change policies on energy prices and bills. DECC, London

9. Albrecht T (2004) Reducing power peaks and energy consumption in rail transit systems by simultaneous train running time control. In: Advances in transport: ninth international conference on computers in railways, Dresden, 17–19 May 2004, pp 885–894

10. IEA (2014) Spreading the net: the multiple benefits of energy efficiency improvements. IEA, Paris

11. González-Gil A, Palacin R, Batty P (2013) Sustainable urban rail systems: strategies and technologies for optimal management of regenerative braking energy. Energy Convers Manag 75:374–388

12. González-Gil A, Palacin R, Batty P, Powell JP (2014) A systems approach to reduce urban rail energy consumption. Energy Convers Manag 80:509–524

13. CBC News (2012) City could pay to turn hybrid buses into diesel buses. http://www.cbc.ca/news/canada/ottawa/story/2012/10/25/ottawa-hybrid-bus-retrofit-pilot-project.html. Accessed 10 Apr 2013

14. Devaux F-O, Tackoen X (2014) Guidelines for braking energy recovery systems in urban rail networks. http://www.tickettokyoto.eu/sites/default/files/downloads/T2K_WP2B_Energy%20Recovery_Final%20Report_2.pdf

15. UITP (ed) (2009) Public transport, the green and smart solution. A new frontier: double market shares by 2025. UITP, Munich

16. Directive 2006/32/EC on energy end-use efficiency and energy services and repealing Council Directive 93/76/EEC

17. Directive 2012/27/EU of the European Parliament and of the Council of 25 october 2012 on energy efficiency, amending Directives 2009/125/EC and 2010/30/EU and repealing Directives 2004/8/EC and 2006/32/EC

18. Landex A. (2012) Annual transport conference, Aalborg University, Aalborg. DTU Transport. http://www.trafikdage.dk/papers_2012/87_AlexLandex.pdf. Accessed 08 Jan 2013

19. Friman M (2004) Implementing quality improvements in public transport. J Public Transp 7(4):49–65

20. Chronopoulos T (2012) Congestion pricing: the political viability of a neoliberal spatial mobility proposal in London Stockholm, and New York City. Urban Res Pract 5(2):187–208

21. LTA (2013) TravelSmart: market research survey results. https://www.lta.gov.sg/content/dam/ltaweb/corp/PublicTransport/files/Travel_Smart_Market_Research_Results.pdf

22. Batty P, Palacin R, González-Gil A (2015) Challenges and opportunities in developing urban modal shift. Travel Behav Soc. doi:10.1016/j.tbs.2014.12.001

23. Beirão G, Sarsfield Cabral JA (2007) Understanding attitudes towards public transport and private car: a qualitative study. Transp Policy 14(6):478–489

24. Fujii S, Kitamura R (2003) What does a one-month free bus ticket do to habitual drivers? An experimental analysis of habit and attitude change. Transportation 30(1):81–95

25. Stradling S, Carreno M, Rye T, Noble A (2007) Passenger perceptions and the ideal urban bus journey experience. Transp Policy 14(4):283–292

26. Kemball-Cook L (2012) Retail therapy provides renewable electricity. Build Eng 87(6):16–17

27. Schlosser N (2013) Transit systems get creative with revenue generation. Metro Mag 109(1):52–54

28. Transport for London (2013) Northern line extension: factsheet 1: funding and finance. https://www.tfl.gov.uk/cdn/static/cms/documents/nl-factsheet-i-web.pdf

29. Pickford J (2013) London project to use risky funding model, Financial Times, 08/04/2013. http://www.ft.com/cms/s/0/fcda4910-9f64-11e2-b4b6-00144feabdc0.html#axzz3Lh06bZEN

30. Redfern H. (2012) Making public transit greener. http://www.metro-magazine.com/blog/transit-dispatches/story/2012/07/making-public-transit-greener.aspx. Accessed 09 Jan 2013

31. Wald M (2011) Saving electricity on a Philadelphia subway line, New York Times edn, 13/06/2011. http://green.blogs.nytimes.com/2011/06/13/batteries-will-save-juice-on-a-philadelphia-commuter-line/. Accessed 09 Jan 2013

32. Poulin (2014) Recovering braking energy: SEPTA's wayside energy storage project, APTA 2014 Rail Conference. Available online https://library.e.abb.com/public/ce6e1725f840073f85257d3f00514387/SEPTA%20Breaking.pdf. Accessed 18 sep 2014

33. ABB (2014) SEPTA stays focused on energy efficiency with second energy storage and unique hybrid system from ABB. http://www.abb.com/cawp/seitp202/186379817a2df9b6c1257d57004c026e.aspx. Accessed 18 Sep 2014

34. Grogan A (2012) Driverless trains: it's the automatic choice! Eng Technol 7(2):54–57

35. Valderrama A, Jørgensen U (2008) Urban transport systems in Bogotá and Copenhagen: an approach from STS. Built Environ 34(2):200–217

36. Wallis I, Bray D, Webster H (2010) To competitively tender or to negotiate—weighing up the choices in a mature market. Res Transp Econ 29(1):89–98

37. ERRAC & UITP (2009) Metro, light rail and tram systems in Europe. http://www.uitp.org/files/ERRAC_MetroLR&TramSystemsinEurope.pdf. Accessed 20 Dec 2012

38. EC (2014) Commission takes action against UK for persistent air pollution problems (Press release). http://europa.eu/rapid/press-release_IP-14-154_en.htm. Accessed 04 Mar 2014

9

The Research of Irregularity Power Spectral Density of Beijing Subway

Dan Lu[1] · Futian Wang[1] · Suliang Chang[2]

Abstract This paper aimed to describe track irregularity of Beijing Subway. A preprocessing data that are from the upward track of Beijing Subway line 10 have been used in the survey. In addition, based on the maximum entropy principle, the spectrum of track irregularity power spectral density which is short for PSD has been drawn using MATLAB and fitted by using a nonlinear least squares method. After all scientific discussion, the results show the characteristic parameters of PSD. Furthermore, by contrasting Beijing Subway with national rail line, a proper function research has been done about track irregularity of Beijing Subway and the characteristics of track irregularity power spectral.

Keywords Beijing subway · Track irregularities · PSD · Fitting

✉ Dan Lu
14125712@bjtu.edu.cn

Futian Wang
ftwang@bjtu.edu.cn

Suliang Chang
773713416@qq.com

[1] State Key Laboratory of Rail Traffic Control and Safety, Beijing Jiaotong University, Beijing, China

[2] The Line Branch of Beijing Subway Operation Limited Company, Beijing, China

Editor: Baoming Han

1 Introduction

Track irregularity is the main cause of the vibration of the locomotive vehicle and the force of the wheel rail. It has an important influence on the safety, stability, comfort, duration of vehicles and track components as well as the environmental noise of the train. When vertical profile irregularity amplitude is too large, the train will be capsized because of upward track. If the situation goes more seriously, it will cause uneven wheel load reduction or even derailing. Hence, grasp the track irregularities can monitor the track quality state, help make reasonable arrangements for the maintenance plan, reduce the costs of maintenance and repair, and ensure the safety and smoothness of the line. Master the irregularity can realize the development and application of the track quality state detection data. It can also improve the utilization rate of detection data, understand the track irregularities of the state, and help improve the reliability of the rail facilities and operating safety. It contributes to the rational allocation of resources and control maintenance costs scientifically. It has vital significance to the repair department which makes repair plans.

British Railways [1] began to study the track irregularity in 1964. Soon afterwards, Britain, Japan, East Germany, the former Soviet Union, India, Czech, the United States, and other countries drew up the spectral density and the correlation function of the respective track irregularity. In 1980s, according to six speed and track state level [2] based on the FRA Division of American Track Safety Standard, the U.S. laid down the track irregularity power spectrum.

China began the research in 1960s–1970s, Changsha Railway Institute and the Academy of Railway Sciences used the "Inertia Reference Method" [3] to measure the track irregularity, combined with ground testing and Inspection Car tracking and other different methods to analyze the track irregularity power spectrum. China Academy of Railway Sciences has studied on China's three main track irregularity power spectrums, and put forward the fitting function of the track irregularity power spectrum of the three main lines. There is a research [4] that makes a feature analysis of the track irregularity of urban rail transit based on track irregularity inspection data from Shanghai Subway. Xu Lei [5] analyzed the track irregularity spectrum of Shenchi—Huanghua Heavy Haul Railway, and compared it to the high- and low-interference spectrum in Germany based on raw data. It shows that there is a good correlation between TQI index and the area of track irregularity spectrum, when using MATLAB programming to analyze the power spectrum and TQI, Therefore, Jianbing Li [6] gave a suggestion that the track irregularity power spectrum should be one of the main factors to control the speed of the track quality.

However, due to different equipments and operation modes, the research and analysis in the National Railway are not suitable for subway. At the same time, the research about the analysis of subway is less so this paper takes the irregularity data from Beijing Subway track inspection car as sample; researching the irregularity PSD of Beijing Subway, the data in the research are from a project which is under way. Meanwhile, it makes the fitting spectrum of the track spectrum density on Beijing Subway and then gets the corresponding characteristic parameters. Thus, we can obtain a scientific evaluation and give a suitable management for railway track irregularity. In addition, we could also put forward reasonable repair advice for Beijing Subway, and provide technical support for the quality of lines.

2 The Characteristic of Beijing Subway

The subway is mainly responsible for the urban traffic, connecting the main passenger distribution points in a city, which makes it different from the local train and the High-speed Rail [7].

(1) The line of Beijing Subway is underground, and its general speed can reach 30–40 km/h while its maximum speed can run up to 80 km/h. The structure of the track is made up of heavy rail, elastic fastener, seamless lines, and monolithic track bed.

(2) Subway line is generally built in prosperous regions of the city and has to change direction frequently due to route planning, terrain, surface object, and buildings. Besides, rail wear is aggravated because the

curve, especially small radius curve, occupies more and more proportion of mileage.

(3) Subway lines are in an underground and humid environment. The track structure corrosion will accelerate the destruction of the track structure because of the leakage of the tunnel.

(4) The work and repair area is narrow in the tunnel and on viaduct. Because of the high traffic density, repair and inspection work cannot be done during the daytime. Workers usually do general repair work at night.

(5) The ballast bed is laid monolithic, so there is a different characteristic. The destruction of monolithic track bed is mainly concrete crack and subsidence.

(6) Subway operation uses the ATO driving mode, which features the same type of vehicles, same axle load, and the same speed in the same section. Because of the heavy traffic and the elasticity of the monolithic roadbed, all the problems that happen in the subway would be complicated and spread fast.

3 Test Data of Beijing Subway

The test data are gathered from track inspection car of the Beijing Subway Company (Number: GW-01). The track inspection car is managed by testing workshop of track inspection of the Beijing Subway Company. All the measuring systems of the track inspection car, including geometric shape detection, rail full section, and the contact rail detection function, are provided by ENSCO. Types of detection data of Beijing track inspection car consist of original wave-data, overloading data, and ALD data. The existing assessment criteria of detection data on track include local gauge and track quality.

The data of track inspection car come from upstream of Beijing Subway line 10 in the situation of 40–50 km/h and 1024 m length in January 4, 2014. The number of raw data is 4096, and the measurement was taken every 0.25 m. The raw data are stored in the binary system of g03 [8] and the data format of which is shown in Table 1.

4 Concept and Function of Track Irregularity PSD

Track vertical irregularity refers that the track is rutted and uneven in the vertical direction along the rail lengthwise direction. It includes roughness on the rail surface, elastic deformation of track, residual deformation, the inconsistency

Table 1 Data format

Item	Unit
Speed	km/h
Mileage	km
Twist warp	mm
Horizontal acceleration	g
Gauge	mm
Standard	mm
Curvature	Rad/km
Normal acceleration	G
Height	mm
Track	mm
Super elevation	mm
ALD	v

of components of gap and the uneven subsidence of roadbed, and so on.

The PSD diagram of track vertical irregularity is a kind of continuous curve which uses the frequency or wavelength as abscissa and the spectral density values (In unit frequency band, it means square value of the irregularity amplitude) as ordinate. It reflects the frequency distribution of the track vertical irregularity power spectrum density on the shaft of the power spectrum diagram. It can also clearly reveal each wavelength composition of random track irregularity waveform. It means that the smaller the area under the spectral line, the better the irregularity. That is to say, the lower the line locates (near the abscissa), the smoother this orbit is.

The PSD diagram of track vertical irregularity clearly indicates the wavelength components of one long rail irregularity and the density of mean square value of each wavelength inside the tiny bandwidth. Different wavelength components can give us all kinds of track irregularities information in both amplitude and wavelength ways. It plays an important role in assessing the state of the track scientifically and judging the track disease.

5 The Calculation of Track Vertical Irregularity Power Spectrum

The spectrum estimation of power spectral varies from theory to theory. It can be divided into the classic power spectrum estimation and modern power spectrum estimation based on the different theories. The classic power spectrum estimation is divided into direct method (periodontal method [9]) and the indirect method [10] (BT). Modern power spectrum estimation method includes parameter model [11] (AR, MA and ARMA model, Prony

spectrum estimation) and non-parametric method [12] (MUSIC method, characteristic vector method).

When comparing the method of literature [13] with other methods, it is easy to see that for both the resolution and the smoothness curve, the maximum entropy spectrum estimation works best.

Maximum entropy, which is based on the known data and information, is to predict the unknown related functions of discrete time delay reasonably without any new hypothesis. A discrete information source can be represented as discrete random variables X, values for $x_1, x_2 \ldots x_n$ and the values of the probability are $p_1, p_2 \ldots p_n$. In the Shannon information theory [13], entropy is the source of each information entropy, the definition of the entropy of discrete random variable X:

$$H(X) = -\sum_{i=1}^{N} P(x_i) \ln P(x_i) = -E[\ln P(x_i)] \qquad (1)$$

In Eq. (1), the $P(x_i)$ is defined as the probability of the event when $X = x_i$; $\ln P(x_i)$ is the definition of information. Information is needed to remove event uncertainty of measurement. E stands for the mathematical expectation; while N and i refer to the number and serial number of event X. For a certain event, the probability of occurrence is 1, so the amount of information and the corresponding entropy are zero. The smaller the probability of an event is, the greater information it has, and the bigger the corresponding entropy value is.

Use the principle of maximum entropy to do the spectral estimation:

$$S(f) = \frac{P_M}{\left| 1 + \sum_{m=1}^{M} a_m e^{-j2\pi mfT} \right|^2} \qquad (2)$$

In Eq. (2): P_M is the prediction error power; m is the order; N is signal sampling points; and a_m $(m = 1, 2 \ldots M)$ is the coefficient of linear prediction filters of M.

The calculation process of track vertical irregularity power spectrum:

(1) Select 4096 data which come from track inspection car in January 4, 2014.
(2) Take a preprocessing for the selection of data, using the method of "3 Times Standard Deviation" to eliminate outliers. The outliers are data that influenced by external factor and they are beyond the normal. It will influence the analysis then.
(3) Through MATLAB programming, use the method of maximum entropy spectral estimation in the pretreatment of the raw data to draw track spectrum.

6 The Curve Fitting Expression of the Track Vertical Profile Irregularity Spectrum Density

6.1 The Fitting Function of Track Vertical Profile Irregularity Spectrum Density

Although the curve fitting of track vertical profile irregularity spectrum density is drew under a large number of real data, it does not have any analytic function. In order to facilitate the description, it is usually expressed by a function which is close to the measured characteristic curve. Based on the fitting formula [14] of the three railway track irregularity spectrums in the U.S., Germany, and China and the analysis of the actual track inspection data, this reference documentation uses programming R and introduces the fitting formula of railway track irregularity power spectrum in our country:

$$S(f) = \frac{AB^2}{f^2(f^2 + B^2)} \tag{3}$$

In Eq. (3), $S(f)$ refers to track irregularities fitting function; f refers to spatial frequency; and A and B refer to spectral characteristic parameter (Table 2).

The actual fitting parameter of Beijing Subway is shown below:

Table 2 Characteristic parameter of the fitting curve of track irregularity spectrum of Beijing Subway

Characteristic parameter	AB^2	B^2
Height	883.4	18.69

6.2 The Fitting Theory of the Nonlinear Least Squares Method

The fitting procedure of the nonlinear least squares method [15] is as follows: the $(f_i, S(f_i))$ $(i = 1, 2..., m)$ includes m data and weight coefficient is $\omega_0, \omega_1, ...\omega_m$, $S(f_i, c)$ refers to a functional model, $c = (A, B)$ is undetermined parameter of fitting function, and $S(f_i, c)$ is a nonlinear function whose coefficient is $c = (A, B)$, make a constructor function $s(f)$ to replace the measured curve approximately. The relationship between the constructor function curve and the measured curve is as below:

$$Q = \sum_{i=0}^{m} \omega_i [S(f_i) - S(f_i, c)]^2 \tag{4}$$

In the formula (4), Q refers to the quadratic sum of the distance between the two curves.

In order to get the minimum of the quadratic function, the condition must be satisfied:

$$\frac{\partial Q}{\partial c_j} = 0 \quad (j = 0, 1, ..., n) \tag{5}$$

6.3 The Fitting Outcome of the Track Vertical Profile Irregularity Spectrum Density Curve

It is not difficult to see from Fig. 1 that the fitting curve trend of the Beijing Subway irregularity and the actual drawing tendency are consistent, so it can describe the irregularity state of Beijing Subway better. According to

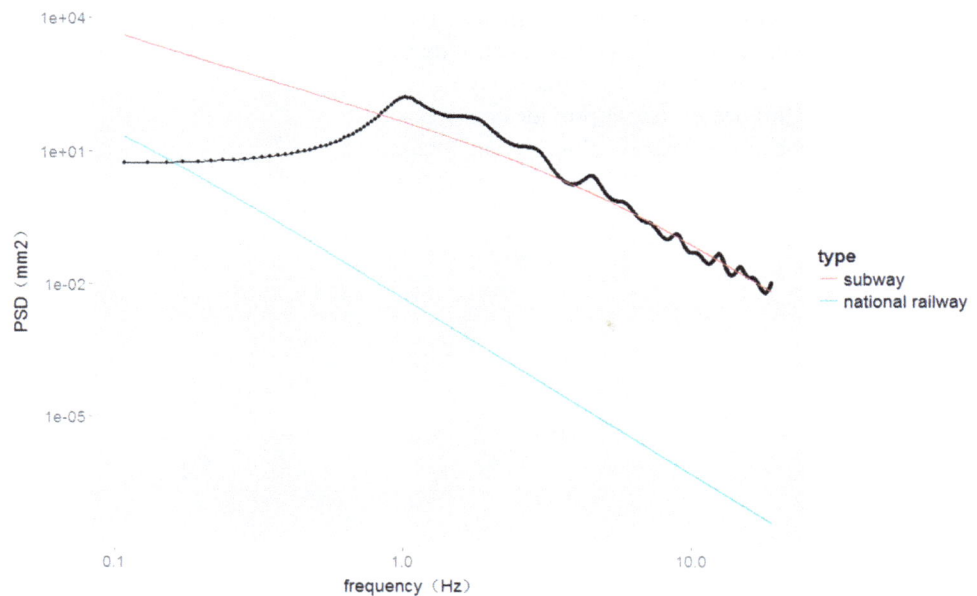

Fig. 1 The comparison of the fitting curve between Beijing subway and national railway

the fitting curve of Beijing Subway and national railway, Beijing Subway irregularity power spectrum curve is mostly above the national railway. The lower the curve position is, the better the track irregularity is. In this case, the national railway state is much better than subway. The frequency curve of track irregularity from 0.1 to 1.0 Hz is smooth. The fluctuations of the curve are more complex when it is beyond 1.0 Hz. This is attributable to unsmooth welding joint. Subway track irregularity power spectrum frequency is obviously increased in the curve amplitude between 0.5 and 1.0 Hz, indicating the need for better rail rolling technology and rail welding technology.

7 Conclusions

This paper introduces the study and significance of the track irregularity PSD. Firstly, the raw data are dealt with 3 Times Standard Deviation to eliminate outliers. Subsequently, it summarizes the operation characteristics of Beijing Subway of line 10, utilizing the maximum entropy method and MATLAB program to obtain the track spectrum estimation based on data that have been dealt, and using the nonlinear least squares method to fit the PSD curve which is proposed for Beijing Subway. Finally, comparing Beijing Subway with the national rail line, an analysis of the trend curve has been done. It points out where the track problems might have. This paper is of guiding significance to the safe operation of the subway lines and can give reasonable suggestions to the maintenance of track, so as to improve the safety of urban rail transit system. Since fitting track spectrum uses only a part of the data, the fitting formula can be a reference to the analysis of Beijing Subway track spectrum, while the fitting process needs to be further polished. At last, in order to analyze the irregularity power spectral density of Beijing Subway, it suggests that different types of data should be made full use of, and it provides more basis for managers to access the quality of orbit and master its changing rules.

References

1. Liu X (2009) Research on simulation and analysis of typical track spectrum. Dissertation, Ji Lin university
2. Zeng H, Jin S, Chen X (2005) Power spectrum density analysis of track irregularity of newly-built railway line for passenger. J Railw Sci Eng 04:31–34
3. Luo L (2006) The control of track irregularity state of wheel-rail system. China Railway Publishing House, Beijing
4. Li Z (2011) Characteristic analysis of track irregularity spectrum of urban. J East China Jiaotong Univ Rail, Transit 05:83–87
5. Xu L, Chen X, Li X et al (2013) Track irregularity spectrum analysis of Shenchi—Huanghua Heavy Haul Railway. J Cent South Univ (Sci Technol) 12:5147–5153
6. Li J, Lian S (2008) Study on the relationship between track irregularity spectrum area and track quality index. J SHIJIAZHAUNG Railw Inst (Nat Sci) 04:9–12
7. Chunhua Yu (2007) The study of urban rail transit track maintenance work's contents and management mode, Urban Rail Transit, pp 72–75
8. White DL (1998) Statistical characterization of vehicle and track interaction using rail vehicle response and track geometry measurements. Dissertation, Virginia Polytechnic Institute and State University
9. Huang Y (2013) Power spectrum estimation based on periodogram. Sci Technol West China 09:1–2
10. Wang S (2011) Research on power spectrum estimation and its application in broadband ADCP signal detection. Dissertation, Ocean university of China
11. Geng X, He Z (2009) Modern digital signal processing and its application. The Press of Tsinghua University, Beijing
12. Lu M, Zhang X (2005) The processing of discrete random signal. The Press of Tsinghua University, Beijing
13. Tian G (2008) The analysis and numerical simulation of railway track irregularity power spectrum. Dissertation, Southwest Jiao Tong University
14. Chen X, Wang L, Tao X et al (2008) Study on general track spectrum for Chinese main railway lines. Railw Sci Technol Res 03:73–77
15. Dong Z (2005) Nonlinear least-square algorithms used for TMA in bearing-only system—the engineering mathematic model and algorithms. Inf Command Control Syst Simul Technol 01:4–8

10

A Streetcar Undesired: Investigating Ergonomics and Human Factors Issues in the Driver–cab Interface of Australian Trams

Anjum Naweed[1] · Helen Moody[2]

Abstract Australia is home to the biggest light rail network and the industry is currently undergoing a renaissance. However, there is littleresearch to indicate the extent to which well-informed human factors and ergonomics practises are being incorporated into tram cab design. A lack of standardised features may create transfer conflicts between cabs, as well as operational issues and concerns for occupational health. The aim of this paper is to improve our understanding of the socio-technical complexity of light rail and to enhance how design standards are informed in this domain. Various human factors methods were used, including observational cab rides, objective force assessments, interviews, and focus groups. Data were collected across two sites and analysed thematically. Analysis of data suggested a substandard level human factors and ergonomics input in the design of the cab and driver interface that violated many key tenets of established design guidelines. These were particularly concerned with the usability of the master controller (i.e. throttle lever) and various issues in the design of the tram driver workspace. Findings also revealed a number of subtle yet significant features associated with delivery of service that created safety-performance conflicts. In conclusion, very little human factors input of tram driving, and the ergonomics considerations of the driver's workplace in general, appear to be going into the design of tram cabs. This may be related to the practice of using non-specific standards for developing trams and/or poorly integrating human factors and ergonomics into their specification processes. Some considerations for future work are given.

Keywords Modern tram · Traffic and transport safety · Light rail · Interdisciplinary transportation research · Cab design · Interface design

1 Background

Australia has over 270 km of tramway and is home to the biggest and oldest urban light rail network in the world [1]. There are also signs that the Australian light rail industry is undergoing a renaissance; new light rail projects are being planned, are under construction, or currently operational in Melbourne, Adelaide, Canberra, Perth, Newcastle, Sydney, and the Gold Coast [2, 3].

Much of the published literature on Australian Trams has tended to explore its history, growth, and design, specifically in terms of its engineering [e.g. 1, 4]. Marketing brochures for modern trams typically advertise innovations in technology, including advanced bogie designs, provision of passenger amenities, and modularised car-body components for collision safety and ease of repair [5]. Some emerging classes have also won design awards [6]; however, very little published research indicates the extent to which good ergonomics and human factors (E/HF) principles have been incorporated into the design of tram cabs. This space typically falls into the design and consultation space of rail manufacturing companies that service the industry. In recent years, the absence of this has been made apparent by its coverage in the media spotlight

✉ Anjum Naweed
anjum.naweed@cqu.edu.au

[1] Appleton Institute for Behavioural Science, Central Queensland University, 44 Greenhill Rd, Wayville, SA 5034, Australia

[2] Injury Prevention and Management, Corporate Health Group, Adelaide, SA, Australia

Editor: Marin Marinov

for reports of work-related injury and stress reducing the drivers' ability to work [7].

Beyond occupational health and safety concerns, there are a number of reasons why the tram system would be important from an E/HF perspective. First, many tram classes operate in Australia's individual states and increasing the number of classes of trams on a single network invariably decreases the range of standardised features. This has the effect of creating inconsistencies in stimulus–response design and may lead to conflicts when transferring between different trams [8], particularly if there has been no consideration to matching and/or retro-fitting frequently used features. Additionally, the lack of standardisation may create variations in the design of the cab interface and impact on driving performance. For example, injuries associated with passengers falling in the saloon are among the most common safety issue [9] and linked with concerns for jerky driving [10]. Whilst some of these issues may be related to human performance and training, it is worth questioning the extent to which they are influenced by problems in tram cab design.

Most of the functional requirements for designing the driver–cab of a tram appear to have been extracted from standards originally written for generic locomotives and driving coaches [e.g. 11]. Although the tram and train driving tasks have synergies, heavy, and light rail are fundamentally different systems, and the resulting designs may not always be suitable. The operation of trams in mixed-road traffic environments (i.e. as streetcars or street moving vehicles) stretches this gap further. Theories that orientate towards systems design such as the joint-cognitive systems movement suggests that drivers are likely to have a very unique dynamic with their vehicles, and an inter-connectedness that means they operate with it as an intimate system [12]. Based on this notion, one may assume that any differences in the features and characteristics of the two environments are likely to create subtle but significant differences in task dynamics. This is one of the reasons why it is important to involve the end-user in participative design processes, so as to ensure that the system, interface and the task work and behave as intended [13].

Norman's six design principles, which still underpin thinking around good design theory, are based on visibility, feedback, affordance, mapping, constraint, and consistency [14, 15]. In terms of tram cab design, these can be considered to correspond with: whether drivers can see the state of the tram, and if controls are positioned in a way that can be easily found and used (visibility); whether drivers know tram state, what it is doing, and what action needs to be performed (feedback); whether the perceived and actual properties of the various control and functions of the tram clearly elucidate their operation (affordance);

whether there is a clear relationship with controls and their effect (mapping); whether there are any clear restrictions to the kind of interactions that can take place (constraints); whether cab-interfaces are designed to have similar operations; and whether systems are learnable and drivers can quickly transfer prior knowledge to new contexts (consistency). One would not expect many of these tenets to be violated or substantially compromised in the design of modern trams.

Industrial ergonomics literature has identified E/HF issues in the design of tram cabs and suggested improvements, particularly for the primary speed/brake throttle controller (also known as the master controller), and "deadman" safety features (i.e. safety devices that operate if the driver is incapacitated) [16]. The general expectation is that once cab design issues have been identified, improvements will filter back into the specifications during the tendering and bidding process for new rolling stock classes. However, there are few publically accessible ways of tracking how, when, and if this happens. In general, new knowledge or insights are also expected to inform standards, but this process does not always translate either.

1.1 Aims

The aim of the research presented into this paper is to improve our understanding of the socio-technical complexity of light rail, and how design standards are informed in this domain. To fulfil the aims of this research, the objective of this paper is to draw on established design theory [14] to provide an account of the key E/HF issues in cab and task design that have been observed in the Australian tram industry. Whilst we share insights that would be of interest to the academic community, our goal is to provide a platform for debate and discussion for academic and industry practitioners, particularly those within the engineering urban rail transit community.

2 Method

2.1 Research Context

The research reported in this article was undertaken at two different sites (i.e. two different light rail networks) in Australia. The work was initiated by tram operators in response to complaints and concerns provided by tram drivers. In one site, this followed a very high incidence of musculo-skeletal injuries with particular incidence of left shoulder/arm injuries, and in other, through reports of cab design issues in the driver–cabs of newly deployed rolling stock.

2.2 Design

Several methods were converged to evaluate and assess the workplace of the tram driver. These fell into a mainstream range of qualitative techniques that are commonly used to evaluate humans at work, and included cab walkthroughs on stationary trams, on-road observations on scheduled services, objective force assessments, interviews, and focus groups.

Converging multiple data collecting techniques has been recommended as an effective means of eliciting knowledge [17] and performed successfully on other rail research [e.g. 18, 19]. Undertaking the work in different sites provided the means to compare common issues across different tram classes and identify the scope of specific E/HF problems. The methodology also included a specific physical ergonomics assessment and a broader human factors investigation. Figure 1 shows an overview of the methodology for each of these components, including individual methods of data collection and the main areas of focus. As Fig. 1 shows, unstructured interviews, cab walkthroughs, and cab rides were common to both components.

The physical ergonomics work assessed the driver's interaction with their cab in terms of how the interface fitted their body and addressed their functional needs. Methods in this component included unstructured interviews, cab walkthroughs (on stationary trams), and cab rides. Unstructured interviews were conducted with groups of drivers in the tram depot, or entirely opportunistically with individual drivers during cab walkthroughs and cab rides. Cab walkthroughs were pre-planned and involved a detailed discussion of the cab with a driver trainer who "walked" through the layout of the various controls, described their functions, and demonstrated how they were operated. The process lasted 30–60 min. Responses to questions were voice recorded, and photos were taken. During the walkthroughs, cabs were also assessed in terms of the push–pull forces and static muscle load required to operate controls. Particular focus was placed on the master controller and "deadman" devices, and measurements were taken using a Mecmesin force gauge. Cab rides were pre-planned "out and back" journeys from the main depot, undertaken during scheduled services (i.e. with passengers) but also out of service. These lasted 30–120 min and included spells of interviewing and silent observation of the driver engaging with their task.

The human factors work examined the driver–cab in terms of the design of the tram-driving task itself and its cognitive demands. The methods in this component also included cab walkthroughs, driving observations, and unstructured interviews, and these were undertaken as previously described. However, this component also included a series of focus groups. These lasted 1.5–2 h,

conducted with 4–6 tram drivers per group, and included a scenario simulation task [20]. This task required each of the participants to create challenging driving scenarios, as a means to overcome conversation-based limitation, but also stimulate situational insight. These methods has been previously applied in the heavy rail industry and have been very effective for identifying specific cognitive task demands, and illustrating how driver's stabilise the task in the face of conflicting goals [20]. Example questions for the general focus group were "What are your thoughts on the [specific class] tram?" and "take me through a challenging part of a route that you drive over." Example questions for the scenario simulation task were "create a really hard stretch of track" and "imagine you have to drive over your route. List the strategies that you would use to navigate it."

2.3 Participants and Recruitment

Participants provided informed consent to take part in the work, but in all cases, contacts at the relevant organisations were required to mediate and facilitate access. Whilst most activities were pre-planned, given the nature of the field-work, this was subject to change.

2.4 Data Collection Decomposition

Our work comprised assessment of two specific tram-types from two different rail manufacturing companies servicing the industry. Our assessments were related to prior incidence of musculo-skeletal injury in one tram, and a concern for the skills transfer issues in the other. In both cases, issues were first raised and reported by tram drivers. A further four tram classes were observed in order to create a basis for cross-comparison, making six tram classes in total. As two of these classes comprised observations of an earlier and later model (of the same class), eight different trams were actually observed across both of the sites.

Figure 2 decomposes the data collection in each of the sites by showing the number and duration of each of the methods and the number of drivers that took part. Six cab rides with a total of 7 h of observation were performed. The unstructured interviews were undertaken with 10 drivers. As shown in Fig. 2, the physical ergonomics assessment was only performed at one site, though this was on two tram classes common to both. The focus groups were also undertaken in one site, though this was the larger network of the two with a much greater variety of tram types.

2.5 Data Analysis

Like the E/HF methods used to collect data, the analysis was largely qualitative. For the ergonomics assessment,

Fig. 1 Overview of the
methodology

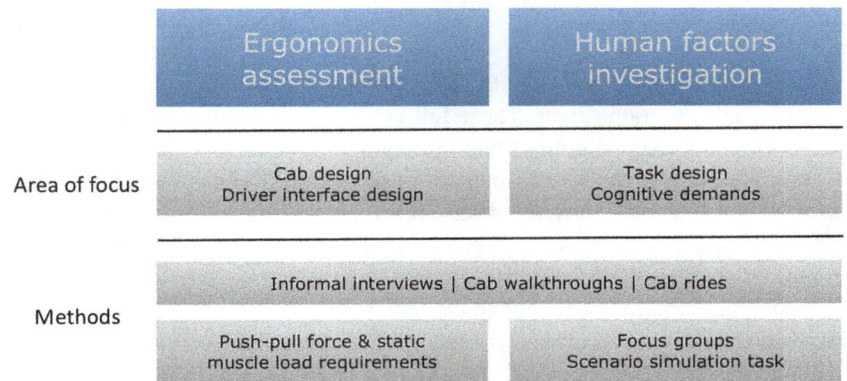

Ergonomics assessment	Human factors investigation
Area of focus	
Cab design Driver interface design	Task design Cognitive demands
Methods	
Informal interviews | Cab walkthroughs | Cab rides	
Push-pull force & static muscle load requirements	Focus groups Scenario simulation task

Fig. 2 Decomposition of
activities in each of the data
collection sites

Site 1	Site 2
Component	
Ergonomics assessment Human factors investigation	Human factors investigation only
Trams	
Walkthroughs: 2 | 2h	Walkthroughs: 6 | 7h
Cab rides: 2 | 1.5h	Cab rides: 4 | 5.5h
Participants	
Unstructured interviews: 4	Unstructured interviews: 6
	Focus groups: 4 (4, 4, 3, 4)*

* Denotes the number of participates in each of the four focus groups

the impact of the controls on postural and muscle fatigue was evaluated by analysing the objective force assessments in 3-dimensional static strength modelling software. Further, the organisations' own statistics associated with the history of injury in the use of their trams was reviewed and the physical loads in relation to injury causing factors were examined. These assessments identified target areas for changes to the physical cab. For the human factors work, data from the focus groups and informal interviews was transcribed and analysed with a number of techniques. Consistent with the aims of the research and scope of this paper, we report specifically on results that were identified through a process of thematic analysis applied to data from both the ergonomics and human factors components. This is in order to generate clear themes that meet the aims of the research and fulfil the objectives of the paper.

The themes and conceptual groupings were derived from phrases and comments from the transcripts, which grounded the findings in the data [21]. The initial findings were refined into overarching groupings and discussed to reach agreement on the codes that were used. This analysis was undertaken in multiple rounds and included a process of constant comparative analysis and cross-data validity check to determine consistency of the analysis and allow for further refining [22, 23, 24]. To achieve this, the data were checked regularly against other sources, such as photos of different cab designs from the various classes, scenarios from the focus groups, and injury statistics. The groupings and emerging themes were then checked with subject matter experts and tram drivers (i.e. end-users) in the participating organisations [25, 26].

3 Results

Before presenting findings from the thematic analyses, we will set the context by briefly describing our observations of the *driver–tram system* in terms of the type of cognitive work involved in the task. After this, we will provide three themes that appeared to evidence very little specification of E/HF considerations in the design of the cab for human use.

The themes comprised design considerations associated with the cab, the driver–cab interface, and the task.

3.1 The Driver–Tram System

The driver and the tram operated in a transportation system that was expressed in terms of how well-traffic moved. There were signals, signs, points (i.e. switches that selected the route), and rules of the road, all of which defined issues of flow and required technical knowledge. Most of the observed trams served as a streetcar in the main; that is, they moved on the road with adjoining vehicles but also on their own rail corridors (i.e. a dedicated light rail corridor section). Some sections of track necessarily had less tram and road vehicle separation, largely as a result of increased passenger densities. The task of tram driving itself was observed to have a number of distinct features. First, it was highly dynamic, both in terms of the number and frequency of tasks needing to be performed. The consensus from participants was that they performed three core tasks: (1) preparing the tram so it was safe before starting the shift; (2) ensuring a smooth journey; and (3) providing good customer service within the time constraints. These goals were closely coupled and expertise was characterised by the ability to regulate them well. Trams could move too fast or too slow with little effort, meaning that they could encroach easily beyond points of no return with relative ease. This included road cross hatchings representing "no mans land" with little clearance for turning trams, going through red lights into road traffic, or driving over points that were set incorrectly.

The complexity in the task was defined by the goals of comfort (braking and accelerating smoothly), time-accuracy (maintaining the stopping pattern), and regulating speeds effectively. Thus, the task incorporated an autonomously driven process on two dimensions: first, the driver controlled the acceleration and braking of the tram, not the speed itself; and second, there was a time requirement to meet with service delivery imposed at higher levels. Analysis of the task revealed six main subtasks, each of which was directly related with one or more of the three overarching goals. These are shown in Fig. 3. All but one (vigilance of the road and its users) was directly related to the three core tasks.

Throttle and braking was complex and drivers needed to accurately estimate the influence of gradients and changing conditions on speed in order to do it efficiently. Tram handling was subject to variation as a result of changes to wheel adhesion brought on by changes in weather. Whilst drivers could deploy sand to improve adhesion, some slip-inducing agents were not always visible (e.g. millipedes). Thus, much like train driving, tram drivers spoke of driving by the "feel of the tram" and by "instinct" highlighting the

role of non-technical skills. The driver in the driver–tram system needed to be attentive, vigilant, and remain highly aware of the environment as it evolved around them. They also needed to think and react very quickly, but as a human, they were susceptible to fatigue and psychological impacts associated with shift work. It was also easy to get distracted by sources related to the task, such as time pressure, as well as non-task-related sources, such as passenger chatter. The final element of the system were the goals that informed decision-making processes. The driver needed to regulate safety against productivity, but in practice, pressures and other motivations influenced this process. Cultural norms within the organisation and environment (e.g. road-user behaviour), and other social influences meant that the way a driver operated a tram in one site was likely to be very different from another.

3.2 Cab Design: Confusion Hath Now Made his Masterpiece[1]

Multiple classes and types of tram operated on the networks represented by both sites. Numerous differences in cab design were observed for trams operating on the same network, attracting potential for conflicts when moving between cabs. In one tram for example, sounds were easily masked by other alarms. In another tram (on the same network), the alarms were very loud, such that they frequently "startled" the driver. For example, participants considered the buzzer associated with vigilance checks on one tram to be overly loud, stress inducing, and resonate in the ears long after it had gone. These issues illustrated problems in the design of auditory icons, inappropriate urgency and hazard matching of cab alarms, and ultimately, design issues with *external* consistency. Although the perception of sound was based on subjective assessment, they appeared to contravene recommended noise design levels (i.e. no greater than 85 dB(A) for any length of time and 15–25 dB(A) above this) for alarms [27]. Problem or difficulties with inconsistency are usually a result of different systems rarely observing the same design standards [14].

Issues were also observed with the design and placement of buttons, such that participants confused some of them during walkthroughs. Figure 4 shows a tram cab where eight buttons were integrated into the armrest of the driver seat. The buttons showed very low discrimination in design—that is, they were identical to feel and touch, which increased head-down activity during driving. They also had no backlighting, which created problems in low light/night driving. To overcome this, some drivers indicated that they "pulsed the [cab] light" to improve

---

[1] Shakespeare, W. (In 2.3. Macbeth).

Fig. 3 Overview of the six subtasks associated with three core tasks in tram driving

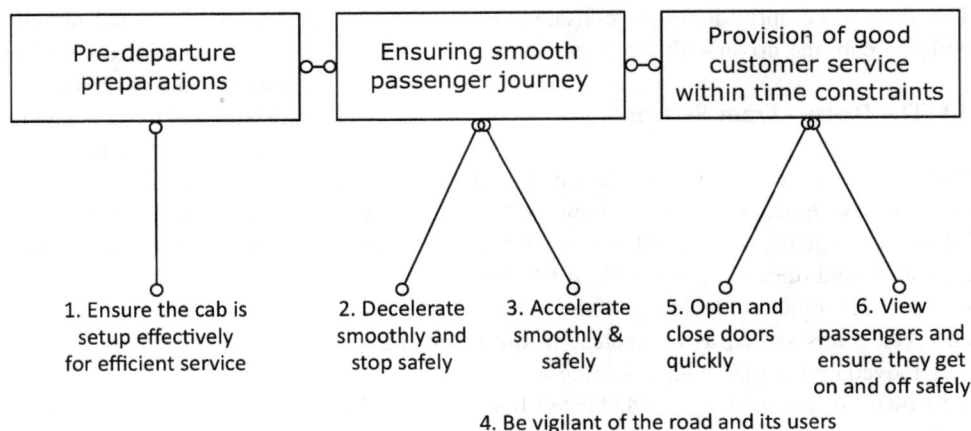

1. Ensure the cab is setup effectively for efficient service

2. Decelerate smoothly and stop safely

3. Accelerate smoothly & safely

5. Open and close doors quickly

6. View passengers and ensure they get on and off safely

4. Be vigilant of the road and its users

visibility whilst others used a torch. Both of these strategies increased task loading, but the problem was exacerbated by light reflecting off the windscreen, which restricted visibility even further. Generally speaking, there was no common design language in the quantity and clustering of buttons. In focus groups, few participants who operated this tram were able to recall the exact location and position of buttons, though some indicated that cab designers had consulted a small reference group of tram drivers as part of the design process. The categories of findings in the cab design theme reflected several departures from Norman's [14] design principles, particularly with visibility, feedback, mapping, constraints, and consistency.

3.3 Interface Design: Action is Eloquence[2]

The master controller was the primary means of controlling the tram and integrated throttle and braking commands into a single lever. These were mapped to the direction of travel (i.e. moving the master controller forward moved the tram forward). Participants described a process of *working* with the master controller to understand how throttle manipulations influenced tram speed. This was a non-technical skill and descriptions supported the joint-cognitive systems view [12]. The master controller is ultimately associated with the safety function of *ensuring a smooth passenger journey* task and subtask, as shown in Fig. 3. The problems observed with the master controller fell into a number of categories and included designs with low correspondence between body and arm posture, variation in reach distances, and forearm pads that did not facilitate contact with all master controller notch positions. For example, the master controller in Fig. 4 was integrated into the right-arm of the driver seat and could only be adjusted by pivoting the armrest up or down at the point it joined the rear of the chair. The way that master controller movements graduated

---

[2] Shakespeare, W. (In 3.2. Coriolanus).

with accelerating and braking also varied between trams, within and between the two sites. Predicting changes in tram speed was therefore less reliable and reported as a contributing factor to work-related stress.

Some master controller designs required application of sustained push force via spring-loading; that is, they were designed to fall back into the brake position if the driver were to let go (i.e. a "deadman" device). One master controller required a static force exertion that exceeded the recommended duration, and as such, was considered to be the main cause for incidence of musculo-skeletal injuries in that tram. During cab rides, we observed frequent changes in the hand positions of participants who operated this type of master controller (see Fig. 5). The same drivers did not show these in conscious demonstrations, suggesting that the process occurred *enactively* [28] (i.e. operated outside of conscious awareness), likely as a strategy to distribute physical loading and fatigue across different muscle groups.

Most rail-based systems use features that perform regular "vigilance" checks. According to standards for generic locomotives and driving coaches, these devices must be capable of stopping the tram in the event that the driver is incapacitated [11]. For this reason, they are usually designed with a contact-point (e.g. a button) that requires frequent contact; the idea being that it acts as a safety measure in the event of physiological breakdown (e.g. from malaise, sleep). In one tram, the maximum holding time for one of these devices was 12 s with 3 s for maximum release. Thus, if the driver held down the button for more than 12 s (without carrying out an action) or released it for longer than 2 s, a buzzer sounded after which the driver had 2 s to carry out an action to prevent the emergency brakes from activating.

The timing given to respond to these devices varied for different trams, between and within the two sites, meaning that the driver had much longer to respond to the device in one type of cab than in another. The mechanism used to

Fig. 4 Buttons that have been integrated into the armrest of the cab. *Left* photo shows whole seat with buttons (*circled*); *right* photo shows a close-up

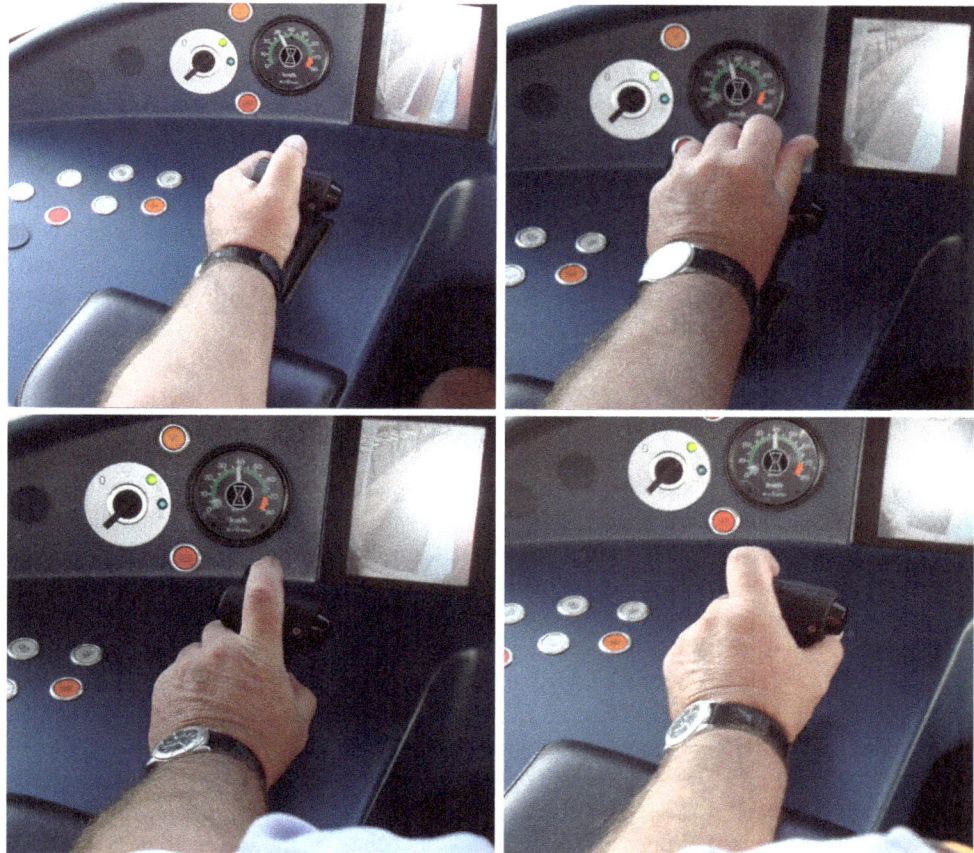

Fig. 5 Matrix of photos showing examples of variation in hand positions observed when operating a master controller that required sustained push force

reset the device also varied. In some it was a foot pedal, in others, a sensitive thumb button/contact, and others had both. In some trams, the driver could not swap between the thumb and foot even though cab design indicated this possibility. In one tram, the system required drivers to touch a sensor on the master controller with their thumb,

but it was reported to have poor responsiveness. During cab rides, it was observed being *tapped* and *swiped*, or *stroked* regularly in clockwise and counter-clockwise motions. This behaviour increased task loading, but also appeared to be a habituated process used to overcompensate for the warning noise (see Cab design theme) and prevent it from sounding. Analysis from the interface design theme identified a number of departures from Norman's design principles [14], namely, feedback, mapping, and consistency.

3.4 Task Design: Time is Out of Joint[3]

Tram driving requires the drivers to wrestle between safety and productivity goals. For some drivers, this generated time pressure. Although this was considered a norm for the task, the experience was intensified by the design of the vigilance task. The focus on time had the effect of taking attention away from other tasks, such as stopping the tram smoothly, opening and closing doors efficiently, and confirming that passengers were getting on and off safely. Further, the design of the task reflected these tensions; as an example, the doors on the trams had sensors, which prevented them from being closed if they were obstructed (e.g. by a person, bag, push-chair). However, the trams also had a *force-door* close button, which could override the passenger detection sensors and close the door. Many drivers confessed to using this button over the regular door close button, in spite of organisational policy.

In newer trams, drivers were presented with a digital time-keeping performance indicator in their interface to show how fast or slow they were moving between stops. This information was updated in real time as a pseudo-estimated time of arrival or ETA feature. Indeed, some participants were observed using these changes in time to regulate the master controller (and therefore tram speed) instead of the information displayed in the speedometer. Thus for some drivers, the time shown in this device was being used to parameterise their speed choices instead of the speeds on the road. Performance penalties for the tram industry were reportedly associated with being too early than for being too late. This was not surprising, given the focus on maintaining flow and minimising impact to other services. At one site, penalties were administered if the driver was more than 1-min early or more than 6-min's late. This was considered to create a "hurry up and wait culture," which emphasised time keeping and anxiety from lack of control in work pacing. This effect is supported by observations in the train system when station dwelling [20], but is of particular concern in the tram context, given that time pressure has been associated with negative emotions, stress, and the increased propensity for risk taking behaviours on roads [23, 29]. The theme of task design revealed a number of breaks from Norman's design principles, namely affordance (i.e. using changes in time from the ETA feature to change speed), mapping, and constraints.

---
[3] Shakespeare, W. (In 1.5. Hamlet).

4 Discussion

The E/HF issues we observed suggested to us that the standards being used to inform the design of older trams, as well as newer ones, may not have been specified well enough for the socio-technical complexity in the system. This was clearly evidenced in the three themes where there were multiple departures and compromises in good-practice design [14]. Most of the trams we observed were modern trams, one of which was undergoing commissioning and acceptance at the time of the research. Thus, they suggested that whilst standards were adequately specified, the process of consultation and end-user design was lacking. They may also have pointed to problems in translating standards into practice. The lack of involvement from E/HF designers and/or good E/HF practises in the design of the trams, particularly in their early stages, is likely to be a contributing factor. However, as designers, managers, and suppliers involved in the tram procurement process were not interviewed as part of the work, this is difficult to substantiate. If these issues are indeed part of the problem, it is important to note that inviting the authors to conduct these assessments is a sign of change. Participants indicated that tram drivers (i.e. end-users) had actually been consulted during the early part of the design process—particularly with respect to data presented in the theme of cab design, and there was consensus to support this. However, given the observed issues with the resulting design (e.g. placement of buttons in the driver seat), the process is likely to have occurred at the cost of expert E/HF input—that is, the tram drivers could have been consulted under the premise that *they* were experts in design *as well* as experts in tram driving.

Many of our observations and findings pointed to important and relevant concerns of the impacts of cab design on work and stress, but there were also data to support this from the perspective of task design. There were strong indications that participants experienced time pressure and anxiety associated with regulating safety goals, such as driving smoothly, letting passengers on and off efficiently, and with their productivity goals, such as keeping to the schedule. These tensions may filter through to the task in the form of suboptimal (i.e. jerky) journeys, and door opening issues; therefore it is useful to monitor these events as a measure for risk. Key observations for E/HF communities, and standards design authorities were as follows:

- Inefficiencies or issues in vigilance device for the cab, interface, and task design spectrum
- Distraction and inattention to the tram-driving task from time pressure and over-emphasis on productivity goals

- Inordinately high forces required to operate the master controller
- Lack of forearm support when operating the master controller
- Inconsistencies in noise, particularly for auditory alarms
- Difficulties for operators to achieve the correct posture due to a combination of poor seating adjustments and console/cabin design.

4.1 Methodological Review

Human Factors integration is concerned with "providing a balanced development of both the technical and human aspects of equipment procurement" and essentially "ensures the application of scientific knowledge about human characteristics through the specification, design and evaluation of systems." [30, p. 6] The key advantage of the methodology used in this paper is that it was driven by a multidisciplinary approach, combining the expertise of mechanical engineers, human factors specialists, and ergonomists.

The methods that were used in the study substantially increased the ability for the investigators to understand the problems. A combination of discussions, generative simulation tasks, and observations of enactive behaviours on- and off-the-job enabled the participants to more easily describe their own knowledge and thought processes, but also overcame the problem of any inaccuracies of self-reporting from memory and decision processes [31]. As a qualitative process, the study provided a richness and depth, particularly given the integration of objective force assessments. However, the methods could be further strengthened by the addition of suitable quantitative methods (e.g. national surveys, simulator work).

4.2 Further Research

For the rail industry, considerations into the redesign of seating, the master controller, and the foot pedal are advocated to achieve more flexibility and correct posture for tram drivers when driving. Work is currently being undertaken in one site involving cab design modifications to determine optimum driver interface with the master controller and foot pedal based on the findings to date with the aid of mock-ups. This will be the subject of further reporting when the testing is completed. Further academic research should also consider the specific make-up of the tram-driving task using task analyses. Towards achieving this aim, more analysis of the data collected in this work may be used to develop a framework for further analysis, specifically in terms of how the task demands

interact with the specific ergonomic and human factors issues.

The findings evidence a general need to investigate any strategies for reducing risk-related stress in this population. This may start by examining the informal strategies to manage and mitigate stress associated with safety risk and performance conflicts. Lastly, it would be useful undertake more work to determine if and how suggested improvement in cab ergonomics filter back into tram classes when specifying bids, tenders, and contracts for new rolling stock.

5 Conclusions

A combination of physical ergonomics assessments and human factors investigations identified issues with design and tram operation in two sites in Australia. The findings suggest a dearth of ergonomics and human factors considerations of the design of the driver interface and the cab. This may be related to the practice of using non-specific standards for developing trams and/or poorly integrating human factors into their specification processes.

Acknowledgments The authors gratefully acknowledge the anonymous operators and contractors that invited the authors to assess their tram cabs. They are also very grateful to Ganesh Balakrishnan for his assistance.

Conflict of interest The authors believe there are no conflicts of interest (financial or otherwise) that impinge on the quality or impartiality of the research in the paper.

References

1. Currie G, Burke M (2013) Light rail in Australia—performance and prospects. Paper presented at the Australasian Transport Research Forum, Brisbane, Australia
2. Australasian Railway Association (2015) Capital metro: Canberra's light rail project in a global context. Author, Canberra
3. Nye B (2015) Light rail in Australasia: the economic, social and environmental case. In: Proceedings of 2015 light rail conference, NSW, AU March 5–6 2015
4. Macdonald A, Coxon S (2011) Towards a more accessible tram system in Melbourne—challenges for infrastructure design. Paper presented at the Australasian Transport Research Forum, Adelaide, Australia
5. Alstom Transport (2011) What will your Citadis be? Author. http://www.alstom.com/transport/products-and-services/trains/tramway-citadis/

6. Good Design Australia (2014) Good design awards 2014. Crowther Blayne, Surfers Paradise

7. Carey A (2012,) Tram cop a low blow as report slams design flaws. The Age (July 14)

8. Wickens CD, Hollands JG (2000) Engineering psychology and human performance, 3rd edn. Prentice-Hall, Upper Saddle River

9. Mitra B, Al Jubair J, Cameron PA, Gabbe BJ (2010) Tram-related trauma in Melbourne, Victoria. Emerg Med Australas 22(4):337–342. doi:10.1111/j.1742-6723.2010.01309.x

10. Middendorp C (2010) Hop on tram for on hell of a scary ride. The Sydney Morning Herald, December 23

11. Union Internationale des Chemins der fer (2009) Driver machine interfaces for EMU/DMU, locomotives and driving coaches—functional and system requirements associated with harmonised Driver Machine Interfaces. Author, Paris, France

12. Woods D, Hollnagel E (2006) Joint cognitive systems: patterns in cognitive systems engineering. Taylor & Francis, Boca Raton

13. Gould JD, Lewis C (1985) Designing for usability: key principles and what designers think. Commun ACM 28(3):300–311

14. Norman D (1988) The design of everyday things. Basic Books, New York

15. Preece J, Rogers Y, Sharp H (2002) Interaction design: beyond human-computer interaction. Wiley, New York

16. Foot R, Doniol-Shaw G (2008) Questions raised on the design of the "dead-man" device installed on trams. Cogn Technol Work 10(1):41–51

17. Cooke NJ (1994) Varieties of knowledge elicitation techniques. Int J Hum Comput Stud 41(6):801–849

18. Naweed A (2014) Investigations into the skills of modern and traditional train driving. Appl Ergon 45(3):462–470. doi:10.1016/j.apergo.2013.06.006

19. Naweed A, Balakrishnan G, Bearman C, Dorrian J, Dawson D (2012) Scaling generative scaffolds towards train driving expertise. In: Contemporary ergonomics and human factors 2012, pp 235–236

20. Naweed A (2013) Psychological factors for driver distraction and inattention in the Australian and New Zealand rail industry. Acc Anal Prev 60:193–204. doi:10.1016/j.aap.2013.08.022

21. Huberman MA, Miles MB (1994) Data management and analysis methods. In: Denzin NK, Lincoln YS (eds) Handbook of qualitative research. Sage, Thousand Oaks, pp 209–219

22. Charmaz K (2006) Constructing grounded theory: a practical guide through qualitative analysis. SAGE Publications Ltd, London

23. Naweed A, Rainbird S, Dance C (2015) Are you fit to continue? Approaching rail systems thinking at the cusp of safety and the apex of performance. Saf Sci. doi:10.1016/j.ssci.2015.1002.1016

24. Naweed A, Balakrishnan G, Bearman C, Dorrian J, Dawson D (2012) Scaling generative scaffolds towards train driving expertise. In: Anderson M (ed) Contemporary ergonomics and human factors 2012: proceedings of the international conference on ergonomics & human factors 2012. CRC Press, Blackpool, p 235–236

25. Powell C (2003) The Delphi technique: myths and realities. J Adv Nurs 41(4):376–382

26. Hsu CC (2007) The Delphi technique: making sense of consensus. Pract Assess Res Eval 12(10):1

27. Edworthy J, Hellier E, Noyes J, Aldrich K, Naweed A, Metcalfe GR (2008) Good practice guide for the design of alarms and alerts. Rail Safety Standards Board, London

28. Branton P (1979) Investigations into the skills of train-driving. Ergonomics 22(2):155–164

29. Cœugnet S, Naveteur J, Antoine P, Anceaux F (2013) Time pressure and driving: work, emotions and risks. Transp Res Part F 20:39–51. doi:10.1016/j.trf.2013.05.002

30. MoD (2000) Human factors integration: an introductory guide. HMSO, London

31. Bisantz A, Roth (2008) Analysis of cognitive work. In: Boehm-Davis DA (ed) Reviews of human factors and ergonomics. Human Factors and Ergonomics Society, Santa Monica, pp 1–43

11

Passenger Stability Within Moving Railway Vehicles: Limits on Maximum Longitudinal Acceleration

J. P. Powell[1] · R. Palacín[1]

Abstract Increasing the acceleration and deceleration of trains within a railway network can improve the performance of the system. However, the risk of passengers losing their balance and falling is also increased. The purpose of this paper is therefore to examine the effect of longitudinal vehicle accelerations on passenger safety and comfort. The literature review brings together two separate disciplinary areas, considering the effects of acceleration on balance from a physiological/kinesiological perspective, as well as looking at the results of previous empirical studies on the levels of acceleration that railway passengers will tolerate. The paper also describes an experiment carried out on the Tyne and Wear Metro, which gathered data on typical acceleration levels to compare against the findings of the literature review. It was found that both the magnitude of the accelerations and their rate of change (jerk) are important. The results also suggest that there may be scope to improve the trade-off between journey times, energy consumption and passenger comfort by fine control of the acceleration/jerk profile. This is particularly relevant to urban rail systems, as they typically feature relatively high acceleration and deceleration. However, the findings for passenger comfort are equally applicable to conventional regional and intercity services.

Keywords Passenger safety · Passenger comfort · Longitudinal acceleration · Braking · Jerk

✉ J. P. Powell
  j.powell2@newcastle.ac.uk

[1] NewRail - Centre for Railway Research, Newcastle University, Stephenson Building, Claremont Road, Newcastle upon Tyne NE1 7RU, UK

Editor: Baoming Han

1 Introduction

An increase in the level of acceleration and deceleration achieved by a train allows a reduction of journey time (and a potential increase in railway system capacity), or alternatively a reduction in energy use for a given journey time. However, higher levels of longitudinal acceleration can compromise passenger comfort, and ultimately safety too if they are sufficient to cause passengers to lose their balance.

This has been highlighted as a significant cause of injury for bus passengers [10], with research suggesting that 'accelerations that are commonly encountered in practice appear to be impossible to endure without support [*such as handgrips*]' [7]. Although bus accelerations are typically higher than trains, passengers in a railway vehicle are more likely to be standing unsupported, or moving around within the vehicle. In Great Britain, the Rail Safety and Standards Board estimate that around 15 % of on-board harm in the railway network in the last 10 years (measured by fatalities and weighted injuries) can be attributed to 'injuries attributable to sudden movements of the train due to lurching or braking' [21].

The purpose of this paper is therefore to examine the effects on railway passengers of the longitudinal accelerations found in regular operation, and the relationship to comfort and safety.

2 Methodology and Paper Structure

The outline methodology of this paper consists of two parts. The literature review in Sect. 3 first examines the biological theory unpinning balance, and then reconciles this with the results of previous empirical studies into the limits of acceleration that passengers will tolerate.

Section 4 describes an experiment undertaken on the Tyne and Wear Metro to measure actual railway vehicle accelerations during regular operation, which are then compared against the findings from the literature, and some conclusions drawn in Sect. 5.

3 Passenger Balance and Stability

3.1 Physiology/Kinesiology

Balance in humans is an unconscious proprioceptive reaction, coordinated by the brain stem, supported by the cerebellum, visual cortex and basal ganglia. Information is obtained from the somatosensory system in the feet, the vestibular system in the inner ear and visual stimuli from the eyes [4]. The somatosensory system detects changes in pressure on the sole of the feet. When there is an imbalance between one foot and the other, it stimulates the muscles in the leg to contract so that the leg stiffens to oppose the increased pressure. The vestibular system consists of the semicircular canals and the otolith organs, and movement of fluid within each of these is detected by cells that stimulate the central nervous system. The semicircular canals provide a static response (effectively measuring position), and so help to stimulate corrective or predictive body movements such a stepping. The otolith organs provide a dynamic response and so control reflex reactions, such as flexing the body to change position. Finally, visual stimuli from the eyes provide an extra frame of reference to help determine position more accurately.

Following the above, three different strategies can be identified for retaining balance under the influence of an external acceleration. Where the acceleration is small, contracting the leg muscles and bending the ankle is sufficient to react against the external acceleration and keep the body balanced; this is known as ankle strategy. If the acceleration magnitude is greater, the body must change position to prevent falling, also bending at the hip. This is known as hip strategy, and requires a longer time for the muscles to actuate. Finally, the applied acceleration may be large enough that one or more steps must be taken to avoid falling; this is the stepping strategy.

Unconscious control of these strategies is a negative feedback system, therefore both the magnitude of the external acceleration and its rate of change (jerk) are important [31]. This implies that both strength and sensing/actuation times of the body's muscles and nervous system must be considered when investigating the case of passengers balancing within a moving vehicle.

Where the jerk is very high, passengers will not have sufficient time to react, and their behaviour can be approximated by a static rigid body. This will topple when the line of action of the resultant force (due to external accelerations acting on it) lies outside of the base of support. This force will act through the body's centre of gravity, and for an average human this is approximately located at 54 % of their height, in line with the front of their knee/ankle joints in a normal standing posture [24, 30].

The minimum time for muscles to react against external forces is typically 0.12–0.13 s [2, 17], and for the body to make larger movements to retain balance takes around 1 s [25]. These figures may be considered to approximate the cases of the ankle and hip strategies respectively.

For a lower jerk level, the maximum tolerable acceleration is greater as the muscles have more time to actuate and resist the force. Where the jerk is very low, the strength of the individual will be the only important human factor, as the acceleration is changing slowly enough for the body to fully react and change posture as required.

Within a given population, there will be significant variation in the ability of individual passengers to balance under the influence of a given level of acceleration and jerk, in accordance with their physiology [4]. This variation means that it is difficult to set universal acceleration/jerk limits for passenger safety. It also means that, depending on their individual reactions to maintain balance, different passengers will have different perceptions of how uncomfortable a particular level of acceleration/jerk is.

3.2 Review of Previous Experimental Work

Empirical research into the levels of longitudinal acceleration that passengers will tolerate can be broadly classified into subjective and objective studies. A review was carried out by Hoberock [13] that included a mixture of both types, partly based on the work of Gebhard [9].

Subjective studies typically use questionnaires and interviews with study participants in order to establish how comfortable different acceleration profiles are for different people. A study by British Railways [16] was carried out to determine the effects on passengers of quasi-static lateral accelerations due to track curvature. For standing passengers, 0.1 g was given the approximate limit that could be attained without discomfort, and around 0.12 g was defined as uncomfortable. Values for seated subjects were somewhat higher, and it was also noted that lower levels of jerk increased the aforementioned acceleration limits. These values were found to have reasonable agreement with previous research in Britain [27] and by South African Railways [22]. The paper also demonstrated that passenger comfort on curves was generally a more limiting case than safety against derailment, and this formed the basis for British track design standards [6].

Hoberock also reported on similar experiments carried out by Japanese National Railways to assess passenger comfort during braking [18, 19], which also considered the effects of jerk. A later study [28] also included an evaluation of whether a given level of deceleration was acceptable to passengers, in addition to the assessments of comfort. It was noted that that the comfort ratings and the acceptability of different decelerations did not always correlate.

This research has been developed further by the Railway Technical Research Institute (RTRI) in Japan. Hiroaki [11] used questionnaires to examine the effect of high jerk values on the acceptability of different levels of acceleration. Curves for the acceptability of different levels of acceleration/jerk were produced, and an example for a group made up of regular commuters is illustrated in Fig. 1. Other curves were presented for occasional regional/intercity travellers, and the acceptability of a given acceleration/jerk level varied significantly depending on the type of passenger and journey being undertaken.

Overall, the variability in the methods of the subjective studies highlighted in this section means that they can only provide a general indication of what can be defined as acceptable or unacceptable levels, especially given the sensitivity of results to individuals' opinions or interpretations. Nonetheless, these studies confirm that both jerk and acceleration influence passenger comfort and stability, and also that unsupported standing passengers facing the direction of the vehicle's acceleration have the lowest tolerance.

Objective studies seek a quantifiable measure of people's reactions to external accelerations, rather than relying on their perception and opinion, and two significant studies were detailed by Hoberock. Hirshfeld [12] reported on a series of experiments intended to determine the effects of longitudinal acceleration on the loss of balance of standing passengers, as part of a wider design programme for the standardised PCC streetcars in the USA. Participants in the study stood on a platform that moved with variable acceleration profiles, and the average value of acceleration at which they either took a step or grabbed a handrail was measured. The study confirmed that different levels of jerk (for the same acceleration) influence the retention of balance, and that unsupported forward-facing passengers were least tolerant, losing their balance at an average of 0.13 g. The combined average for all unsupported standees was 0.165 g, increasing to 0.23 g with an overhead strap for support and 0.27 g with a vertical grab rail.

The experiments carried out by Browning [3] had similar objectives and methodology, although as part of a programme looking at the design of moving pedestrian walkways. The results were categorised in terms of the observed movement of the participants by an expert panel and presented in terms of acceleration against rise time (where jerk equals acceleration divided by rise time). Curves were produced for the approximate limits for each of the observed movement categories, and these are illustrated in Fig. 2.

Based on both the subjective and objective studies, Hoberock's principal conclusion was that it is difficult to set conclusive limits on acceleration and jerk, as passenger's reactions strongly depend on the individual concerned. A range was nonetheless suggested for maximum permissible accelerations of 0.11–0.15 g as an outline guide, with jerk limited to 0.30 g/s.

A more recent review [8] proposed 'large movement' in Browning's results to be approximately equivalent to Hirshfeld's case of passengers either stepping or requiring external support. The review also included acceleration values at which seated passengers start to be dislodged from their seats, based on the results of Abernethy et al. [1]. A limit for transverse (forward- or backward-facing) seats was given as 2.45 m/s^2, well above the guidelines for standing passengers, but a lower limit of 1.4 m/s^2 was given for longitudinal (side-facing) seats.

RTRI have also carried out further experimental studies that combine subjective and objective approaches [15], investigating the jerk limits required for high deceleration levels to be acceptable to passengers. The two graphs in Fig. 3 illustrate the data points and fitted curves for acceptability (left) and the ability of passengers to retain their balance (right) with four jerk levels.

A different type of experiment was carried out by Kamper et al. [14], in which the postural stability of a small group of wheelchair users with tetraplegia or paraplegia was examined under the influence of quasi-static accelerations typically found in road vehicles. 95 % of the participants were able to retain balance within the wheelchair at an acceleration of 0.126 g, and the average at which balance was lost was around 0.22 g. Balance retention was improved with a lower level of jerk.

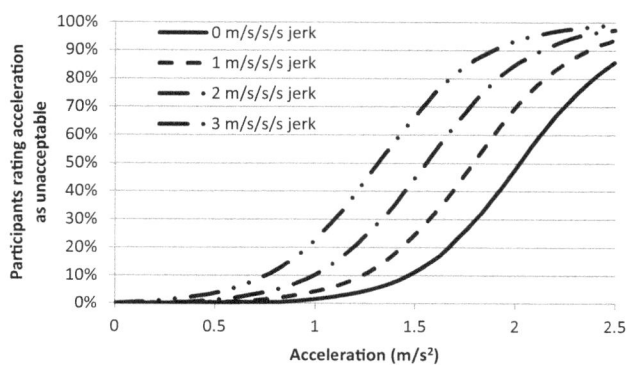

Fig. 1 Acceptability of acceleration/jerk levels [11]

Fig. 2 Browning's results, as summarised by Dorn [8]

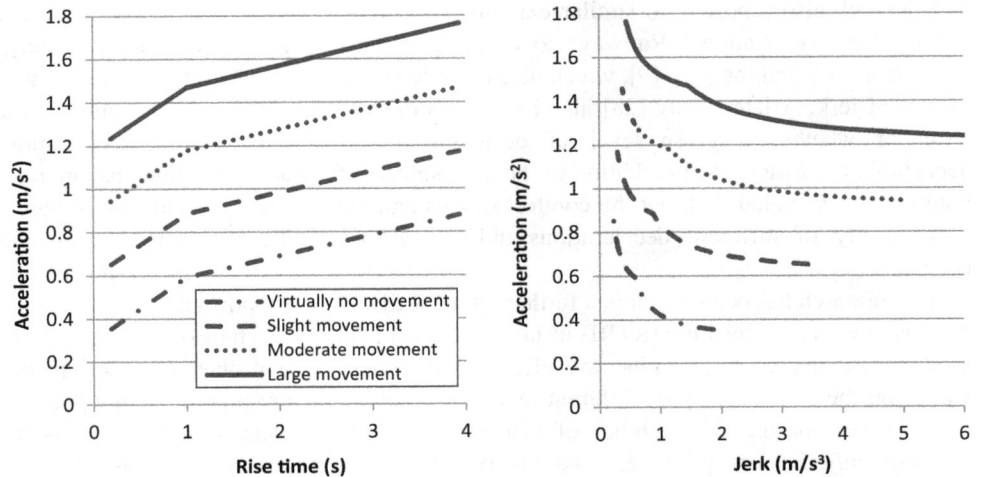

Fig. 3 Subjective and objective results for forward-facing standing passengers [15]

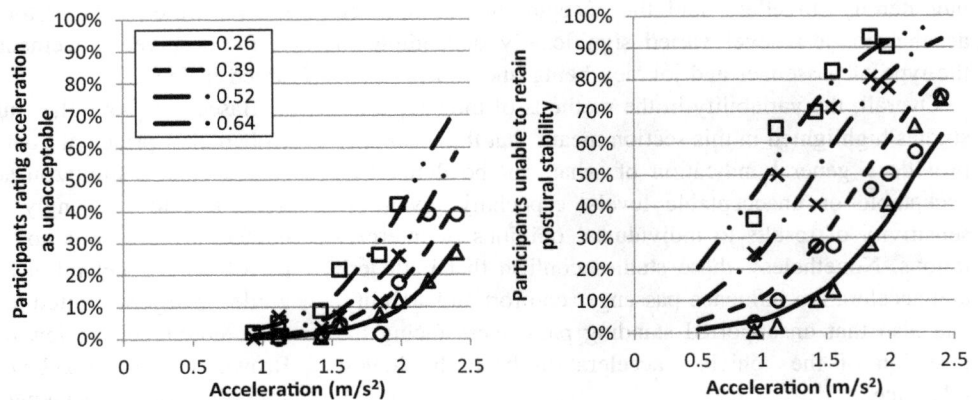

Finally, Sari [23] considered the likelihood of passengers walking within railway vehicles falling under the influence of low frequency (0.5–2 Hz) lateral oscillations. Although not directly applicable to this paper, as the accelerations are transient and the passengers are already in motion, it nonetheless provides a useful point of reference for comparison. A range of accelerations between 0.1 and 2.0 m/s^2 were tested across the range of frequencies. As may be expected, balance was generally lost at a lower acceleration level than the results for standing or seated passengers highlighted in this paper. Likewise, it was reported that results also varied with frequency of the oscillations, effectively the rate of change of these accelerations.

3.3 Current Practice

Although guideline figures for the safe limits of longitudinal accelerations in railway vehicles are used when specifying rolling stock, the source of the values is often unclear and can vary significantly [5, 20, 29]. Table 1

provides some examples from main line and light rail vehicles in Great Britain, from data provided by vehicle manufacturers/operators.

Note that some values given are estimates and should be taken as representative rather than exact. The values are the absolute maxima, and the average values achieved during braking in regular service are typically rather lower—this is illustrated further in Sect. 4 for the Tyne and Wear Metro. The operators also noted that emergency track brakes (indicated by an asterisk * in Table 1) are used as a last resort, as experience has shown their use carries a high risk of passenger injury.

3.4 Findings

Passenger safety becomes an issue when the acceleration/jerk levels require passengers to take one or more steps to retain balance (the stepping strategy), as this introduces the risk of falling. Passenger comfort is a more subjective measure, but may be considered quantitatively as how

Table 1 Example maximum accelerations for railway vehicles in Great Britain

Vehicle	Maximum acceleration (m/s^2)		
	Traction	Service brakes	Emergency brakes
Class 390 Pendolino (intercity EMU)	0.37	0.88	1.18
Class 156 Super Sprinter (regional DMU)	0.75	0.7–0.8	0.7–0.8
Class 323 (suburban EMU)	0.99	0.88	1.18
London Underground 1992 tube stock	1.3	1.15	1.4
Tyne and Wear Metrocar	1.0	1.15	2.1 (*)
Manchester tram (Ansaldo T-68)	1.3	1.3	2.6 (*)
Sheffield Supertram (Siemens-Düwag)	1.3	1.5	3.0 (*)
Croydon tram (Bombardier FLEXITY)	1.2	1.3	2.73 (*)
Nottingham tram (Bombardier)	1.2	1.4	2.5 (*)

close a particular individual is to their own limit of balance. This correlation is not exact however.

The RTRI studies [11, 15] provide an overview of the acceptability of acceleration/jerk for a population, while the results from Browning [3] provide some insight into the effects on individuals. There is a distinct change in the response at a rise time of around 1 s, and it is proposed that this is due to the unconscious change from ankle strategy to hip strategy. For low values of jerk, or high values that correspond to a rise time of less than 0.12 s, the acceleration value becomes independent of the jerk. It is instead related only to the strength of individual passengers for low values of jerk, or their posture and location of their centre of gravity for high values.

There is considerable variation between the perceptions and stability of different individuals however. This can be observed in the scatter in the results of Fig. 3, and more generally by the differences in the findings of the studies reviewed. It is therefore not possible to set precise passenger limits for longitudinal acceleration for passenger safety. Nonetheless, it can be concluded that previously suggested guidelines of 1.1 to 1.5 m/s^2 are reasonable—this is reflected in the current railway practice illustrated by Table 1. The values also suggest that passengers are likely to be more tolerant of discomfort on short metro-type service by comparison with intercity/regional services.

4 Measurement of Accelerations in Service

4.1 Introduction to Experimental Work

An experiment was carried out on the Tyne and Wear Metro to measure the accelerations actually experienced by passengers in service, with the aim of comparing the findings of Sect. 3 against a mixture of subjective (qualitative) and objective (quantitative) data.

The Tyne and Wear Metrocars are towards the upper end of the range of accelerations given in Table 1. The Metro infrastructure consists of old railway alignments converted to Metro use, a new tunnel through Newcastle upon Tyne city centre (built specifically for the Metro), and sections of track shared with current heavy rail services. It therefore also features a range of curvature and gradient values typical of railway systems, including the highest values likely to be found on railway infrastructure. Although on-street tram systems can feature more extreme accelerations, passenger behaviour in trams is likely to be more similar to buses than railway vehicles. The Metro therefore provides a good case study for the purposes of this paper.

4.2 Experimental Methods

An afternoon peak time diagram was chosen for acceleration measurements, running empty from South Gosforth depot to Regent Centre, then in passenger service from Regent Centre to Pelaw and Pelaw to Monkseaton, before returning empty back to the depot.

The equipment consisted of three triaxial accelerometers with a range of ±18 g and a multi-channel data acquisition system with a sample rate of 128 Hz. This was set up within the depot in the B carriage of Metrocar 4007, on the double seat adjacent to the innermost door set (door 5). One accelerometer was glued to the overhead grab rail, one glued to the seat back, and one placed on a metal plate on the seat, sited so that passenger interference would be minimal. These are illustrated in Fig. 4. The accelerometers were calibrated by rotating through ±90° before final placement, effectively a 2 g inversion. The track in the depot where the equipment was installed was close to straight and level, minimising any offset in the readings (due to track geometry) during calibration.

During the test, a log was kept of arrival and departure times at each station, any other significant events (such as signal checks) and approximate passenger loadings. Notes

Fig. 4 Experimental apparatus in situ on Metrocar 4007

were also taken throughout the journey to qualitatively describe the comfort level at different locations.

4.3 Results

Figure 5 illustrates an example set of results for the accelerations measured in the longitudinal, lateral and vertical directions during the tests. The profile illustrated here includes section of running on old main line railway alignments, the tunnel designed specifically for Metro trains and track shared with heavy rail services.

Given that passengers may stand facing in any direction within the vehicle, the resultant of the lateral and longitudinal acceleration was calculated for each measured point and filtered to remove high frequency vibration in order to give the maximum quasi-static acceleration in the horizontal plane. For a static body, a vertical acceleration changes the effective weight (but not mass) of the body, changing the point at which it will topple. This mechanism was assumed to also apply to the cases of ankle and hip strategy, and the resultant horizontal plane acceleration was modified accordingly. The corresponding jerk for each measured point was then obtained by dividing the change

in quasi-static acceleration between adjacent points by the sample time.

The wide variation in assessments of passenger comfort between different studies has already been noted. Therefore, the acceleration/jerk pairs from the measured data were compared against the proposed curves in both Figs. 2 and 3. There were several locations where the measured acceleration/jerk pairs were highlighted as problematic by these methods, and these matched up well with the subjective observations on comfort recorded during the test. These locations were therefore analysed to look for patterns that might suggest how to improve passenger comfort.

4.4 Discussion

The majority of the cases where acceleration/jerk levels were outside the proposed limits were found to be when the train was stopping at a station, as it came to a standstill. By definition, the jerk approaches infinity at the moment speed equals zero when stopping, and therefore a lower acceleration magnitude is necessary. It is common driving practice to reduce the braking effort demanded as the train comes to

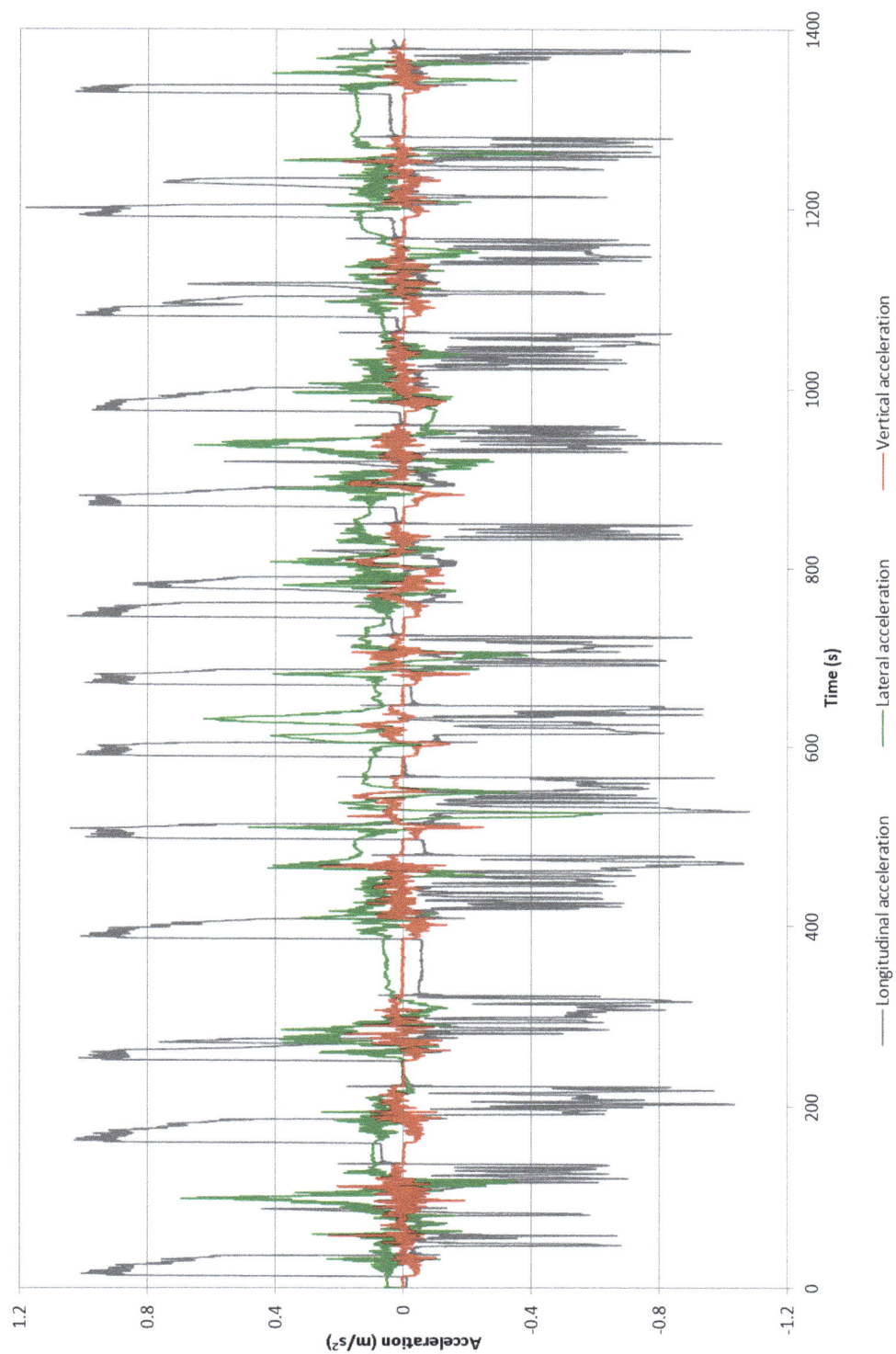

Fig. 5 Sample acceleration profile

a standstill to address this. Sone and Ashiya [26] provide an example of how pure electric braking (rheostatic or regenerative, using three phase AC traction motors) can achieve superior passenger comfort in this respect.

The worst case measured was when traction power was cut off at a relatively low speed while accelerating, in order to meet a speed restriction. The Tyne and Wear Metrocars have camshaft resistance control of DC traction motors, and although the first couple of camshaft steps limit the jerk when power is first applied, power is cut off abruptly by contactors when the combined power/brake controller is returned to neutral. As in the previous case, this results in a large jerk. However, the camshaft control also means that driver has less control over the level of tractive effort by comparison with braking, and for relatively low speeds the train's acceleration will be close to its maximum level. This situation occurs far less often than station stops, but passengers are less likely to be able to predict and anticipate it, and the subjective observations on the Metro suggest it is also the most likely reason for passengers making large movements to correct their balance.

There were also a few cases where the train changed immediately from accelerating to braking, or where the driver was moving the controller frequently between discrete brake notches while decelerating. This can be observed in Fig. 5, where the negative longitudinal accelerations during braking show significantly more variation than the positive longitudinal accelerations. The locations at the cases highlighted generally also coincided with specific infrastructure features that influence the acceleration and jerk levels, such as sharp curves and their associated transitions, which increased the lateral acceleration and jerk. Automatic Train Operation (ATO) can mitigate this variation and provides a more consistent control of braking effort, improving passenger comfort.

Overall, resultant quasi-static accelerations were routinely observed approaching 1.4 m/s². The majority of these cases were found to be acceptable by both the qualitative observations during the test and the quantitative data collected—the locations identified as problematic were not necessarily the locations with a high acceleration level, but in all cases did involve a high value of jerk.

5 Conclusions

Passenger tolerance to longitudinal accelerations varies significantly between different individuals, according to their physiology and psychology. The acceptability of a given level of acceleration depends strongly on the rate of change of the acceleration (jerk).

Acceleration and jerk limits are typically given as single figures in rolling stock specifications. However, this paper suggests that there may be scope to improve the trade-off between journey times, energy consumption and passenger comfort by fine control of the acceleration/jerk profile. ATO and electric braking (using three phase AC traction motors) are likely to be a prerequisite for the level of control required however.

This research is particularly applicable to urban rail systems, as they typically operate vehicles at higher acceleration and deceleration levels than conventional regional or intercity passenger trains. However, passengers on regional or intercity services are likely to expect a higher level of comfort than urban rail passengers, and control of the jerk then becomes important in this respect.

Acknowledgments The authors would like to thank Dr Kazuma Nakai of RTRI for his assistance with the Japanese research and helpful suggestions for the paper, and Nexus for the opportunity to carry out experimental work on the Tyne and Wear Metro (the work described here was carried out when the Metro was managed as an integrated system, prior to train operations being split from infrastructure and contracted separately).

References

1. Abernethy CN, Jacobs HH, Plank GR, Stoklosa JH, Sussman ED (1980) Maximum deceleration and jerk levels that allow retention of unrestrained, seated transit passengers. Transp Res Rec 774:45–51
2. Allum JHJ (1983) Organization of stabilizing reflex responses in tibialis anterior muscles following ankle flexion perturbations of standing man. Brain Res 264(2):297–301
3. Browning AC (1972) The tolerance of the general public to horizontally accelerating floors, with special reference to pedestrian conveyors. Technical Report TR71105, Royal Aircraft Establishment, Ministry of Defense
4. Carpenter RHS (2003) Neurophysiology, 4th edn. Arnold, London
5. Cole C (2006) Longitudinal train dynamics. In: Iwnicki S (ed) Handbook of railway vehicle dynamics. CRC/Taylor & Francis, Boca Raton, pp 239–277
6. Cope GH (1993) British railway track: design, construction and maintenance, 6th edn. Permanent Way Institution, Barnsley
7. De Graaf B, Van Weperen W (1997) The retention of balance: an exploratory study into the limits of acceleration the human body can withstand without losing equilibrium. Hum Factors: J Hum Factors Ergon Soc 39(1):111–118
8. Dorn MR (1998) 'Jerk, acceleration and the safety of passengers' Technology for Business Needs. Birmingham, 24th–26th November
9. Gebhard JW (1970) Acceleration and comfort in public ground transportation. Transportation Programs Report 002, The Johns Hopkins University
10. Halpern P, Siebzehner MI, Aladgem D, Sorkine P, Bechar R (2005) Non-collision injuries in public buses: a national survey of a neglected problem. Emerg Med J 22(2):108–110

11. Hiroaki S (1995) Ability to withstand sudden braking. Railw Res Rev 52(5):18–21
12. Hirshfeld CF (1932) Disturbing effects of horizontal acceleration. Bulletin No. 3, Electric Railway President's Conference Committee
13. Hoberock LL (1976) A survey of longitudinal acceleration comfort studies in ground transportation vehicles. Research Report 40, Council for Advanced Transportation Studies
14. Kamper D, Parnianpour M, Barin K, Adams T, Linden M, Hemami H (1999) Postural stability of wheelchair users exposed to sustained, external perturbations. J Rehabil Res Dev 36(2): 121–132
15. Koji O, Hiroharu E, Hiroaki S, Hiroshi S (2007) Ride evaluation during high deceleration braking. Railw Res Rev 64(7):26–29
16. Loach JC, Maycock MG (1952) Recent developments in railway curve design. ICE Proc: Eng Div 1(5):503–541
17. Maki BE, Fernie GR (1988) A system identification approach to balance testing. In: Pompeiano O, Allum JHJ (eds) Progress in brain research. Elsevier, Oxford, pp 297–306
18. Matsudaira T (1960) Dynamics of high speed rolling stock, Railway Technical Research Institute, Quarterly Reports, (Special Issue), pp 57–65
19. Matsui S (1962) Comfort limits of retardation and its changing rate for train passengers. Jpn Railw Eng 3(11):25–27
20. Profillidis VA (2006) Railway management and engineering, 3rd edn. Ashgate, Aldershot
21. RSSB (2014) Annual Safety Performance Report 2013/14. Rail Safety and Standards Board
22. SAR (1948) Superelevations and maximum permissible speeds on curves. Research Circular No. 25.027, South African Railways
23. Sari MH (2012) Postural stability when walking and exposed to lateral oscillations. Doctoral thesis, University of Southampton
24. Shumway-Cook A, Woollacott MH (2001) Motor control: theory and practical applications. Lippincott Williams & Wilkins, Philadelphia
25. Simoneau M, Corbeil P (2005) The effect of time to peak ankle torque on balance stability boundary: experimental validation of a biomechanical model. Exp Brain Res 165(2):217–228
26. Sone S, Ashiya M (1998) An innovative traction system from the viewpoint of braking, International Conference on Developments in Mass Transit Systems. London, 20th–23rd April
27. Thompson JT (1939) Railway track-work for high speeds. J ICE 10(3):405–407
28. Urabe S, Nomura Y (1964) Evaluations of train riding comfort under various decelerations, Railway Technical Research Institute, Quarterly Reports 5(2):28–34
29. Vuchic VR (2007) Urban transit systems and technology. John Wiley & Sons, Hoboken
30. Watkins J (1983) An introduction to mechanics of human movement. Boston, [Mass.]; Lancaster: MTP Press
31. Zigmond MJ (1999) Fundamental neuroscience. Academic, San Diego

12

Analysis of Wheel-Roller Contact and Comparison with the Wheel-Rail Case

Binbin Liu[1] · Stefano Bruni[1]

Abstract Full-scale roller rigs are recognized as useful test stands to investigate wheel-rail contact/damage issues and for developing new solutions to extend the life and improve the behaviour of railway systems. The replacement of the real track by a pair of rollers on the roller rig causes, however, inherent differences between wheel-rail and wheel-roller contact. In order to ensure efficient utilization of the roller rigs and correct interpretation of the test results with respect to the field wheel-rail scenarios, the differences and the corresponding causes must be understood a priori. The aim of this paper is to derive the differences between these two contact cases from a mathematical point of view and to find the influence factors of the differences with the final aim of better translating the results of tests performed on a roller rig to the field case.

Keywords Wheel-rail contact · Wheel-roller contact · Creepage · Contact patch · Roller rig

1 Introduction

The contact between wheel and rail is one of the most important features of the railway system, and this contact pair has attracted great attention since the beginning of railway engineering. Unfortunately, the problems involved in the wheel-rail contact interface have not been

✉ Binbin Liu
 binbin.liu@polimi.it

[1] Dipartimento di Meccanica, Politecnico di Milano, Via La Masa 1, 20156 Milan, Italy

Editor: Xuesong Zhou

completely solved due to the complexity of the problem. Many attempts have been made from both theoretical and experimental points of view. Moreover, field experiments on wheel-rail contact mechanics and dynamics are often challenging due to the difficulties in adequately controlling the test conditions [1]. Roller rigs are a good alternative in this case, thanks to their high controllability and flexibility, and have been used as experimental tools in railway application over a long time. A variety of roller rig designs have been introduced, targeting different research aims, more detail on this topic can be found in [2–4]. A. Jaschinski et al. [2] performed a comprehensive survey for both full-scale and scaled model roller rigs on the application to railway vehicle dynamics. Zhang et al. [4] reviewed the development history of the roller rig for railway application and performed a detailed comparison between rollers and track in terms of geometry relationship with wheel, creep coefficient, stability, vibration response and curve simulation. Allen [5] and Yan [6] documented in detail on the errors caused by scaled roller rigs for the study of the dynamic behaviour of railway bogies. Keylin et al. [1] derived explicit analytical expressions for comparing contact patch dimensions and Kalker's coefficients for a wheel moving on a roller and on a tangent track, based on Hertz and Kalker's linear theory. Taheri et al. [7] compared the contact patch formed by a single wheelset when coupled to a roller and to a tangent track under the assumptions of the Hertz's theory. Zeng [8] compared the geometry contact characteristics of the wheel-rail and that of the wheel-roller based on a three-dimensional contact searching method.

Nevertheless, a systematic analysis of wheel-roller contact and the differences with respect to the wheel-rail case are missing in the literature. It should be noted that the roller rig test will never completely replace the field test due to inherent differences caused by the replacement of the rail by rollers in a roller rig system. Therefore, it is very important to know the differences between these two systems and the corresponding reasons in order to efficiently perform wheel-rail contact study on a roller rig and to correctly interpret the test results and to compensate for deviations between the roller rig and a real track. This analysis should cover in particular

- the geometry of wheel-rail/wheel-roller contact;
- differences in the formation of the contact patch;
- factors affecting the creepages in a test performed on a roller and their effect on tangential contact forces.

The aim of this paper is to provide a thorough examination of the differences between these two contact cases from a mathematical point of view and to find the influence factors of the differences for better translating the test results on the roller rig to the field test. To this aim, a new approach for solving the normal contact problem for the wheel-roller couple is proposed, and the expressions of the creepages and spin are obtained for the wheel-roller couple. The results of this new approach are presented comparing the case of the wheelset running on a standard track and on rollers, and the differences between these two cases are discussed in the light of their effects on surface damage and degradation occurring in the wheels and the rails.

2 Full-Scale Roller Rigs for Tests on a Single Wheelset

Among all of the roller rigs existing in laboratory, the full-scale roller rig for a single wheelset test is one of most similar systems to a real wheelset-track system from both dynamics and contact mechanics points of view. The mechanical layout for a roller rig of this type is shown in Fig. 1. It consists of a roller with two wheels driven by a

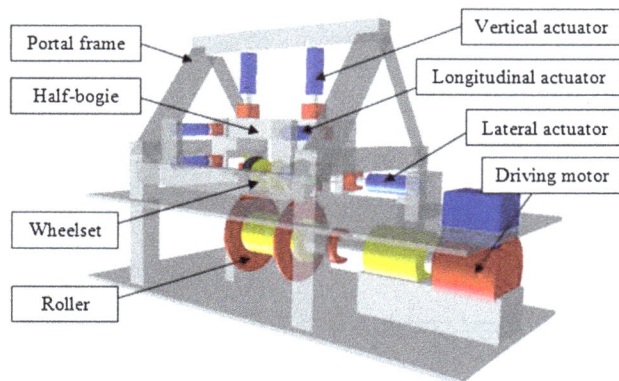

Fig. 1 Layout of a full-scale roller rig for a single wheelset test

motor. A full-scale wheelset is mounted on the top of two rollers with real rail profiles and connected through primary suspensions to a transversal beam representing one half of the bogie frame. Compared with the roller rig for test on a complete vehicle, the high controllability and flexibility of the single wheelset roller rig make it possible to obtain adequate data on the dynamics of the system and on wheel-rail contact under various conditions which are essential for investigating the adhesion and creep of the wheel over the rail. This configuration of the roller rig allows to perform studies of wheel-rail interaction and also tests concerning the dynamic behaviour of a wheelset/bogie, see [9, 10]. However, this paper only concentrates on wheel-rail contact.

3 Contact Formulation for Wheel-Rail and Wheel-Roller Systems

From a mathematical point of view, the contact problem can be solved for both wheel-rail and wheel-roller contact according to the following four steps [11, 12]. The first step is to solve the geometrical problem, in which the locations of the contact points on the contacting bodies are determined. This is followed by solving the normal contact problem, in which the shape and size of the contact patch formed in the contact interface due to body deformation and the corresponding pressure distribution over the contact patch are determined. The third step is to deal with the kinematic problem, in which normalized kinematic quantities, the so-called creepages, are determined. These quantities measure the relative velocities between the contacting bodies at the contact points. In the final step, the tangential problem is solved; this concerns the prediction of tangential stresses at the contact interface which is generated by friction and creepages within the contact zone [11, 13, 14]. All these steps need to be dealt with differently for the case of wheel-roller contact compared to the wheel-rail case, as described in the next section.

3.1 Geometrical Problem

The contact geometry analysis deals with the contact point searching problem between the contacting bodies, i.e. wheel-rail and wheel-roller pairs in this case. The location of contact depends on the dynamic conditions as well as material properties of the contact pair if body deformation is considered. There are many approaches for the detection of the contact points for wheel-rail contact as documented in [11]. Most of the approaches available in the literature assume that the yaw angle of the wheelset against the track is very small and negligible when solving the geometric contact problem so as to form the so-called bi-dimensional methods [14]. This assumption largely simplifies the calculation, leading to fast solutions that can be implemented in rail vehicle online dynamics simulation. However, the replacement of the rail by a roller makes the geometric contact problem more complicated, and the traditional bi-dimensional methods may not be applicable any longer due to the considerable yaw influence on the contact location in the case of the wheel-roller contact. To deal with this problem, a three-dimensional model is needed. Some existing approaches for wheel-roller geometry contact analysis can be found in [4, 8, 15].

It is clear that the geometric contact condition between the wheel and rail/roller is the same for zero yaw angle conditions assuming the contacting bodies to be rigid. The comparisons on the geometry contact relationship between the wheel-rail and wheel-roller contact pairs are available in the literature, for instance [4, 8, 12]. Therefore, no further discussion is needed here, but one typical case study is shown in Fig. 2 for the completeness of this study. The calculation conditions are as follows: profile combination is new S1002/UIC60 with 1:20 rail inclination, the radius of the roller is 1 m and the yaw angle of the wheelset is 60 mrad.

It is interesting to note in Fig. 2 that two-point contact occurs when the wheelset is shifted by 5 mm approximately in lateral direction for wheel-rail contact, but this value is slightly different on the roller rig due to the curvature of the rollers. Obviously, the differences can be decreased by increasing the radius of the roller, but the dimension of the roller is limited in practice considering the increased cost and difficulties related with the manufacturing and installation of the rig. It should be also noted that 60 mrad is a quite large yaw angle for the wheelset and that smaller differences are found for smaller yaw angles.

3.2 Normal Problem

The well-known Hertzian theory [16] is widely used for solving the normal contact problem in rail vehicle dynamics simulation for its simplicity and calculation efficiency. However, Hertzian theory is valid based on half-space assumption and elliptic contact condition. In order to obtain more realistic contact information for the purpose of comparison between wheel-rail and wheel-roller contact (especially in terms of shape and size of the contact patch and of pressure distribution), the use of a more advanced contact model is required. The most elaborate contact model to date can be established by finite element method [17, 18] which is quite complicated and time consuming. The same problem can be dealt with using the boundary element method as done, e.g. by Prof. Kalker's algorithm CONTACT [19] and by the model proposed by Knothe and Le The in [20]. The so-called approximate contact methods represent a trade-off between efficiency and accuracy in the solution of the normal problem and therefore are generally considered as best suited for both local contact analysis and for online dynamics simulation. Well-known approximate models include the Kik–Piotrowski model based on virtual penetration concept [21–23], the STRIPES model proposed by Ayasse and Chollet [24] and Linder's model [25]. Some interesting surveys of the existing approximate methods and the comparison and analysis among them can be found in [22, 26].

The Kik–Piotrowski model has been chosen as the basis for this study. Some modifications have been introduced to extend the original method to deal with both wheel-rail and wheel-roller contact conditions. The Kik–Piotrowski model is a fast and non-iterative method to calculate normal contact problem. An outline of this method will be given in this section, for more details the reader is referred to [21–23]. The idea of this method is presented in Fig. 3. When

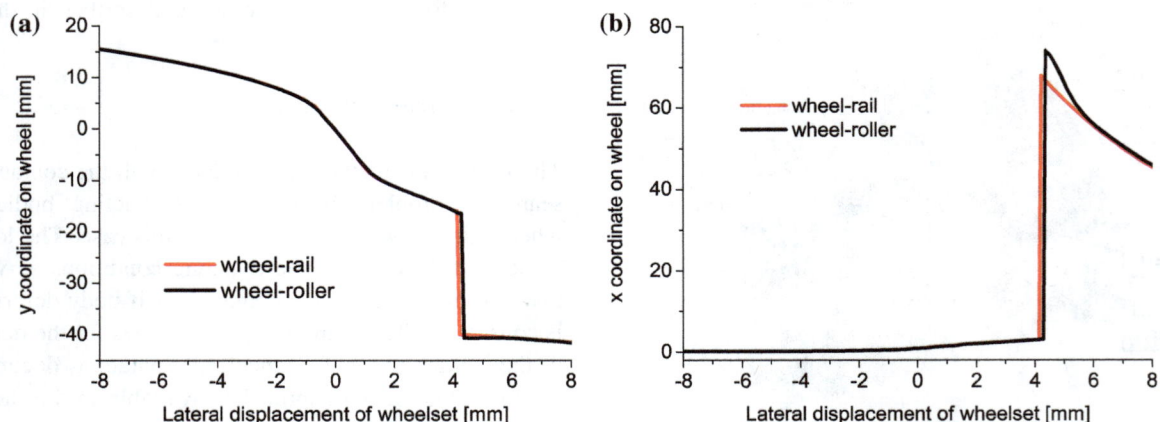

Fig. 2 Comparisons of the contact location on wheel in lateral (**a**) and longitudinal (**b**) directions with 60 mrad yaw angle

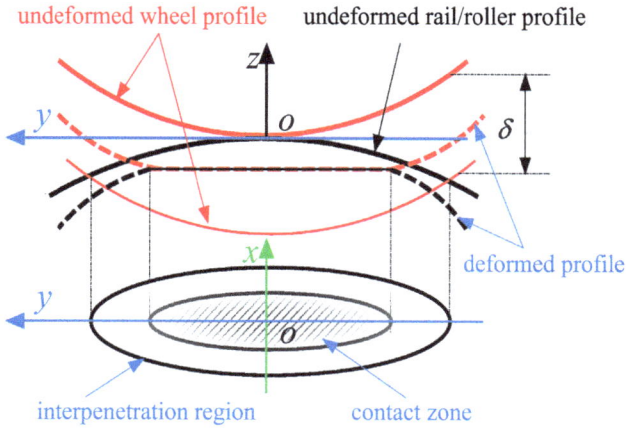

Fig. 3 Contact zone determined with virtual interpenetration method, adapted from [21]

the undeformed surfaces of wheel and rail/roller, touching in the geometrical point of contact O which is determined from geometric contact analysis, are shifted towards each other by a distance δ, called the penetration, they penetrate and intersect on a closed line, whose projection on the x-y plane is called the interpenetration region. On the basis of some similarity of shapes of the contact zone and interpenetration region, the contact zone is determined by scaling the interpenetration depth $\delta_o = \varepsilon\delta$ with an approach scaling factor of $\varepsilon = 0.55$ and the resulting interpenetration region is taken as the real contact zone.

In order to solve the problem numerically, a coordinate system $Oxyz$ representing the contact reference system is defined firstly with the x-axis pointing along the rolling direction of the wheelset, and the y-axis parallel to the wheel axle. The undeformed surface with the same x, y coordinates in contact reference system is assumed as

$$z(x,y) = f_{yz}(y) + \left(\frac{1}{R_w} + \frac{1}{R_r}\right)\frac{x^2}{2},\qquad(1)$$

where subscript yz stands for the function defined in $y-z$ plane and R_w and R_r are the principal radii of the wheel and rail/roller, respectively, at the geometrical point of contact in rolling direction. For wheel tread and rail top contact R_r goes to infinite, while this is not the case for the wheel-roller contact case. The separation of profiles $f_{yz}(y) = z_{yz}^w(y) + z_{yz}^r(y)$ is obtained from the sum of the cross-sections $z_{yz}^w(y)$ of the wheel rolling surface and $z_{yz}^r(y)$ of the rail surface by $x = 0$ in the contact plane. To proceed with the presentation of the method, the interpenetration function of the profiles is defined in the contact plane as

$$g_{yz}(y) = \begin{cases} \delta_0 - f_{yz}(y) & \text{if } f_{yz}(y) \le \delta_0 \\ 0 & \text{if } f_{yz}(y) > \delta_0 \end{cases}\qquad(2)$$

where δ_0 is the virtual interpenetration. The width of the contact patch can be determined by solving Eq. (2). It should be mentioned that the contact shape can be corrected by adjusting the interpenetration function, cf. [22, 23], but no shape correction is applied in this study for simplicity.

The contact zone is determined by the x coordinates of its leading and trailing edges described by formula (3) in the original method based on the assumption that the wheel-rail contact problem is stated in terms of two bodies of revolution with their axes laying in the same plane. The same assumption was made by Linder [25].

$$x_{xz}^l(y) = -x_{xz}^t(y) \approx \sqrt{2R_w g_{yz}(y)}\qquad(3)$$

where subscript xz means that the function is defined in the x-z plane and superscripts l and t indicate the terms associated with the leading and trailing edges of the contact patch, respectively.

In order to determine the contact boundary for wheel-roller contact, the following modifications of Eq. (3) are proposed. Firstly, the contact patch is partitioned into strips paralleling with the x-axis towards to the rolling direction of the wheel. Hence, the profile functions are converted to discrete forms by strips y_i ($i = 1...n$). Then, the extremities of each strip can be determined by solving Eq. (4) instead of Eq. (3). The two solutions of Eq. (4) for each strip correspond to the coordinates of the leading and trailing edges of that strip. All the coordinates comprise the boundary of the contact zone.

$$z_{xz}^w(x, y_i) - z_{xz}^r(x, y_i) = g_{yz}(y_i)\qquad(4)$$

where $g_{yz}(y_i)$ is the interpenetration function at the i-th strip in contact patch, $z_{xz}^w(x,y_i)$ and $z_{xz}^r(x,y_i)$ are the rolling circle of the wheel and roller with the radius of $R_{cw}(y_i)$ and $S_{cr}(y_i)$, respectively, over the i-th strip, which are expressed by Eqs. (5) and (6).

$$z_{xz}^w(x, y_i) = R_{cw}(y_i) - \sqrt{R_{cw}^2(y_i) - x^2}\qquad(5)$$

$$z_{xz}^r(x, y_i) = S_{cr}(y_i) - \sqrt{S_{cr}^2(y_i) - x^2}\qquad(6)$$

where $x \in [-R_{cw}(y_i), R_{cw}(y_i)]$ is the longitudinal coordinate of the rolling circle in the contact plane. It should be noted that $S_{cr}(y_i)$ is a straight line in the case of tangent track.

Moreover, it is assumed that the normal pressure is semi-elliptical in the direction of rolling and has the following expression:

$$p(x_j, y_i) = \frac{p_0}{x_l(0)}\sqrt{\left(x_{xz}^l(y_i)\right)^2 - x_j^2(y_i)}.\qquad(7)$$

Following the same procedure documented in Ref. [21], the normal contact problem can be solved numerically for both wheel-rail and wheel-roller contact system.

It should be pointed out that the influence of the wheelset's yaw angle on the normal contact solution is essential for wheel-roller contact and can be included based on the similar idea proposed here, but it will not be discussed further in the current paper for simplicity.

3.3 Kinematical Problem

With reference to Fig. 1, it can be seen that the wheelset on the roller rig has the same degrees of freedom as on a track, except the constraint in longitudinal direction. Furthermore, the two rollers fixed on the same axle can only rotate around its axle. To accomplish the kinematic analysis of a wheelset on a pair of rollers, a convenient set of reference frames should be introduced as shown in Fig. 4. The wheelset reference frame is denoted by $O_w X_w Y_w Z_w$ attached to the wheelset's centre of mass so that axis $O_w Y_w$ coincides with the wheelset's axis of rotation, the $O_w Z_w$ axis points upwards and the $O_w X_w$ axis completes the right-handed coordinate system. Similarly, a roller reference

frame is introduced and denoted by $O_{ro} X_{ro} Y_{ro} Z_{ro}$ attached to the roller's centre of mass which is defined as the inertial frame. Two contact reference frames $O_{cl} X_{cl} Y_{cl} Z_{cl}$ and $O_{cr} X_{cr} Y_{cr} Z_{cr}$ are introduced at the contact interfaces between the left-hand and right-hand wheels of the wheelset and rollers at the wheelset central position.

It is assumed that the wheelset reference frame $O_w X_w Y_w Z_w$ is obtained from the inertial frame by performing two successive rotations. The axes of the reference frames are parallel before rotation, and the first rotation is made about the z-axis by an angle ψ called yaw angle (positive in counter-clockwise direction) followed by a second rotation about the x-axis by an angle φ called roll angle. Therefore, the transformation matrix A^{w2i} connecting the wheelset frame to the inertial frame is expressed as follows:

$$A^{w2i} = \begin{bmatrix} \cos\psi & -\sin\psi & 0 \\ \sin\psi\cos\varphi & \cos\psi\cos\varphi & -\sin\varphi \\ \sin\psi\sin\varphi & \cos\psi\sin\varphi & \cos\varphi \end{bmatrix} \quad (8)$$

Since the angles of rotation are generally small in railway dynamics, the small angle approximation can be applied, so that the transformation matrix reduces to

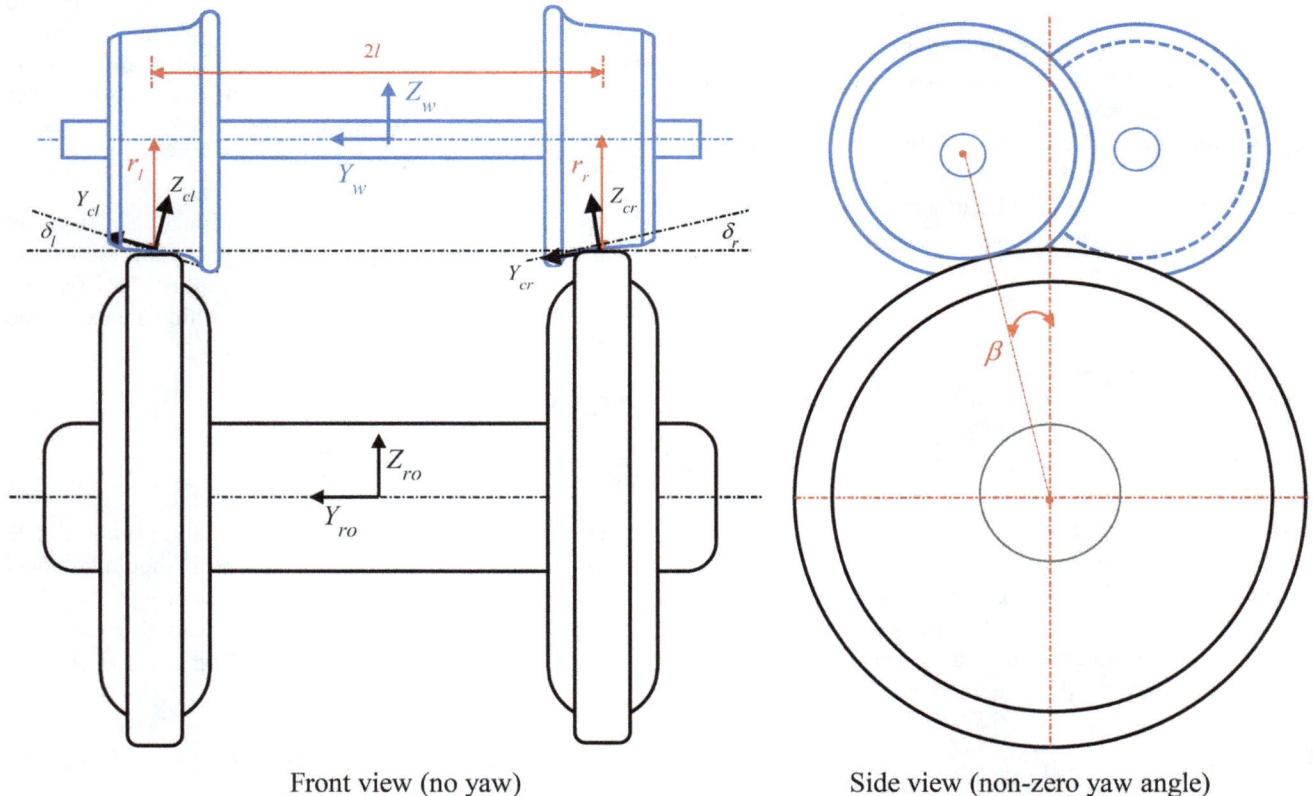

Front view (no yaw)   Side view (non-zero yaw angle)

Fig. 4 Reference frames defined in the roller rig system

$$A^{\mathrm{w}2i} \approx \begin{bmatrix} 1 & -\psi & 0 \\ \psi & 1 & -\varphi \\ 0 & \varphi & 1 \end{bmatrix}. \tag{9}$$

The position vectors of the contact points on the wheel and roller can be defined in the inertial frame as follows:

$$r_{\mathrm{w}i} = R_{\mathrm{w}} + A^{\mathrm{w}2i}\bar{u}_{\mathrm{w}i} \quad (i = l, r) \tag{10}$$

where the position vector of the origin of the wheelset reference frame in the inertial frame is expressed as

$$R_{\mathrm{w}} = \begin{bmatrix} 0 & y & z \end{bmatrix}^{\mathrm{T}} \tag{11}$$

and the position vectors of the contact points in the wheelset reference frame can be expressed in the following forms: for the left-hand wheel

$$\bar{u}_{\mathrm{w}l} = \begin{bmatrix} r_l\sin\beta & l & -r_l\cos\beta \end{bmatrix}^{\mathrm{T}} \tag{12}$$

and for the right-hand wheel

$$\bar{u}_{\mathrm{w}r} = \begin{bmatrix} -r_r\sin\beta & -l & -r_r\cos\beta \end{bmatrix}^{\mathrm{T}}, \tag{13}$$

where r_i $(i = l, r)$ represents the radius of the left-hand and right-hand wheels, respectively, l is the half distance between the contact points on the left-hand and right-hand wheels and β is the shift angle of the contact point on the roller with respect to the vertical plane of the inertial frame caused by a non-zero yaw angle ψ. It is assumed that this angle is the same for the left-hand and right-hand side on the roller and can be approximated by Eq. (14), since it is very small in ordinary circumstances.

$$\beta = \frac{l\psi}{r_0 + s_0} \tag{14}$$

In Eq. (14), r_0 and s_0 denote the radii of the wheel and roller at the central position, respectively.

Taking the derivative of Eq. (11), the velocity vector of the contact point located on the wheel with respect to the inertial frame is obtained as

$$v_{\mathrm{w}i} = \dot{R}_{\mathrm{w}} + \omega_{\mathrm{w}} \times u_{\mathrm{w}i} \quad (i = l, r), \tag{15}$$

where $u_{\mathrm{w}i} = A^{\mathrm{w}2i}\bar{u}_{\mathrm{w}i}$ is the position vector of the point of contact on the wheel defined in the inertial frame which is determined from Eqs. (9), (12) and (13) for the left-hand and right-hand wheels, respectively, as follows:

$$u_{\mathrm{w}l} = A^{\mathrm{w}2i}\bar{u}_{\mathrm{w}l} \approx \begin{bmatrix} r_l\sin\beta - l\psi \\ r_l\psi\sin\beta + l + r_l\varphi\cos\beta \\ l\varphi - r_l\cos\beta \end{bmatrix}$$
$$\approx \begin{bmatrix} r_l\beta - l\psi \\ l + r_l\varphi \\ l\varphi - r_l \end{bmatrix} \tag{16}$$

and

$$u_{\mathrm{w}r} = A^{\mathrm{w}2i}\bar{u}_{\mathrm{w}r} \approx \begin{bmatrix} -r_r\sin\beta + l\psi \\ -r_r\psi\sin\beta - l + r_r\varphi\cos\beta \\ -l\varphi - r_r\cos\beta \end{bmatrix}$$
$$\approx \begin{bmatrix} -r_r\beta + l\psi \\ -l + r_r\varphi \\ -l\varphi - r_r \end{bmatrix} \tag{17}$$

and ω_{w} is the absolute angular velocity vector at the point of contact defined in the inertial system as

$$\omega_{\mathrm{w}} = \begin{bmatrix} 0 \\ 0 \\ \dot{\psi} \end{bmatrix} + A^{\mathrm{w}2i}\begin{bmatrix} \dot{\varphi} \\ \Omega_{\mathrm{w}} \\ 0 \end{bmatrix} = \begin{bmatrix} \dot{\varphi}\cos\psi - \Omega_{\mathrm{w}}\sin\psi\cos\varphi \\ \dot{\varphi}\sin\psi + \Omega_{\mathrm{w}}\cos\psi\cos\varphi \\ \dot{\psi} + \Omega_{\mathrm{w}}\sin\varphi \end{bmatrix}$$
$$\approx \begin{bmatrix} \dot{\varphi} - \Omega_{\mathrm{w}}\psi \\ \dot{\varphi}\psi + \Omega_{\mathrm{w}} \\ \dot{\psi} + \Omega_{\mathrm{w}}\varphi \end{bmatrix} \tag{18}$$

with $\Omega_{\mathrm{w}} = V/r_0$ the rolling angular velocity of the wheelset.

Substituting Eqs. (11), (17) and (18) into Eq. (15), the velocity vectors of the contact point on the wheelset in the inertial frame are obtained as follows. For the left-hand wheel:

$$v_{\mathrm{w}l} = \dot{R}_{\mathrm{w}} + \omega_{\mathrm{w}} \times u_{\mathrm{w}l}$$
$$\approx \begin{bmatrix} -l\dot{\psi} - r_l(\dot{\varphi}\psi + \dot{\psi}\varphi + \Omega_{\mathrm{w}}) \\ \dot{y} - l(\dot{\varphi}\varphi + \dot{\psi}\psi) + r_l(\dot{\varphi} + \dot{\psi}\beta - \Omega_{\mathrm{w}}\psi) \\ \dot{z} + l\dot{\varphi} + r_l(\dot{\varphi}\varphi - \Omega_{\mathrm{w}}\beta) \end{bmatrix} \tag{19}$$

and for the right-hand wheel:

$$v_{\mathrm{w}r} = \dot{R}_{\mathrm{w}} + \omega_{\mathrm{w}} \times u_{\mathrm{w}r}$$
$$\approx \begin{bmatrix} l\dot{\psi} - r_r(\dot{\varphi}\psi + \dot{\psi}\varphi + \Omega_{\mathrm{w}}) \\ \dot{y} + l(\dot{\varphi}\varphi + \dot{\psi}\psi) + r_r(\dot{\varphi} - \dot{\psi}\beta - \Omega_{\mathrm{w}}\psi) \\ \dot{z} - l\dot{\varphi} + r_r(\dot{\varphi}\varphi + \Omega_{\mathrm{w}}\beta) \end{bmatrix}. \tag{20}$$

Similarly, the velocity vector of the contact point on the roller in the inertial frame can be expressed as

$$v_{\mathrm{ro}i} = \omega_{\mathrm{ro}} \times u_{\mathrm{ro}i} \quad (i = l, r), \tag{21}$$

where ω_{ro} is the angular velocity of the roller with the following form:

$$\omega_{\mathrm{ro}} = \begin{bmatrix} 0 \\ \Omega_{\mathrm{or}} \\ 0 \end{bmatrix} = \begin{bmatrix} 0 \\ -\dfrac{V}{s_0} \\ 0 \end{bmatrix} \tag{22}$$

and $u_{\mathrm{ro}i}$ stands for the position vector of the contact point in the inertial frame. For the left-hand roller, the expression of this vector is

$$u_{\mathrm{ro}l} = \begin{bmatrix} -s_l\sin\beta & l & s_l\cos\beta \end{bmatrix}^{\mathrm{T}} \tag{23}$$

and for the right-hand roller the expression is

$$u_{\mathrm{ror}} = [s_r\sin\beta \quad -l \quad s_r\cos\beta]^{\mathrm{T}}. \tag{24}$$

Hence, the velocity vector of the point of contact is obtained. For the left-hand roller,

$$v_{\mathrm{rol}} = \omega_{\mathrm{ro}} \times u_{\mathrm{rol}} = \begin{bmatrix} 0 \\ -\dfrac{V}{s_0} \\ 0 \end{bmatrix} \times \begin{bmatrix} -s_l\sin\beta \\ l \\ s_l\cos\beta \end{bmatrix}$$

$$= \begin{bmatrix} -\dfrac{V}{s_0}s_l\cos\beta \\ 0 \\ -\dfrac{V}{s_0}s_l\sin\beta \end{bmatrix} \approx \begin{bmatrix} -\dfrac{V}{s_0}s_l \\ 0 \\ -\dfrac{V}{s_0}s_l\beta \end{bmatrix} \tag{25}$$

and for the right-hand roller:

$$v_{\mathrm{ror}} = \omega_{\mathrm{ro}} \times u_{\mathrm{ror}} = \begin{bmatrix} 0 \\ -\dfrac{V}{s_0} \\ 0 \end{bmatrix} \times \begin{bmatrix} s_r\sin\beta \\ -l \\ s_r\cos\beta \end{bmatrix}$$

$$= \begin{bmatrix} -\dfrac{V}{s_0}s_r\cos\beta \\ 0 \\ \dfrac{V}{s_0}s_r\sin\beta \end{bmatrix} \approx \begin{bmatrix} -\dfrac{V}{s_0}s_r \\ 0 \\ \dfrac{V}{s_0}s_r\beta \end{bmatrix}. \tag{26}$$

Thus, the velocity differences between the wheel and roller at each point of contact in the inertial frame can be calculated as follows. For the left side

$$\Delta v_l = v_{\mathrm{w}l} - v_{\mathrm{rol}}$$

$$= \begin{bmatrix} -l\dot\psi - r_l(\dot\varphi\psi + \dot\psi\varphi + \Omega_{\mathrm{w}}) + \dfrac{V}{s_0}s_l \\ \dot y - l(\dot\varphi\varphi + \dot\psi\psi) + r_l(\dot\varphi + \dot\psi\beta - \Omega_{\mathrm{w}}\psi) \\ \dot z + l\dot\varphi + r_l(\dot\varphi\varphi - \Omega_{\mathrm{w}}\beta) + \dfrac{V}{s_0}s_l\beta \end{bmatrix} \tag{27}$$

for the right side:

$$\Delta v_{\mathrm{r}} = v_{\mathrm{wr}} - v_{\mathrm{ror}}$$

$$= \begin{bmatrix} l\dot\psi - r_{\mathrm{r}}(\dot\varphi\psi + \dot\psi\varphi + \Omega_{\mathrm{w}}) + \dfrac{V}{s_0}s_{\mathrm{r}} \\ \dot y + l(\dot\varphi\varphi + \dot\psi\psi) + r_{\mathrm{r}}(\dot\varphi - \dot\psi\beta - \Omega_{\mathrm{w}}\psi) \\ \dot z - l\dot\varphi + r_{\mathrm{r}}(\dot\varphi\varphi + \Omega_{\mathrm{w}}\beta) - \dfrac{V}{s_0}s_{\mathrm{r}}\beta \end{bmatrix} \tag{28}$$

and the difference of angular velocity is

$$\Delta\omega = \omega_{\mathrm{w}} - \omega_{\mathrm{ro}} = \begin{bmatrix} \dot\varphi\cos\psi - \Omega_{\mathrm{w}}\sin\psi\cos\varphi \\ \dot\varphi\sin\psi + \Omega_{\mathrm{w}}\cos\psi\cos\varphi + \dfrac{V}{s_0} \\ \dot\psi + \Omega_{\mathrm{w}}\sin\varphi \end{bmatrix}$$

$$\approx \begin{bmatrix} \dot\varphi - \Omega_{\mathrm{w}}\psi \\ \dot\varphi\psi + \Omega_{\mathrm{w}} + \dfrac{V}{s_0} \\ \dot\psi + \Omega_{\mathrm{w}}\varphi \end{bmatrix}. \tag{29}$$

To determine the creepages and spin, the velocity differences obtained above must be resolved in the contact plane where they are defined. It is assumed that the contact frames are connected to the wheelset frame by the following transformation matrices for the left and right wheels, respectively.

$$A^{\mathrm{w2cl}} = \begin{bmatrix} \cos\beta & 0 & \sin\beta \\ -\sin\beta\sin\delta_l & \cos\delta_l & \cos\beta\sin\delta_l \\ -\sin\beta\cos\delta_l & -\sin\delta_l & \cos\beta\cos\delta_l \end{bmatrix}$$

$$\approx \begin{bmatrix} 1 & 0 & \beta \\ 0 & 1 & \delta_l \\ -\beta & -\delta_l & 1 \end{bmatrix} \tag{30}$$

and

$$A^{\mathrm{w2cr}} = \begin{bmatrix} \cos\beta & 0 & -\sin\beta \\ -\sin\beta\sin\delta_r & \cos\delta_r & -\cos\beta\sin\delta_r \\ \sin\beta\cos\delta_r & \sin\delta_r & \cos\beta\cos\delta_r \end{bmatrix}$$

$$\approx \begin{bmatrix} 1 & 0 & -\beta \\ 0 & 1 & -\delta_r \\ \beta & \delta_r & 1 \end{bmatrix}, \tag{31}$$

where $\delta_i(i = l, r)$ denotes the contact angle.

Hence, the transformation matrices connecting the inertial frame to the contact frame can be obtained for the left and right side wheels by the following operation. For the left side

$$A^{\mathrm{i2cl}} = A^{\mathrm{w2cl}}A^{\mathrm{i2w}} = A^{\mathrm{w2cl}}(A^{\mathrm{i2w}})^T$$

$$\approx \begin{bmatrix} 1 & \psi & \beta \\ -\psi & 1 & \delta_l + \varphi \\ -\beta & -\delta_l - \varphi & 1 \end{bmatrix} \tag{32}$$

and for the right side:

$$A^{\mathrm{i2cr}} = A^{\mathrm{w2cr}}A^{\mathrm{i2w}} = A^{\mathrm{w2cr}}(A^{\mathrm{i2w}})^T$$

$$\approx \begin{bmatrix} 1 & \psi & -\beta \\ -\psi & 1 & -\delta_r + \varphi \\ \beta & \delta_r - \varphi & 1 \end{bmatrix} \tag{33}$$

Therefore, the velocity differences between the wheel and roller in the contact plane are obtained as

$$\begin{aligned} \Delta v_{\mathrm{c}i} &= A^{\mathrm{i2c}i}\Delta v_i \\ \Delta\omega_{\mathrm{c}i} &= A^{\mathrm{i2c}i}\Delta\omega_i \end{aligned}. \tag{34}$$

Now, the creepages can be obtained by definition as follows. The longitudinal creepages on the left and right wheels are

$$\xi_{lx} = \frac{\Delta v_{\mathrm{c}lx}}{V} \approx -\frac{l\dot\psi}{V} - \frac{r_l\dot\psi\varphi}{V} - \frac{r_l}{r_0} + \frac{\dot y\psi}{V} + \frac{s_l}{s_0} + \frac{\dot z\beta}{V} + \frac{l\dot\varphi\beta}{V}$$

$$\xi_{rx} = \frac{\Delta v_{\mathrm{c}rx}}{V} \approx \frac{l\dot\psi}{V} - \frac{r_r\dot\psi\varphi}{V} - \frac{r_r}{r_0} + \frac{\dot y\psi}{V} + \frac{s_r}{s_0} - \frac{\dot z\beta}{V} + \frac{l\dot\varphi\beta}{V}$$

$$\tag{35}$$

the lateral creepages are

$$\xi_{ly} = \frac{\Delta v_{\text{cly}}}{V} \approx \frac{\dot{y} - \frac{s_l}{s_0}\psi V + r_l\dot{\psi}\beta + r_l\dot{\varphi} + l\delta_l\dot{\varphi} + \dot{z}\delta_l + \dot{z}\varphi}{V}$$

$$\xi_{ry} = \frac{\Delta v_{\text{cry}}}{V} \approx \frac{\dot{y} - \frac{s_r}{s_0}\psi V - r_r\dot{\psi}\beta + r_r\dot{\varphi} + l\delta_r\dot{\varphi} + \dot{z}\delta_r - \dot{z}\varphi}{V}$$

(36)

and the spin creepages are

$$\xi_{lz} = \frac{\Delta\omega_{\text{clz}}}{V} \approx -\frac{\delta_l}{r_0} + \frac{\dot{\psi}}{V} - \frac{\dot{\varphi}\beta}{V} + \frac{\Omega_{\text{or}}\delta_l}{V} + \frac{\Omega_{\text{or}}\varphi}{V}$$

$$\xi_{rz} = \frac{\Delta\omega_{\text{crz}}}{V} \approx \frac{\delta_r}{r_0} + \frac{\dot{\psi}}{V} + \frac{\dot{\varphi}\beta}{V} - \frac{\Omega_{\text{or}}\delta_r}{V} + \frac{\Omega_{\text{or}}\varphi}{V}$$

(37)

It can be seen from the expressions above that the radius of the roller and the shift angle (function of yaw) contribute to the differences in terms of creepages and spin with respect to wheel-rail contact condition. The corresponding expressions for wheel-rail contact condition can be obtained by setting $s_0 = \infty$ and $\beta = 0$. Moreover, the longitudinal and lateral creepages can be simplified further by assuming that the contacting bodies remain in contact at all times which means the z components vanish in the expressions (35) and (36).

3.4 Tangential Problem

The common method to solve the wheel-rail tangential contact problem is represented by the FASTSIM algorithm [13], also due to Kalker. This method was originally developed for elliptic contact condition, but can be extended to cover a more general geometry of the contact patch. The difficulty is to determine the flexibility parameter that is required by this method. To overcome this, Kik and Piotrowski proposed a method to define an equivalent ellipse for each separate contact zone by setting the ellipse area equal to the non-elliptic contact area and the ellipse semi-axes ratio equal to length to width ratio of the patch. The flexibility parameter is determined by equating the two solutions obtained from the linear complete theory and from the simplified theory for elliptical contact area and pure longitudinal, lateral and spin creepages. In addition, there are two options with respect to the choice of the flexibility parameter, namely considering one single weighted mean flexibility parameter or three flexibility parameters one for each creepage component. According to [27], the single flexibility parameter will reduce the agreement of FASTSIM to the exact theory. Therefore, three flexibility parameters are used in the current study.

From the main assumption of the linear theory which neglects slip in the contact zone, the tangential stress distribution is derived in the form:

$$\begin{cases} \tau_x(x,y) = \dfrac{\xi_x}{L_1} - \dfrac{y\xi_z}{L_3}(x - x_{xz}^l) \\ \tau_y(x,y) = \dfrac{\xi_y}{L_2}(x - x_{xz}^l) + \dfrac{\xi_z}{2L_3}\left(x^2 - (x_{xz}^l)^2\right) \end{cases}$$

(38)

where $\xi_i (i = 1 - 3)$ are the longitudinal, lateral and spin creepages, and $L_i (i = 1 - 3)$ denotes the flexibility parameter for each creepage component.

The stresses stated in Eq. (38) cannot exceed the so-called traction bound. Slip occurs in the region where the tangential stresses predicted by Eq. (38) are greater than the traction bound. The formulation for the traction bound used in this paper is obtained by applying Coulomb's friction law locally with a constant friction coefficient, i.e. $\mu p(x_i, y_j)$. The tangential forces are obtained from the numerical integration of the stresses over the contact patch.

4 Results and Discussions

The effects of roller rig testing in the experimental investigation of wheel-rail contact have been addressed in Sect. 3 under four different points of view. In reality, all of these factors interact with one another, thereby it is essential to investigate their combined influence on the contact solution. To this end, a set of cases with various contact positions and radii of roller have been chosen to quantify the influence. The calculation parameters listed in Table 1 are used throughout the simulations.

For simplicity, the track irregularities are neglected and no wheelset velocity component is considered except in the rolling direction. The creepages are calculated according to expressions (35)–(37) as presented in Table 2. According to Eq. (35), the last three terms represent the additional

Table 1 Calculation parameters

Parameter type	Value
Wheel profile	New S1002
Rail/roller profile	New UIC60
Rail/roller inclination	1:40
Track/roller gauge	1435 mm
Wheel flange back spacing	1360 mm
Tape circle to flange back distance	70 mm
Wheel radius	460 mm
Roller radius	0.5 m/1.0 m
Young's modulus	210 MPa
Poisson's ratio	0.3
Friction coefficient	0.35
Normal force	80 kN
Velocity	72 km/h

Table 2 Parameters defining the case studies

No.	y (mm)	Longitudinal ξ_x			Lateral ξ_y			Spin ξ_z (m^{-1})		
		Rail	Roller $s_0 = 1$ m	Roller $s_0 = 0.5$ m	Rail	Roller $s_0 = 1$ m	Roller $s_0 = 0.5$ m	Rail	Roller $s_0 = 1$ m	Roller $s_0 = 0.5$ m
1	0	0	0	0	0	0	0	0.075	0.109	0.143
2	3	−0.0017	−0.0022	−0.0027	0	0	0	0.197	0.287	0.377

contribution of the roller rig to the longitudinal creepage with respect to wheel-rail contact case. It is clear that the major difference in the longitudinal creepage is caused by the variation of the roller head circumferential velocity across its profile. The lateral creepage is zero when the yaw angle is assumed to be zero based on Eq. (36) under the considered contact condition. It can be seen from Eq. (37) that the additional contribution of the roller rig to the spin creepage is coming from the last three terms in the expression that represents, respectively, the effect of the wheelset yaw angle and of the angular velocity of the roller. These additional terms explain the remarkable increase of the spin for the roller rig case which is shown in Table 2. The contact estimation results are presented in two groups for normal contact solution and tangential contact solution, respectively.

4.1 Normal Contact Solution

The solutions of the normal contact problem in terms of the shape and area of the contact patch and the corresponding pressure distribution within the contact region are obtained by the method proposed in Sect. 3.2 for wheel-rail and wheel-roller contact conditions, respectively. The calculation results for the case studies listed in Table 2 are presented in Fig. 5, and the results of wheel-rail contact and of wheel-roller contact obtained at the same contact position are presented in the same figure for comparison.

It can be seen from Fig. 5 that these two simulation cases correspond to a highly non-elliptic contact condition in the first case and nearly elliptic contact condition in the second case. In Fig. 5a, b, it is observed that the length of the contact patch in longitudinal direction decreases for the roller rig with respect to the rail due to the finite radius of roller, this effect being more visible for the smaller value of the roller radius. On the contrary, the width of the contact patch is slightly increased in the case of wheel-roller contact. The maximum contact pressure over the contact zone is increased by a decrease of the roller radius as the same load is spread across a smaller contact area, see Fig. 5c, d. The change of the contact patch also affects the semi-axis ratio and consequently affects the creep coefficient and the creep forces. The differences caused by the

roller rig in terms of contact area and maximum contact pressure should be taken into account when the roller rig is used for contact deterioration mechanism studies such as wear and rolling contact fatigue. To quantify the difference involved in the normal contact solution, a statistical summary of the results is presented in Table 3.

It can be concluded from Table 3 that the differences are increasing with the decrease of radius of the roller and the agreement between wheel-roller and wheel-rail contact is better for the approximately elliptic contact condition, i.e. case 2. It should be mentioned that the results reported here are dependent on the particular parameters assumed in this study, but the analysis approach and the conclusions are generally applicable to any roller rig of this kind.

4.2 Tangential Contact Solution

The corresponding tangential contact solutions for the cases introduced in Sect. 4.1 are presented in Fig. 6 in terms of the stress distribution and division of the contact patch into a stick region and a slip region.

It can be observed from Fig. 6 that the pattern of the stress distribution over the contact patch is similar for wheel-rail and wheel-roller contact conditions in both cases considered. However, the relative percentage of the slip region over the whole contact area is slightly larger in the case of wheel-roller contact. The resultant tangential creep forces in longitudinal and lateral directions are calculated by integration over the contact area, and are presented in Table 4 together with the differences caused by the roller rig test.

It can be seen from Table 4 that the resultant longitudinal force produced on the roller rig differs significantly from the same quantity in the wheel-rail contact case when the contact patch is highly non-elliptic, i.e. for $y = 0$ mm, and the difference increases as the radius of roller decreases, whereas the differences are relative small for approximately elliptic contact condition, i.e. for $y = 3$ mm, especially when the lateral component of the tangential force is concerned.

To evaluate the contact surface damage situation the frictional power at the contact patch is calculated for each case study, and the results are summarized in Table 5.

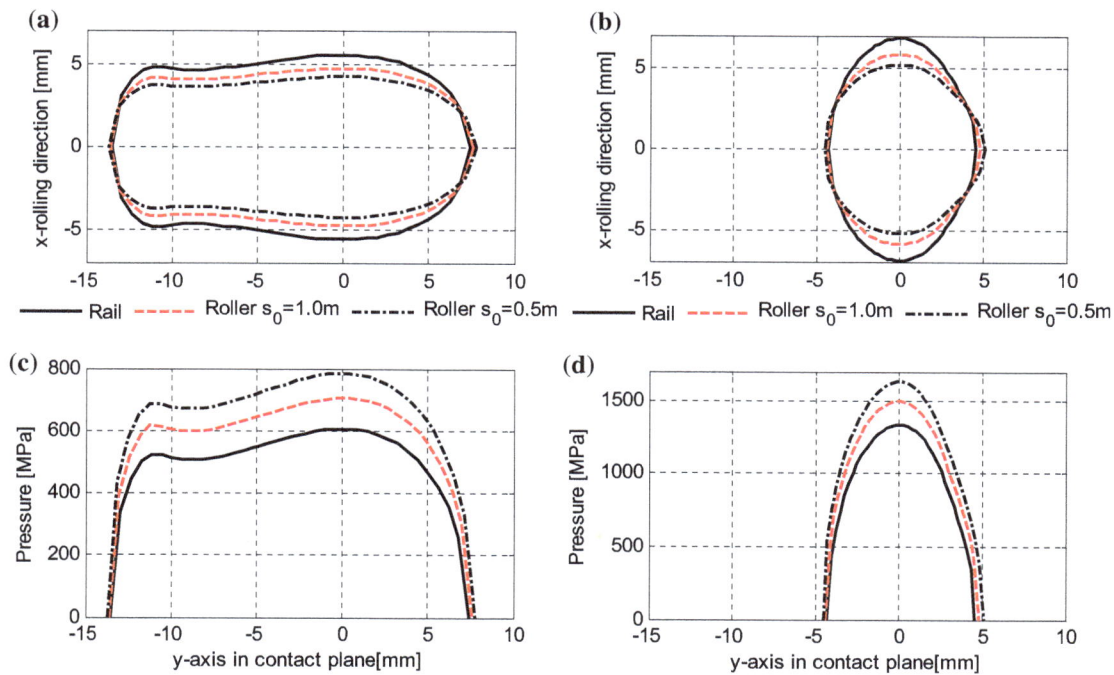

Fig. 5 Contact patch (*top*) and the pressure along its *y*-axis (*bottom*) for cases 1 (*left column*) and 2 (*right column*)

Table 3 Summary of normal contact solution

y (mm)	Contact area A_c (mm^2)			A_c.difference (%)		Max. pressure P_m (MPa)			P_m.difference (%)	
	Rail	Roller $s_0 = 1$ m	Roller $s_0 = 0.5$ m	Roller $s_0 = 1$ m	Roller $s_0 = 0.5$ m	Rail	Roller $s_0 = 1$ m	Roller $s_0 = 0.5$ m	Roller $s_0 = 1$ m	Roller $s_0 = 0.5$ m
0	201	176	159	−12.4	−20.9	598	682	754	14.0	26.1
3	94	84	78	−10.6	−17.0	1336	1498	1632	12.1	22.2

It is clear from Table 5 that the frictional power increases considerably in the case of wheel-roller contact with respect to wheel-rail contact under the same condition, and the differences are particularly relevant for a smaller radius of the roller. It should be noted that the increase of frictional power implies an accelerated manifestation of wear and fatigue effects in the contact pair. This accelerated effect caused by the roller should be taken in proper account by wheel-rail surface damage/deterioration studies performed using roller rigs. It is worth mentioning that this accelerated effect is desirable for wheel material comparison/optimization concerning wear, because the roller rig is capable of reproducing wear patterns within a much shorter time compared to field testing. The test results from the roller rig can be used to examine and document differences in hardening, profile development, polygonalization and possible crack formation of the wheel under test [28]. More details on the wear test on the roller rig can be found in references [28, 29].

5 Conclusions

This paper investigated the differences between wheel-rail contact and wheel-roller contact, with the final aim of assessing the extent to which the results obtained on a roller rig can be extended to the case of a wheelset running on a real track.

A systematic description and comparison on the methodology for solving the contact problem at the wheel-rail and wheel-roller interfaces have been done in terms of the geometric contact problem, normal contact problem, kinematic problem and tangential contact problem. A modified Kik–Piotrowski model has been proposed to deal with the wheel-roller contact problem for zero yaw angle contact conditions.

Simulation results have pointed out the differences implied by a test performed on a roller rig compared to wheel-rail contact in terms of size and shape of the contact patch and distribution of the normal and tangential stresses.

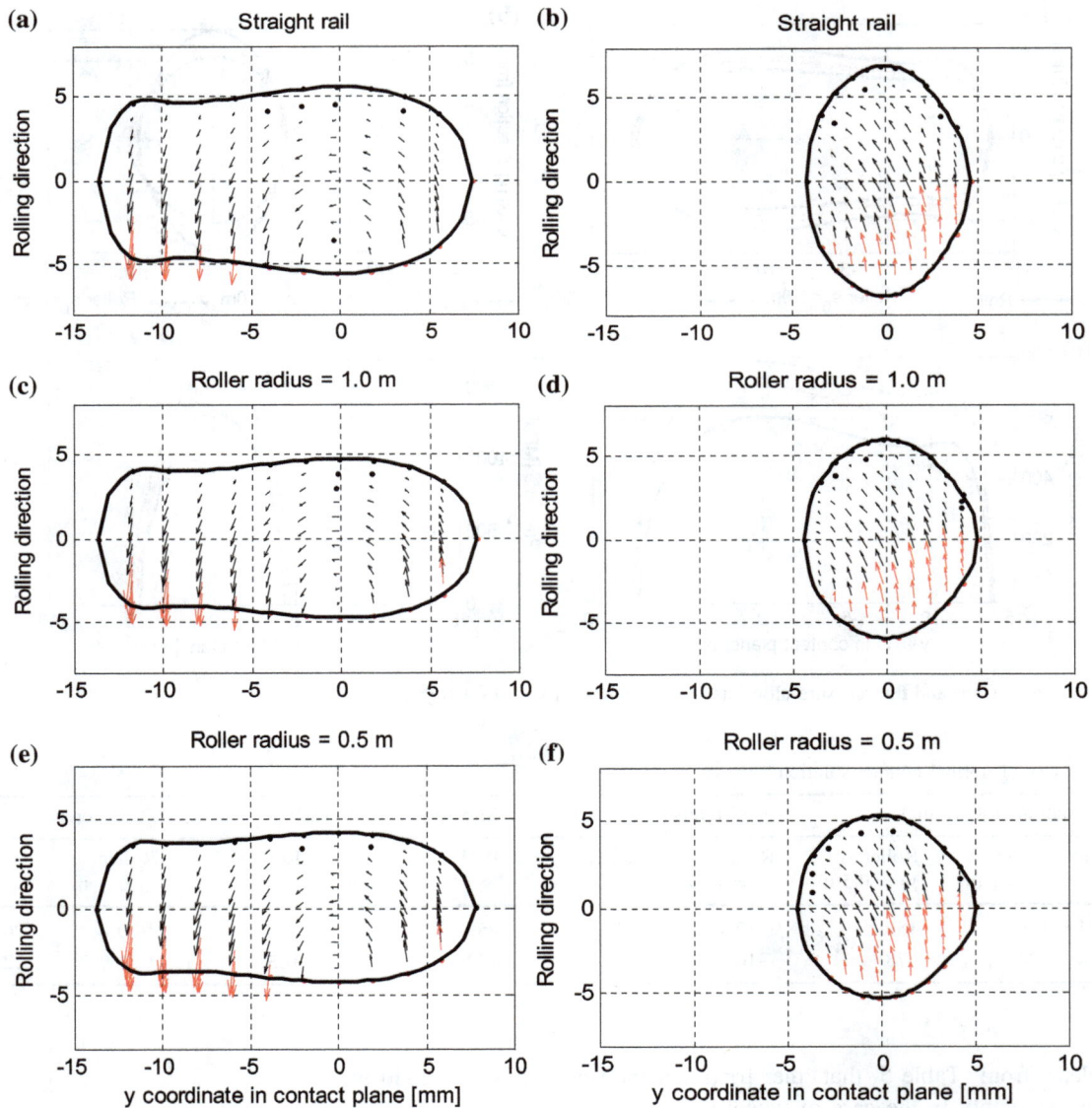

Fig. 6 Stress distribution over the contact patch formed by wheel-rail (*top*), wheel-roller with a radius of 1 m (*middle*) and wheel-roller with a radius of 0.5 m (*bottom*) for cases 1 (*left column*) and 2 (*right column*). *Black* and *red arrows* represent the stress vector in the stick and slip regions, respectively

Table 4 Summary of tangential contact solution

y (mm)	F_x (kN)			F_x difference (%)		F_y (kN)			F_y difference (%)	
	Rail	Roller $s_0 = 1$ m	Roller $s_0 = 0.5$ m	Roller $s_0 = 1$ m	Roller $s_0 = 0.5$ m	Rail	Roller $s_0 = 1$ m	Roller $s_0 = 0.5$ m	Roller $s_0 = 1$ m	Roller $s_0 = 0.5$ m
0	−4.96	−5.76	−6.16	16.13	24.19	−3.24	−3.46	−3.45	6.79	6.48
3	16.46	17.03	17.87	3.46	8.57	−5.70	−5.45	−5.65	−4.39	−0.88

This analysis provides a useful framework for interpreting the results of tests performed on a roller rig, e.g. wear and/ or rolling contact fatigue tests and for extending the results to the real behaviour of the wheelset in the field. The accelerated effect on wheel surface deterioration during the test on the roller rig is preferable for material optimization

Table 5 Frictional power

y (mm)	Frictional power P_f (Nm/s)			P_f difference (%)	
	Rail	Roller $s_0 = 1$ m	Roller $s_0 = 0.5$ m	Roller $s_0 = 1$ m	Roller $s_0 = 0.5$ m
0	72.41	116.59	200.94	61.01	177.50
3	608.50	847.42	1119.02	39.26	83.90

study, while it is not the case for reproducing the wear process of the wheel in service. In this second case, the above-mentioned effect must be taken into account when translating the results of the test to the field case.

Further work will focus on the development of a contact model which is capable of taking into account the influence of yaw angle for both normal and tangential problems.

References

1. Keylin A, Ahmadian M (2012) Wheel-rail contact characteristics on a tangent track vs a roller rig. In: Proceedings of the ASME 2012 rail transportation division fall technical conference, Omaha
2. Jaschinski A, Chollet H, Iwnicki S, Wickens AH, Würzen JV (1999) The application of the roller rigs to railway vehicle dynamics. Veh Syst Dyn 31:345–392
3. Meymand SZ, Craft MJ, Ahmadian M (2013) On the application of roller rigs for studying rail vehicle systems. In: Proceedings of the ASME 2013 rail transportation division fall technical conference, Altoona
4. Zhang W, Dai H, Shen Z, Zeng J (2006) Roller rigs. In: Iwnicki S (ed) Handbook of railway vehicle dynamics. Taylor & Francis Group, Boca Raton, pp 458–504
5. Allen PD (2006) Scaling testing. In: Iwnicki S (ed) Handbook of railway vehicle dynamics. Taylor & Francis Group, Boca Raton, pp 507–525
6. Yan M (1993) A study of the inherent errors in a roller rig model of railway vehicle dynamic behaviour. ME thesis, Manchester Metropolitan University
7. Taheri M, Ahmadian M (2012) Contact patch comparison between a roller rig and tangent track for a single wheelset. In: Proceedings of the 2012 joint rail conference, Philadelphia
8. Zeng Y, Shu X, Wang C, Yu W (2013) Study on three-dimensional wheel/rail contact geometry using generalized projection contour method. In: Zhang W (ed) proceedings of the IAVSD 2013 international symposium on dynamics of vehicles on roads and tracks, Qingdao
9. Bruni S, Cheli F, Resta F (2001) A model of an actively controlled roller rig for tests on full size wheelsets. Proc Inst Mech Eng Part F: J Rail Rapid Transit 215:277–288
10. Liu B, Bruni S (2015) A method for testing railway wheel sets on a full-scale roller rig. Veh Syst Dyn 53(9):1331–1348
11. Shabana AA, Zaazaa KE, Sugiyama H (2007) Railroad vehicle dynamics: a computational approach. Taylor & Francis Group, LLC, Boca Raton
12. Bosso N, Spiryagin M, Gugliotta A, Somá A (2013) Mechatronic modeling of real-time wheel-rail contact. Springer, London
13. Kalker JJ (1982) A fast algorithm for the simplified theory of rolling contact. Veh Syst Dyn 11:1–13
14. PomboJ Ambrósio J, Silva M (2007) A new wheel-rail contact model for railway dynamics. Veh Syst Dyn 45(2):165–189
15. Yang G (1993) Dynamic analysis of railway wheelsets and complete vehicle systems. Doctoral dissertation, Delft University of technology, Delft
16. Hertz H (1882) Über die berührung fester elastische körper. J Für Die Reine U Angew Math 92:156–171
17. Damme S, Nackenhorst U, Wetzel A, Zastrau B (2002) On the numerical analysis of the wheel-rail system in rolling contact. In: Popp S (ed): system dynamics and long-term behaviour of railway vehicles, track and subgrade. Lecture Notes in applied mechanics, vol 6, Springer-Verlag, Berlin, pp 155–174
18. Vo KD, Zhu HT, Tieu AK, Kosasih PB (2015) FE method to predict damage formation on curved track for various worn status of wheel/rail profiles. Wear V322–323:61–75
19. Kalker JJ (1990) Three-dimensional elastic bodies in rolling contact. Solid mechanics and its applications. Kluwer Academic Publishers, Dordrecht
20. Knothe K, Le The H (1984) A contribution to calculation of contact stress distribution between elastic bodies of revolution with non-elliptical contact area. Comput Struct 18(6):1025–1033
21. Kik W, Piotrowski J (1996) A fast, approximate method to calculate normal load at contact between wheel and rail and creep forces during rolling. In: Zobory I (ed.), proceedings of 2nd mini-conference on contact mechanics and wear of rail/wheel systems, Budapest
22. Piotrowski J, Chollet H (2005) Wheel–rail contact models for vehicle system dynamics including multi-point contact. Veh Syst Dyn 43(6–7):455–483
23. Piotrowski J, Kik W (2008) A simplified model of wheel/rail contact mechanics for non-Hertzian problems and its application in rail vehicle dynamic simulations. Veh Syst Dyn 46(2):27–48
24. Ayasse J, Chollet H (2005) Determination of the wheel rail contact patch in semi-Hertzian conditions. Veh Syst Dyn 43:161–172
25. Linder Ch (1997) Verschleiss von Eisenbahnrädern mit Unrundheiten. Diss. Techn. Wiss. ETH Zürich, Nr. 12342
26. Sichani MS, Enblom R, Berg M (2014) Comparison of non-elliptic contact models: towards fast and accurate modelling of wheel–rail contact. Wear 314:111–117
27. Vollebregt EAH, Wilders P (2011) FASTSIM2: a second-order accurate frictional rolling contact algorithm. Comput Mech 47:105–116
28. Ullrich D, Luke M (2001) Simulating rolling-contact fatigue and wear on a wheel/rail simulation test rig. World Congress on Railway Research, Cologne
29. Braghin F, Bruni S, Resta F (2001) Wear of railway wheel profiles: a comparison between experimental results and a mathematical model. In: Ture H (ed.), 17th international symposium on dynamics of vehicles on roads and tracks, pp 478–489

13

Examining Job Accessibility of the Urban Poor by Urban Metro and Bus: A Case Study of Beijing

Chun Zhang[1] · Joyce Man[2,3]

Abstract Using data from Beijing, this paper evaluates job accessibility for people living in affordable housing to job centers by public transit, including urban metro and bus. By comparing the middle and low-income group who mainly use public transit and higher income group who mainly travel by car, results show an accessibility gap for different modes of transportation as travel by public transit takes nearly double the amount of time as travel by car. While commuting time is closely linked to the location of the provided affordable housing, it is also dependent on the quality of local public transit service. Areas with substantial travel time differences between public transit and car travel reveal the weaknesses of public transit provision. Furthermore, average commuting time by both public transit and car from areas of affordable housing built after 2004 is much longer than that from previously built areas implying that low-income groups are being driven to more disadvantaged locations with time changes. In contrast to the classical job-housing mismatch hypothesis in U.S. cities, the mismatch model in Chinese cities is that while major job opportunities are still concentrated in the central city, affordable housing residents who rely on urban metro and bus are being moved further afield into distant suburban areas. The paper will provide the implication for affordable housing and transportation planning in Chinese cities in the future. Improving job accessibility by further establishment of urban metro system for this demographic will promote the urban economy and provide social welfare for the disadvantaged.

Keywords Accessibility · Job-housing mismatch · Urban metro · Affordable housing · Beijing

1 Introduction

Job accessibility has been a hot research topic for urban planning professions and policy makers for a long time, especially with the rise of the New Urbanism movement and smart growth policy [27, 31]. It is believed that better job accessibility will not only calm down traffic [15] and reduce greenhouse gas emissions [1], but also improve local employment and social inequality for the disadvantaged groups [4, 19]. In this sense, better job accessibility is desired by both urban planners and policy makers.

However, the detachment between employment and housing has been on the rise across the globe, though especially in U.S. cities in the past few decades [8, 9, 10, 36]. This detachment trend is simultaneous with large-scale urbanization and urban sprawl at the end of twentieth century. Both spatial and non-spatial factors play their part in determining job accessibility. For example, empirical study supports that land-use pattern decides about one-third of the driving distance between the workplace, home, and other destinations [7]. In this sense, the balance of land mix use between job and housing will reduce distance commuting and increase job accessibility [15]. Non-spatial factors also play an important role, for example, discriminatory housing policy toward the Blacks in the neighborhoods in certain American cities [14,

✉ Chun Zhang
zhangc@bjtu.edu.cn

[1] School of Architecture and Design, Beijing Jiaotong University, Beijing, China

[2] School of Public and Environmental Affairs, Indiana University Bloomington, Bloomington, USA

[3] Lincoln Institute Center for Urban Development and Land Policy, Beijing, China

Editor: Prof. Haishan Xia

22, 29]. Because of racial, economic, or even linguistic factors, African-American residents of the inner city are unable to find credible jobs in the city core, but have to commute a long way to labor-intensive jobs located in far suburban areas [24, pp. 95–131; 26].

Similarly to the U.S. cities, the job-housing balance in the cities of developing countries [19, 29, 36], such as China, has also been decreasing in this era of rapid urban expansion. This paper selected Beijing as a case study, for it shows a unique job accessibility pattern carrying strong post-socialism characters as the national capital and the second biggest city [34].

In recent years since the 1990s, large amounts of affordable housing units have been developed in the urban fringe far from the city center [13]. At the same time, however, the majority of job opportunities are still concentrated in the urban center [40]. And although the level of car ownership keeps increasing, it is still not high, which means that the urban poor are still mostly dependent on public transit [38].

This paper selects Beijing as a case study and evaluates the accessibility of jobs for the urban poor who live in affordable housing by calculating the average public transit time from their homes to commercial and industrial job centers. This paper also compares the accessibility of affordable housing projects from before and after 2004 in order to show if the average public transit time has increased or decreased in the wake of the comprehensive plan of 2004.

Due to the limited levels of car ownership in Chinese cities, this paper implies that the calculated job accessibility of the urban poor holds implications not only for urban employment but also for the level of social inequality for disadvantaged groups. Evaluating the accessibility of affordable housing projects will also hopefully lead urban planners and policy makers to reflect on the selection on current affordable housing project sites.

The following parts of this paper will be structured as follows: The second part will review accumulated research on the definition and measurement of accessibility, the benefit of improving accessibility, and related empirical studies in Chinese cities. The third part will give a brief introduction to the background and data of the Beijing case study. The fourth part will explore the spatial distribution of jobs across city, job accessibility across different travel modes, different locations, and in different time periods before and after 2004. The last part will give a conclusion regarding job accessibility of the urban poor in Beijing and its unique features in comparison to that of U.S. cities.

2 Literature Review

There are many ways to define and measure accessibility [7], and this paper selects the traditional and simple definition and measurement based on gravity model and

network model. Then, this paper reviews the literature on the benefit of improving job accessibility and increasing job-housing balance, and also on the accessibility-related domestic studies on Chinese cities.

2.1 Accessibility and Measurement

Accessibility can be defined as the ability and ease to move from one place to access urban facilities in another place and overcome friction such as distance and travel cost [17]. This concept was first brought into the urban planning field in 1959 to measure the potential of interaction [18]. Different from the original idea in the transportation field, the concept of job accessibility in the urban planning field places more emphasis on the relationship between accessibility and the urban land-use pattern [33]. This concept continues to draw attention from both scholars and decision makers, who regard it as an important indicator of good urban planning.

The traditional measurement of accessibility is the gravity model, which is brought up based on distance decreasing rules in geography. The model argues that potential gravity exists among the urban land. In order to access from one place to another driven by this potential, the traveler needs to overcome some friction [7]. Then, accessibility can be measured by aggregate relation to these places in terms of distance, time, or cost. The weight of each place can also be calculated in this way. A longer distance, time, or higher cost indicates that people have less opportunity to work in this area.

Specifically, the simple expression of gravity model is

$$A_i = \sum_j E_j f(C_{ij}), \tag{1}$$

where A_j is the accessibility score for people living in zone i, E_j is the number of employment opportunities in zone j, $f(C_{ij})$ is the impedance function associated with the cost of travel, and C for travel between zone i and zone j, For a metropolitan region with Z zones, $i, j = 1, 2, ..., N$.

The gravity model successfully explains how individuals choose which job centers to work at. However, accessibility in real life highly depends on the transportation network rather than just the spatial distance itself [17]. Thus, the introduction of the real transportation network is an improvement on the original gravity model.

The rules of network analysis come partly from graph theory, focusing on the connectivity among nodes. The simplest network can be depicted as a line between two nodes, A and B and they generate two links: A–B and B–A. When adding more nodes to the network, the links among all the nodes should be calculated based on *Law of the Network* as

$$S = N(N - 1), \tag{2}$$

where S is the network size (number of links) and N is the number of nodes.

Then, the accessibility can be measured using impendence in a real road or public transit network. By introducing the network analysis, in Eq. (1), the equation will be improved from geographic distance to real travel distance, time, or cost based on the transportation network.

2.2 The Benefit of Better Job Accessibility

Accumulative research shows that there are many environmental and social benefits of improving job accessibility. New urbanism advocates argue that by improving the job accessibility, not only traffic but also energy consumption and urban sprawl will all be reduced [15, 31]. In addition, studies show that it will relieve traffic jam [12] and cut down the time on road [36]. For example, a case study in San Francisco found that if job opportunities increase by 10 %, the commuter traffic will reduced by 3.29 % [10].

Recently, better job accessibility proved to have extra environmental and economical benefits. By reducing traffic, it will help to maintain better air quality [1, 7, 15]. Besides by reducing vehicular commuting time, public health will also be improved [36]. Furthermore, the average cost of commuting will be cut down due to reduced gas consumption.

2.3 Job Accessibility Study in Chinese Cities

In the traditional socialist cities, job accessibility was never a problem such as in big, modern Chinese cities. According to the planning system of the planned economy era, employment and housing used to be perfectly balanced within the Work Unit compound [11]. Living and working within the same compound wall is depicted vividly as a "spatial bond" [34]; thus, the job accessibility used to be pretty high in the planned economy era. The character of such a job-housing relationship led to minimal domestic research on the issue of job accessibility until the 1990s.

With the acceleration of economic reform in 1980s, the traditional Work Unit began to collapse and the job-housing relationship began to fundamentally change [11]. The termination of state-provided housing in 1998, a milestone in Housing Reform, triggered the process of large-scale residential relocation to suburban areas [20]. Nonetheless, the economic heart of the city still remained in the city center, inducing a change of the job-housing relationship and decrease in job accessibility.

For example, Liu and Wang found that job-housing spatial mismatch began to form and impacted the

everyday commuting behavior based on the questionnaire in Beijing, and job accessibility began to decrease [41]. Zhao and Lu argue that the traditional job-housing relationship still has an institutional legacy on job accessibility in the current transition period of Beijing [38]. In a case study of a southern city, Guangzhou, it was found that with the disappearance of Work Unit and the commercialization of housing, the commuting distance became longer, and the commuting spatial structure was changed [25]. Similarly, Zhou noted the excess commuting phenomena based on the TAZ analysis in Xi'an as a western city in China [43]. Based on the cases above, job accessibility turned worse since the 1990s in most of the Chinese cities with the trend of quick urbanization and job-housing relation changes.

3 Method and Data

3.1 Study Area

Beijing is the national capital in the north of China and has a population of over 17 million [3]. While China has been successful in reducing poverty in recent decades, the urban poor in Chinese cities still faces the usual difficulties that come with limited income. In reality, the gap between the urban rich and the urban poor is not only based on salary, but also on the urban resources supported or provided by the municipal government. These urban resources include transportation facilities, education facilities, hospitals et al., and this paper will focus on the job accessibility of the urban poor. It is worthy to note that the way people travel has evolved tremendously from planned economy period to the transition period.

Pre-reform Beijing was defined by the development of self-contained work units, or *danwei*. People work and live within their *danwei*, and all activity (shopping, recreation and education) occurs within the boundaries of these *danwei* [37]. However, with the collapse of the traditional work unit, the relationship between occupation and housing has fundamentally changed. On one hand, people began to find jobs outside of the walls of their respective *danweis*, releasing the spatial bond. On the other hand, the privatization of the housing market and inner city redevelopment has driven people into large housing projects in the outer edges of the urban fringe, while the job centers remain in the center of the city. This has resulted in the low job accessibility and job-housing spatial mismatch apparent in post-reform Chinese cities, the result of the detaching of jobs and housing.

Along with the increased work-home commute, private car ownership in Beijing has also increased dramatically.

By 2008, there were approximately 3.25 million cars in Beijing, and this number has been rising steadily by 15 % each year [39]. While huge amounts of investment are supporting the rapid growth of the road network, another half of all transportation investment goes toward building public transit systems including bus, mass rapid transit, and rail. By 2010, there were a total of 678 bus lines and a total length of 174.2 thousand km of road covered by the bus system in Beijing. At the same time, the urban metro system grew at a dramatic speed in the first decade of twenty-first century. Based upon subway Line 1 and Line 2, which were built in 1969, the subway system has expanded to a sprawling 14 lines and a total length of 323 km by 2010. Despite the investment in the improvement of public transit, a great amount of maintenance fees goes to subsidize the bus and subway ticket prices, making Beijing one of the cheapest public transit cities in China.

Despite the substantial investments on both urban metro and bus, the spatial mismatch between housing and employment still plagues the urban poor. However, the occurrence of spatial mismatch is different from the U.S. cities: (1) the affordable housing projects are mostly located in the urban fridge [13] and some of these projects are far away from public transit. (2) Main job centers are still concentrated in the central city. (3) The rate of car ownership is still low and commuters rely more on public transportation [39]. In consideration of the above characteristics, the job accessibility of low income can be measured as the commuting time that the urban poor living in the urban fridge have access to central city job opportunities by public transit.

3.2 Data Source and Method

This paper has selected the Beijing urban area within the 6th ring roads as the study area including the administrative districts of Dongcheng, Xicheng, Chaoyang, Haidian, Fengtai, and Shijingshan (Fig. 1). The land, job, and transportation data were collected between 2001 and 2004, while the affordable housing data were collected from 1999 to 2010 [44].

First, the housing data come from the affordable housing projects, which partly represent where the urban poor lives. Second, this paper combines the land-use map in 2004 and economic data in 2001 to identify the job centers. Specifically, it extracts the location of commercial land as the potential job centers. By overlapping the job density within the zip zones based on the economic basic unit survey conducted in 2001, the intensity of the concentration of jobs at each center can be calculated. Third, the transportation data between housing and job come from the public transit network and road network map compiled in 2004, representing those who travel by public transit and private car, respectively. Based on the housing,

employment, and transportation network data, the OD matrix has been generated to measure the network distance or travel time between job and housing. Then, the job accessibility of the poor living in affordable housing is measured as the average single-way commuting time to all the various potential job centers (Fig. 2).

The method outlined above is based on the following hypotheses: (1) The residents of the affordable housing units are identified as low income.[1] (2) Low-income groups tend to rely on public transit, while those in higher income groups tend to drive their own cars. (3) The commuting time by public transit is calculated by Bus time = walking to the nearest stop (A) + waiting time (B) + travel time on public transit (C) + walking to the destination (D); With commuting time by car as Car time = driving time (E) + parking time (F). For simplicity's sake, this paper supposes A, B, D, and F to be 5 min. (4) That the average speed for a public bus is 16.6 km/h [42], while the average speed for a private car is 40 km/h (BJTRC [2].

4 Discussion and Results

Based on the data and hypothesis above, most of the employment opportunities are still concentrated in the city center. Here, job accessibility can be measured by the average commuting time to all the potential job opportunities. By comparing the commuting time between using public transit and driving, the job accessibility gap between low-income group and higher income group can be observed. Furthermore, the comparison of the commuting times calculated with a starting point of affordable housing locations built before and after 2004 will show the change in the job accessibility of the urban poor.

4.1 Distribution of Employment Opportunities

According to the employment data of 2001, despite the fact that most of the affordable housing projects are located in the suburban area outside of the third ring road, most of the employment opportunities are highly concentrated in the city center. It especially shows that the job density is much higher in the inner city than the suburban area and also much higher in the northern than in the southern part of city (Fig. 3).

Specifically, three main job-rich areas are identified as their job density is over 300 jobs per hectare: Guomao in the east, Jinrongjie in the west, and Anzhen in the northeast. There is nearly no affordable housing located within

---

[1] Although this statement is debatable, it is generally believed that the affordable housing residents are at least middle and low income due to the qualification requirements of affordable housing applications.

Fig. 1 Case study of Beijing
with affordable housing projects
and commercial job centers

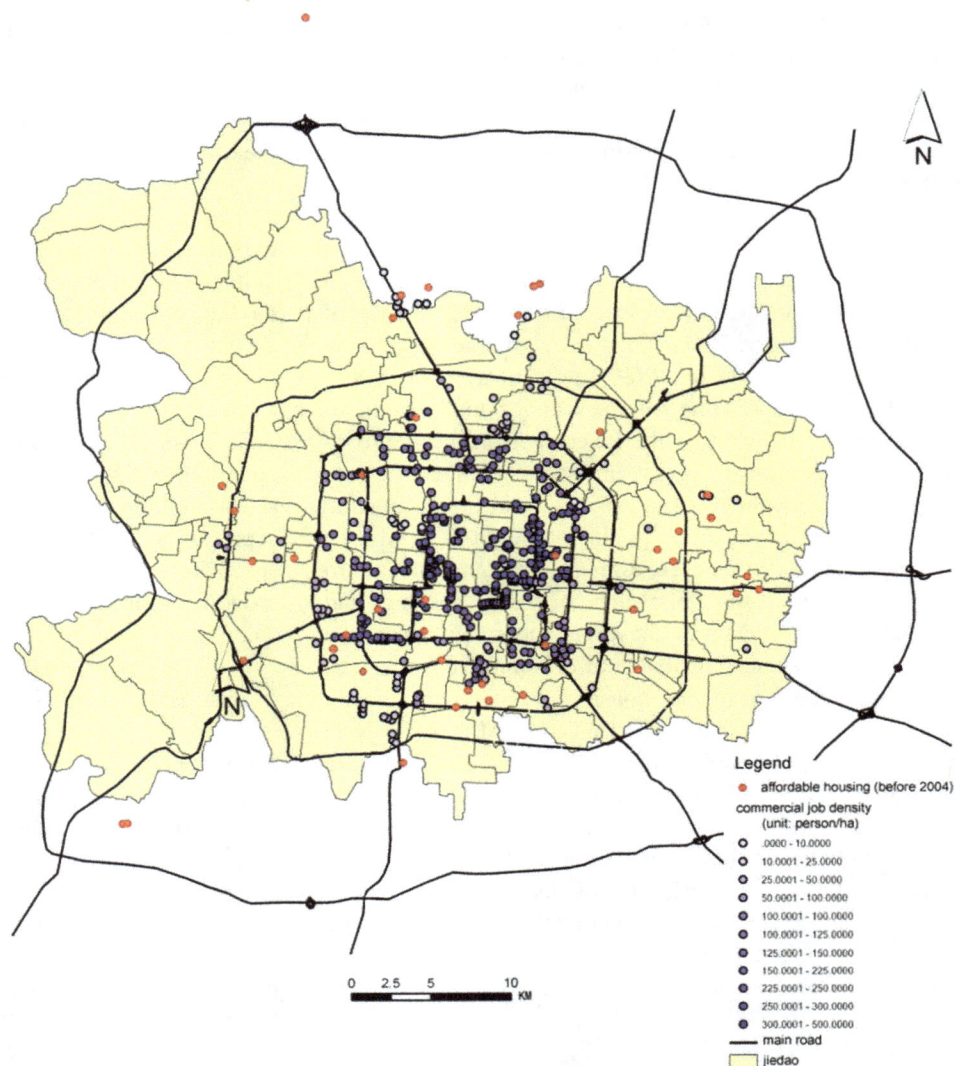

Legend

● affordable housing (before 2004)
commercial job density
(unit: person/ha)

○ .0000 - 10.0000
○ 10.0001 - 25.0000
○ 25.0001 - 50.0000
◐ 50.0001 - 100.0000
◐ 100.0001 - 100.0000
◑ 100.0001 - 125.0000
● 125.0001 - 150.0000
● 150.0001 - 225.0000
● 225.0001 - 250.0000
● 250.0001 - 300.0000
● 300.0001 - 500.0000
— main road
▭ jiedao

the job-rich areas. On the contrary, in the suburban area zip code zones, where the affordable housing is located, the job density is usually as low as 20–50 jobs per hectare.

Basically, there were no sub-centers located in the suburban area as of 2001, a fact which clearly reveals job-housing mismatch problem for affordable housing residents. In this way, the main commuting direction for most affordable residents is toward the city center. As discussed in the accumulated literature, the highly concentrated distribution of employment opportunity shows the strong character of the post-socialist city.

4.2 Travel Mode Difference of Job Accessibility

According to the reach method and data preparation, the measurement of accessibility can be simplified to create the OD matrix from the all the affordable housing projects to all the potential commercial job centers.

First, the service area analysis shows the accessible range difference between who travel by bus and by car. Taking an affordable housing development, *Fengtishidai* in Fengtai district, as an example, 34.9 km^2 of commuting area can be reached within 30 min of travel by public transit (Fig. 4, left), while as large as 454.8 km^2 can be reached by car (Fig. 4, right). In this case, the commuting area is much larger for car drivers than for public transit users.

Second, based on the service area analysis, the accessibility for each affordable housing project can be measured by using the average commuting time to all potential commercial job opportunities. The OD matrix of commuting time based on the bus network is generated using the origins of 73 affordable housing areas to the destinations of 418 commercial job centers. For example, the highlighted case of affordable housing area *Wangjingxincheng* shows that residents need an average of 56.5 min to travel to all the commercial job centers (Fig. 5, left). This average

Public Transit lines/Streets

Land Use Map

Economic Data

Bus Network/ Road network

Commercial Land

Job Density (Zip code zone)

Affordable housing

Job centers

Network Analysis

O-D Time Cost Matrix

Accessibility Value

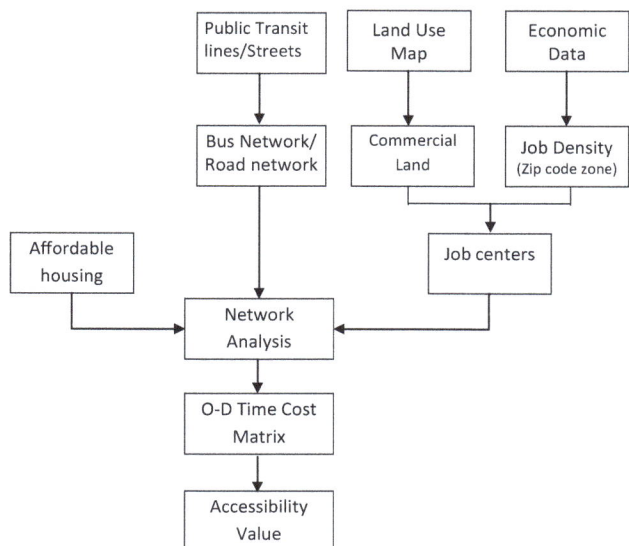

Fig. 2 Network analysis approached to job accessibility

Fig. 3 The commercial employment density of Beijing in 2001

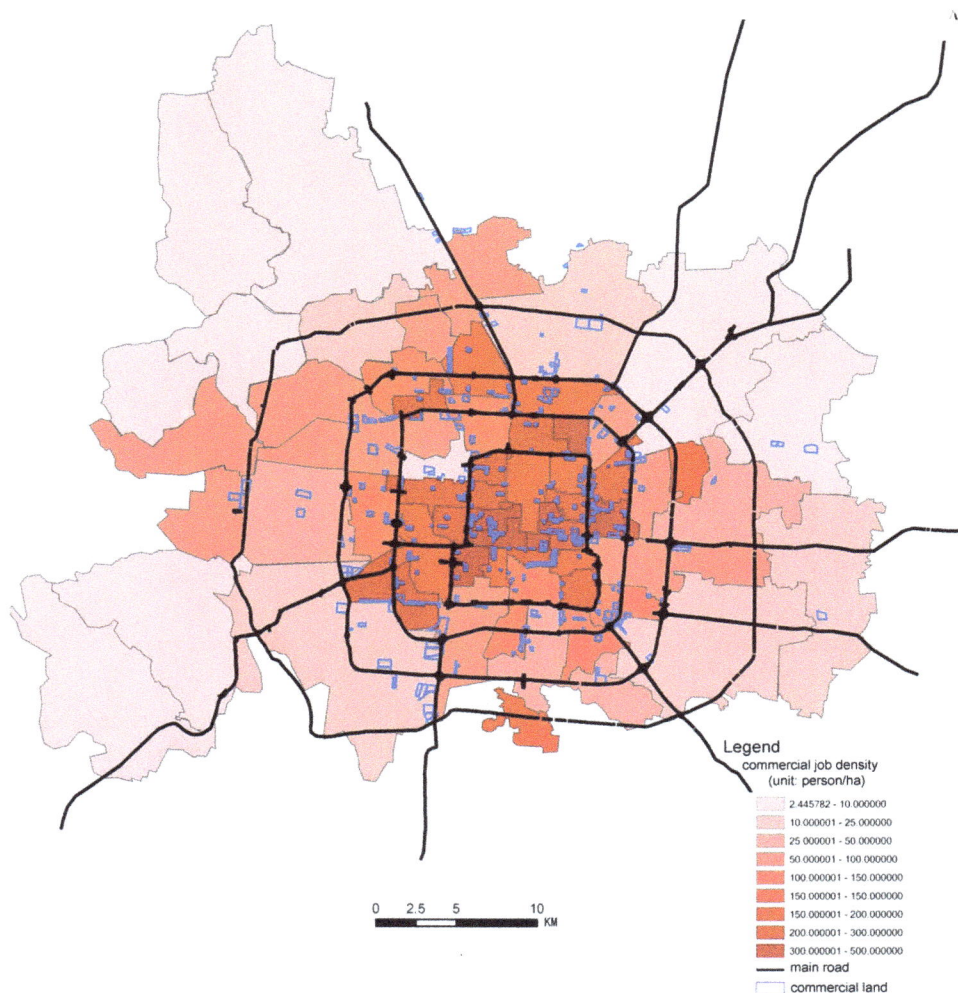

commuting time can be an index to measure job accessibility. Generally speaking, it takes 59.4 min on average for the affordable housing residents to travel to commercial job centers, which is higher than the one-way commuting time for all the residents in Beijing [41].

Third, a similar OD matrix based on the road network from the same origins and destinations is used to measure the job accessibility by car. Again, using the affordable housing area of *Wangjingxincheng* as an example, the average commuting time by private car is only 26.7 min which is significantly shorter than the average travel time by public transit. For car users, the average commuting time from all affordable housing sites to commercial job centers as a whole is only 28.9 min; nearly one half of the average commuting time by public transits (Fig. 5, right). This conclusion has similar results to the empirical studies conducted in the U.S. cities, in which different accessibility levels among different income groups do exist [16].

Legend
commercial job density
(unit: person/ha)

2.445782 - 10.000000
10.000001 - 25.000000
25.000001 - 50.000000
50.000001 - 100.000000
100.000001 - 150.000000
150.000001 - 150.000000
150.000001 - 200.000000
200.000001 - 300.000000
300.000001 - 500.000000
main road
commercial land

0 2.5 5 10 KM

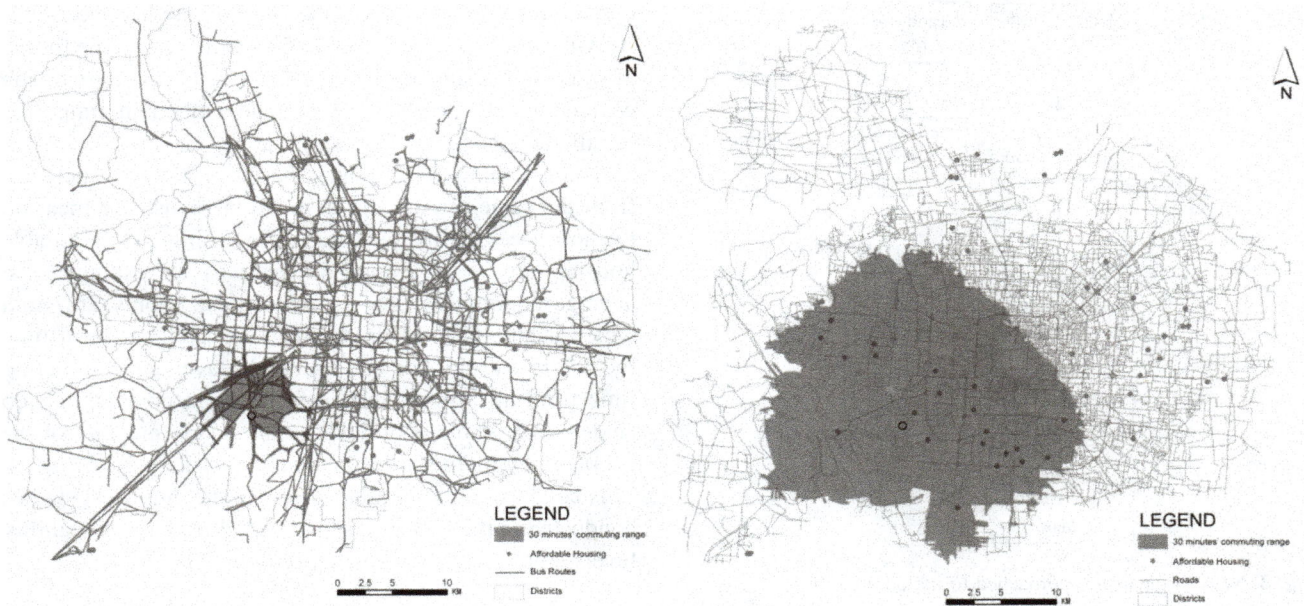

Fig. 4 The accessible commuting area within 30 min by bus (*left*) and car (*right*)

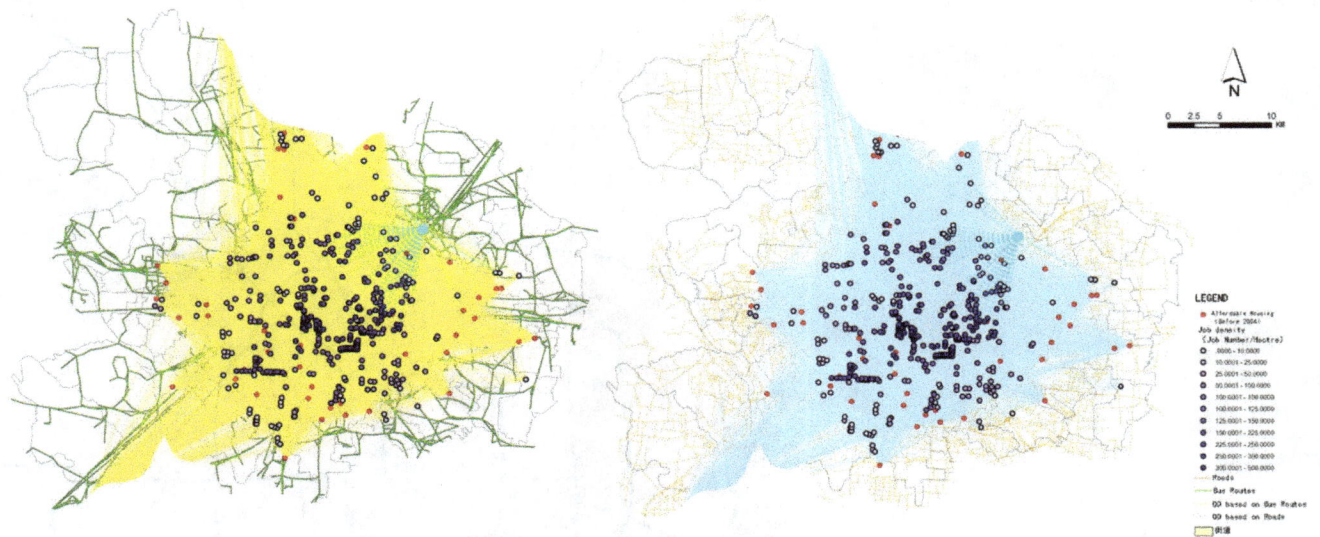

Fig. 5 OD matrix of commuting time based on bus network (*left*) and road network (*right*)

4.3 Spatial Difference of Job Accessibility

As discussed above, the commuting time using different travel methods reveals the difference of accessibility between different income groups. On average, traveling by public transit takes nearly double the amount of time than traveling by car. As Fig. 6 on the left shows, the average commute by bus is concentrated in the range of 46–60 min, while the average commuting time by car is concentrated in the range of 21–30 min. However, in addition to varying commuting time by different travel methods, the spatial

difference of accessibility is also an important issue for the urban low-income groups.

The spatial difference of accessibility is large where the commuting time gap between public transit and car is large. This difference reveals an insufficient supply of public transit services where traveling by bus needs much longer than travel by car (Fig. 6, right). For example, some of the considered cases of affordable housing projects closely follow this model. The affordable housing projects of *Huilongguan* in the north, *Chaoyang New Town* and *Dingfu Garden* in the northeast, and *Tiancunluobei* in the

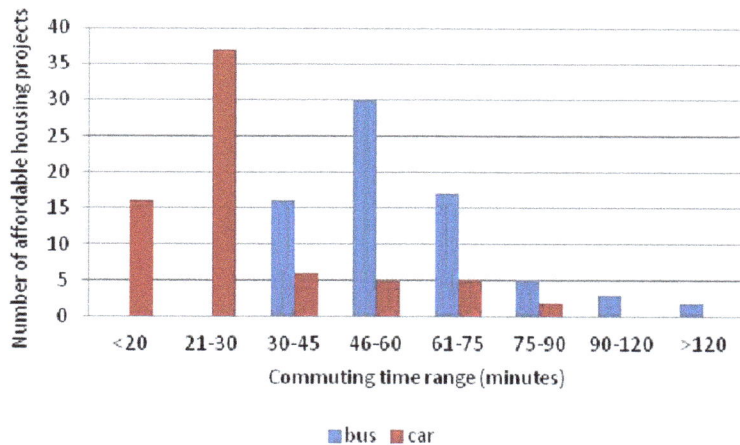

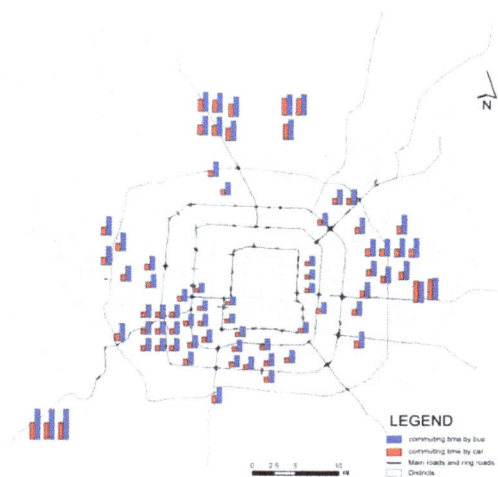

Fig. 6 Commuting time gap between bus and car

west will all need more public transit investment to improve local job accessibility in the future.

From the spatial distribution aspect, the job accessibility of affordable housing is influenced by both the distance to city center and local public transit services. With an increase in distance to city center, both average commuting time by public transit and car are not directly increased, which alludes to the interaction of other local factors (Fig. 7). These results are similar to those concluded on Western cities. Outside of geographical distance, local factors such as land use and public facilities might also influence accessibility [17]. For example, although *Huanghuixiaoqu* and *Jiandongyuan* in Dongzhou district are far from the city center, the commuting time gap between bus and car is not large due to the convenient and efficient BTR line and the proximity to subway Line 8.

The discussion above suggests that the limited job accessibility for some affordable housing residents might not be solely due to the location factor, but also the lack of local public transit facilities. That suggests that low-income groups might be driven to disadvantaged places without convenient public transit facility, especially for the area without urban metro or BRT connection. In American cities such as Los Angeles and Detroit, it has been observed that areas where low-income groups concentrate and have a lack of adequate bus services are prone to experience low-employment decay [23].

4.4 Time Changes for Job Accessibility

Previous sections discussed the job accessibility of all affordable housing sites built between 1999 and 2010 to commercial job centers in 2001, this part will compare the job accessibility for the sites of affordable housing projects before and after 2004.

Generally speaking, a longer commuting time regardless of the transportation method is needed for residents living in affordable housing areas built after 2004 compared to those living in areas built before 2004. If traveling by public transit, the average commuting time increases from 56.5 min before 2004 to 74.7 min after 2004. If traveling by car, the average commuting time increases from 28.9 min before 2004 to 42.8 min after 2004 (Fig. 8).

The average commuting time needed by residents who travel by public transit shifts to right, red bar from the left, blue bar (Fig. 8, upper right). Before 2004, the average commuting time was concentrated between 46 and 60 min. With the average commuting time's rightward shift, the commute to work for those living in affordable housing has increased with time into the range from 75 to 90 min. This implies that more than a half of affordable housing residents need to travel over 90 min for a one-way commute.

Similarly, the average commuting time by car also shifts rightward (Fig. 8, right down). Compared to bus travel time, commuting time by car is spread out over a greater time range. The amount of time for those living in affordable housing projects to go to work obviously increased, especially for those who need to drive 31–40, 41–50, and above 60 min.

Again similar to those affordable housing areas built before 2004, the commuting time by both public transit and car from affordable housing areas built after 2004 does not directly increase as the distance from the city center increases, but is rather more turbulency. For example, the relationship between commuting time and the distance to the city center is more irregular after 2004 than before. As for public transit in locations where it is insufficiently supplied, commuting time by bus will obviously require a longer time than commuting time from other affordable housing areas with a similar distance to the city center

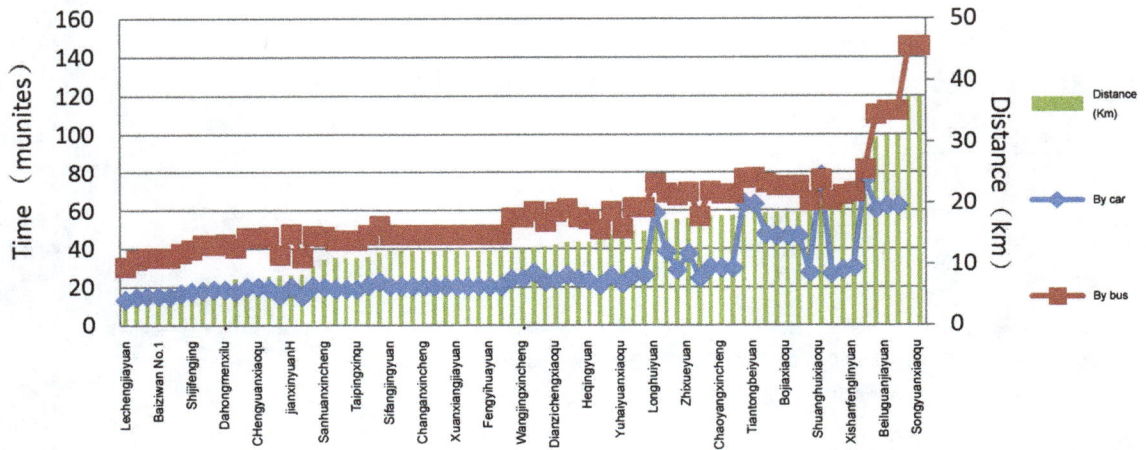

Fig. 7 Distance to city center and commuting time by bus and car for affordable housing (pre 2004)

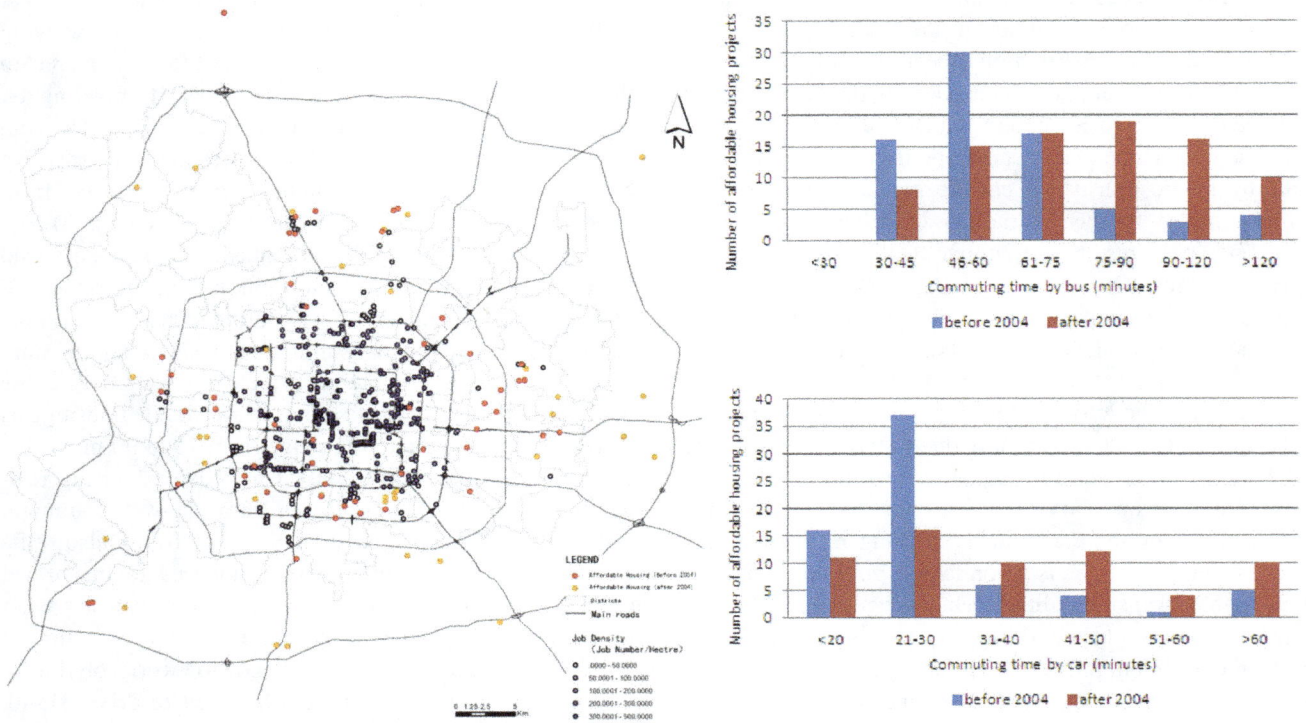

Fig. 8 Commuting time changes for affordable housing before and after 2004

(Fig. 9), such as the affordable housing projects near *Chaoyang New Town* in the northeast and *Tiantongyuan* in the north.

In conclusion, from aspect of job accessibility, the affordable housing areas built after 2004 are in worse locations than those built before 2004, which has forced the urban poor into more disadvantaged locations. That mostly because the selection of affordable housing sites moves to more remoting place far away from the city center after 2004 in Beijing. Actually in most Chinese cities, although there is no housing discrimination policy toward the minority low-income groups [5, 6], these urban low-income groups are still relegated into affordable house in the undesirable locations with limited job accessibility in the long run.

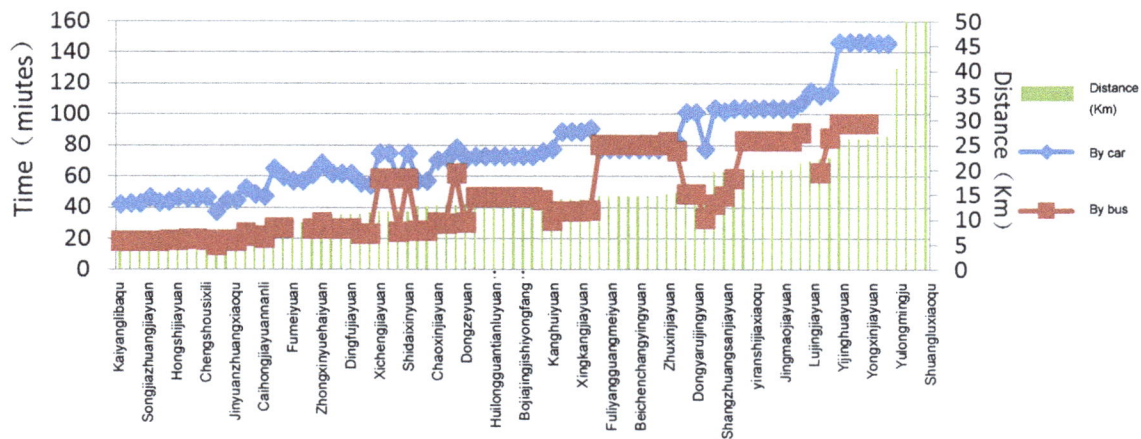

Fig. 9 Distance to city center and commuting time by bus and car for affordable housing (after 2004)

5 Conclusion

This paper uses Beijing as case study to explore the job accessibility for low-income groups living in affordable housing areas. A comparison of the average commuting time between using public transit and driving car shows that travel by public transit requires nearly twice the amount of time as driving. This difference in travel methods also reflects the job accessibility gap between different income groups, namely poorer low-income group who take public transit and richer car owners. From the spatial perspective, local public transit is insufficient in areas where travel by both urban metro and bus needs much more time than travel by car. Furthermore, the average commuting time by bus increased from 56.5 min before 2004 to 74.7 min after 2004, while the average commuting time by car increases from 28.9 min before 2004 to 42.8 min after 2004. The fact that the residents of new affordable housing areas built after 2004 need to travel longer times after 2004 than before suggests that they are being continuously driven into more disadvantaged locations with limited job accessibility.

Meanwhile, the case study of Beijing shows its unique features of job-housing spatial mismatch as compared to the classical model seen in American cities. In Beijing's case, the affordable housing locations near the city center have better job accessibility, while the locations in the far out suburban areas are weaker in job accessibility. This is different from the classical spatial mismatch hypothesis in which job opportunities move to suburban area, while the low-income groups stay in the inner city [21]. The job opportunities in Chinese cities such as Beijing still concentrated in the inner city without spreading outwards, while the affordable housing areas for low-income groups are being built further and further from the city center. Considering the lack of public transit around these suburban areas, this spatial pattern might induce the low accessibility for low-income groups in Chinese cities.

This unique Chinese model for spatial mismatch reveals the limited accessibility due to income differences. Compared to Western cities, job accessibility in China is based on income factors rather than racial factors [4, 19]. The limited job accessibility for low-income groups forces them into worse housing locations farther and farther away from job-rich areas. Furthermore, this new spatial mismatch in Chinese cities severs the "spatial bond" which tied housing and employment together within the same Work Unit compound as seen before the economic reform. This broken "spatial bond" also decreases job accessibility for low-income groups.

Exploring job accessibility and its change for low-income individuals can provide implications for future transportation developments and urban land use. For example, in the selection of better locations for future affordable housing projects, urban planners could make sure to make it easier for individuals to access job opportunities, improve local public transit services, or make the last mile home more convenient. Especially for big Chinese cities, establishing well-connected urban rail system for affordable housing residents might be a good way to improve their job accessibility. As the advocators of New Urbanism and Smart Growth proposed, improving the accessibility of low-income groups have many benefits for the entire city, as it will not only reduce private car use and promote urban employment, but also improve social equality for low-income groups. This can then lead planning professionals and decision makers who care about the disadvantaged in the city to provide them with equal footing to employment opportunities.

Acknowledgments Supported by the China Fundamental Research Funds for the Central Universities (2014JBZ020) and the China Ministry of Education Social Science Founding Youth Program (13YJCZH240).

References

1. Armstrong M, Sears B (2001) The new economy and jobs/housing balance in Southern California. Retrieved 1 April 2012
2. Beijing Bus (2012) Report on Development of Beijing Bus in Recent Years. Retrieved 3 Dec 2012
3. Beijing Statistical Bureau (2001) Beijing statistical yearbook. China Statistic Press, Beijing
4. Blumenberg E, Ong P (1998) Job accessibility and welfare usage: evidence from Los Angeles. J Policy Anal Manag 17(4):639–657
5. Boustan L, Margob R (2009) Race, segregation, and postal employment: new evidence on spatial mismatch. J Urban Econ 65:1–10
6. Brooks R (2002) Covenants and conventions. Northwestern University School of Law
7. Cervero R (1989) Jobs-housing balance and regional mobility. J Am Plan Assoc 55(2):136–150
8. Cervero R (1991) Jobs/housing balance as public policy. Urban Land 50(10):10–14
9. Cervero R (1996) Jobs-housing balance revisited: trends and impacts in San Francisco Bay area. J Am Plan Assoc 62:492–511
10. Cervero R, Sandoval O et al (2002) Transportation as a stimulus of welfare-to work: private versus public mobility. J Plan Educ Res 22(1):53–63
11. Chai, Y, Zhang, C (2009) Geographical approach to Chinese Cities' Danwei: a key to understand the transition of Urban China. Int Urban Plan 24 (5): 2–6. [柴彦威, 张纯. 地理学视角下的城市单位:解读中国城市转型的钥匙. 国际城市规划, 2009,24 (5):2–6]
12. Downs A (1992) Stuck in traffic: coping with peak hour traffic congestion. Brookings Institution, Washington, DC
13. Deng FF, Huang Y (2004) Uneven land reform and urban sprawl in China: the case of Beijing. Prog Plan 61:211–236
14. Ellwood D (1986) The spatial mismatch hypothesis: Are there teenage jobs missing in the ghetto? In: Freeman R, Holzer H (eds) The black youth employment crisis. University of Chicago Press, Chicago
15. Ewing R (1996) Best development practices: doing the right thing and making money at the same time. Planners Press, Chicago
16. Grengs J (2010) Job accessibility and the modal mismatch in Detroit. J Transp Geogr 18(1):42–54
17. Handy SL, Niemeier DA (1997) Measuring accessibility: an exploration of issues and alternatives. Environ Plan A 29(7):1175–1194
18. Hansen W (1959) How accessibility shapes land use. J Am Inst Plan 25:73–76
19. Holzer H (1991) The spatial mismatch hypothesis: What has the evidence shown? Urban Stud 28(1):105–122
20. Huang, Y. (2005). From work-unit compounds to gated communities: housing inequality and residential segregation in transitional Beijing. Restructuring the Chinese city: changing society, economy and space. L. J. C. M. a. F. L. Wu. London, Routledge
21. Kain J (1968) Housing segregation, negro employment, and metropolitan decentralization. Q J Econ 2:175–197
22. Kain J (1992) The spatial mismatch hypothesis: there decades later. Hous Policy Debate 3(2):371–460
23. Kawabata M (2003) Job access and employment among low-skilled autoless workers in US metropolitan areas. Environ Plan A 35(9):1651–1668
24. Ley D (1993) Social geography of the City. Harper & Row Press, New York
25. Liu W, Yan X, Xie L (2012) Employment and residential mobility and its spatial structure change based on the 3 years survey analysis. Geogr Res 31(9): 1685–1697. [刘望保,闫小培,谢丽娟.转型时期广州居民职住流动及其空间结构变化——基于3个年份的调查分析. 地理研究, 2012, 31(9):s1685–1697]
26. Mattingly D (1999) Job search, social networks, and local labor market dynamics: the case of paid household work in San Diego, California. Urban Geogr 20(1):46–74
27. New Urban News (2010) http://www.newurbannews.com. Retrieved 6 Apr 2010
28. Omer I (2006) Evaluating accessibility using house-level data: a spatial equity perspective. Comput Environ Urban Syst 30:254–274
29. Raphael S (1998) The spatial mismatch hypothesis and black youth joblessness: evidence from the San Francisco Bay Area. J Urban Econ 43:79–111
30. Shen Q (2000) A spatial analysis of job openings and access in a US metropolitan area. J Am Plan Assoc 67(1):53–68
31. Smart Growth America (2007) What is smart growth? Retrieved 13 Dec 2007
32. Taylor B, Ong P (1955) Spatial mismatch or automobile mismatch? An examination of race, residence, and commuting in US metropolitan areas. Urban Stud 32(9):1537–1557
33. Wachs M, Kumagai TG (1973) Physical accessibility as a social indicator. Soc Econ Plan Sci 7:437–456
34. Wang E, Song J, Xu T (2011) From "spatial bond" to "spatial mismatch": an assessment of changing jobs housing relationship in Beijing. Habitat Int 35:398–409
35. Warner SBJ (1978) Streetcar suburbs: The process of growth in Boston 1870–1990. Harvard University Press, Cambridge
36. Weitz J, Schindler T. (1997) Are Oregon's communities balanced? A test of the jobs-housing balance policy and the impact of balance on mean commute times. Unpublished manuscript. Department of Urban Studies and Planning, Portland State University
37. Zhang C, Chai Y (2014) Un-gated and integrated work unit communities in post-socialist urban China: a case study from Beijing. Habitat Int 43:79–89
38. Zhao P, Lu B (2010) Exploring job accessibility in the transformation context: an institutionalist approach and its application in Beijing. J Transp Geogr 18:393–401
39. Zhao P, Lu B et al (2009) Consequences of governance restructuring for quality of urban living in the transformation era in Beijing: a view of job accessibility. Habitat Int 33:436–444
40. Liu B, Shen F (2008) Study on the characters of job-housing spatial structure in Beijing metropolitan area. Hum Geogr 120 (4):40–47 (**In Chinese**) [刘碧寒, 沈凡卜. 北京都市区就业—居住空间结构及特征研究. 人文地理, 2008, 120(4):40–47]
41. Liu Z, Wang M (2011) The influence of job-housing mismatch on commuting behavior of residents in Beijing: based on the anaylsis of accessibility and commuting time. Acta Geogr Sin 65:191. [刘志林, 王茂军.北京市职住空间错位对居民通勤行为的影响分析——基于就业可达性与通勤时间的讨论. 地理学报, 2011, 66(4):457–467]
42. Sun F, Wang Z (2003) Statistical analysis of the average speed of buses in Beijing. Automot Eng 25(3):219–242. [孙逢春, 王震坡等. 北京市公共汽车平均车速统计分析. 汽车工程, 2003, 25(3):219–242]
43. Zhou J, Zhang C, Chen XJ, Huang W, Yu P (2014) Has the legacy of Danwei persisted in transformations? the jobs-housing balance and commuting efficiency in Xi'an. J Transp Geogr 40:64–76
44. Lu B, Zhang C, Chen TM (2013) Study on changes in job accessibility for the urban low-income: a case study of Beijing. City Plan Rev 37(1):56–63. [吕斌, 张纯, 陈天鸣. 城市低收入群体的就业可达性变化研究一以北京为例. 城市规划, 2013, 37(1):56–63]

14

A Study of the Feasibility and Potential Implementation of Metro-Based Freight Transportation in Newcastle upon Tyne

Alex Dampier[1] · Marin Marinov[2]

Abstract This paper discusses the concept of using a metropolitan railway network to transport freight directly to a city centre from the surrounding businesses. Specifically we look in depth at the Tyne and Wear Metro system, situated in Newcastle upon Tyne, to determine if such a scheme would be feasible. Through research into the modes of transport available, along with a review of literature and case studies, it was found that the current method of transporting the majority of freight by road is unsustainable and damaging to both the environment and local communities. Other options for the transportation of freight have been reviewed, and results showed that a modal shift will be necessary in the near future. The system was then modelled using software provided by the Department for Transport, which demonstrated that the implementation of such a scheme would provide vast accident savings, a reduction in the number of casualties on the road, and a monetary saving as a result of the lower casualty rate. The conclusion was reached that the scheme is viable; however, further research and study are necessary before implementation.

Keywords Urban freight · Metro · Metropolitan rail networks · COBALT · Defra traffic calculator

✉ Marin Marinov
marin.marinov@ncl.ac.uk

[1] Mechanical and Systems Engineering School, Newcastle University, Newcastle upon Tyne, UK

[2] NewRail, Mechanical and Systems Engineering School, Newcastle University, Newcastle upon Tyne, UK

Editor: Jing Teng

1 Introduction

The transportation of freight throughout the country has a huge effect on Britain's economy, and both large and small businesses alike rely daily on the collection and transportation of goods. In 2010, 1489 million tonnes of goods were lifted in the UK [17, 18], 82 % of which were transported by road. Rail, in comparison, had just 5 % of total modal share [45]. After a peak in 2007, registered numbers of HGVs have been falling, but due to an increase in the size of preferred lorry, the volume of goods transported remains largely unchanged [21, 22].

Given that the vast majority of goods are delivered within a 50 km radius of the journey start point [35], there has been much research undertaken into the potential impact of using light rail systems such as inner city metros and trams to transport freight within our urban areas [5]. This paper aims to establish whether such a system could be implemented using the Metro network in Newcastle upon Tyne, to deliver goods which originate from the local area directly into the heart of the city. A potential scheme has been developed, and an in-depth study of the town of Killingworth to the north of Newcastle has been completed, to assess the impact that such a scheme could have.

The outline of the proposed scheme is as follows:

(1)   All goods bound for the city centre are taken by each business to a micro-consolidation centre at Palmersville, the nearest Metro station to Killingworth. This facility could simply be based on the platform, utilizing the existing infrastructure. All businesses selected for the trial are a maximum of 3 miles away from Palmersville, so the use of a conventionally powered vehicle would not have a large effect on emissions. In time, it would be desirable to transfer

this initial drayage leg to an ultra-low emissions vehicle (ULEV), to cut CO_2 emissions.

(2)   Goods are condensed into one trainload, and loaded onto a modified Metro train, repurposed for the transportation of freight. This may be made easier by the use of standardised 'Bento Boxes', to allow for easy differentiation between goods from each company, and also to increase manoeuvrability. The train would be dedicated to freight transportation, thus minimising the disruption to passengers.

(3)   The train departs from Palmersville Metro station and arrives at a track situated between Jesmond Metro station and Manors Metro station, used only late at night by Nexus to transport trains back to the maintenance depot. This track holds great potential for the building of a distribution centre, as it is only a short distance from the city and is unused for the vast majority of the day.

(4)   Goods would be sorted and taken the last mile to businesses through the use of bicycle couriers and electric vans, using a scheme similar to that used by GNewt Cargo (Sect. 2.6).

This scheme would reduce the numbers of vehicles on the road, reduce the CO_2 and particulate matter emissions due to the full pay load and higher efficiency of trains, and reduce inner city noise and congestion. The scheme may present disadvantages such as added cost due to the 'double handling' of freight, and it may take longer to transport the freight using a multimodal system than via direct road transport, so it may not be suitable for time-sensitive goods.

The system proposed is not envisaged to handle heavy haul rail freight operations on the Tyne and Wear metro network. Instead it will focus on the transportation of small medium size parcels, low density high value goods and recyclable material. To address the technical aspect of rolling stock, the existing metro fleet should be redesigned to transport such freight. Hence all trains running on the Tyne and Wear metro network will behave like the current metro fleet for passengers.

To understand this study fully, there is a need to understand the reasons behind the dominance of road transport, and to evaluate the modal choices available to logistics companies. The feasibility of such a system relies on its ability to compete with other transport modes, and its capability to overcome problems such as a lack of 'door-to-door' service and the need for investment.

The remainder of the paper is as follows: Sect. 2 is a comprehensive review of the current situation regarding the transportation of freight, both in the UK and globally. It includes case studies relevant to the discussion, along with a review of the benefits and weaknesses presented by each mode of transport, in an attempt to determine the reasoning behind the overwhelming choice of road to transport freight. Section 3 outlines the use of software named COBALT, to assign a monetary value to the reduction of accidents through removal of vehicles from the roads surrounding Newcastle. Section 4 presents the results from this modelling, with Sect. 5 presenting observations about the implementation of such a system, and Sect. 6 coming to a conclusion on the study as a whole. Section 7 then outlines any further work that is recommended on the subject.

2 Current Transport Situation

2.1 The Fall of Rail Freight

According to Headicar [31], "freight haulage by road overtook the volume carried by rail in 1995", and although it has seen a resurgence in recent years, it has not regained its modal share. When one looks at the scale of goods transportation by road, Headicar also tells us that "just under one-fifth of all road vehicle miles are represented by freight movements". It is therefore clear that the transportation of goods by road contributes largely to the volume of traffic on the roads, and as a result also contributes to congestion, harmful emissions and noise pollution. Marinov [35] suggests that the reason for the surge in popularity of road transport was the beginning of mass production in the late twentieth Century; rather than taking directly from the market share of freight by rail, road took on the need for increased volumes of transportation.

Woodburn [63] reports that returning to the volumes carried by rail at the height of the industry's dominance would be a very difficult task, due to the many changes in transport infrastructure that have occurred in recent years. The Institute of Mechanical Engineers has predicted that by 2030 there will have been a 60 % increase in the volumes of freight being carried and an increase of 114 % in tonne-kilometres [33]. It goes on to say that "this growth cannot be met by the current infrastructure available", concluding that "we are outgrowing our multi-functional network". If we are to achieve the targets set out by the most recent European Commission White Paper that "30 % of road freight over 300 km should shift to other modes such as rail or waterborne transport by 2030, and more than 50 % by 2050", [8], the current rail network must undergo substantial amounts of renovation, requiring significant investment.

If such a modal shift was achieved, it would have the potential to have a significant impact on many aspects of the UK's transport network. Just some of the statistics [33] regarding freight transportation by rail include:

- CO_2 emissions from rail freight are nearly ten times lower than those from HGVs.

- UK businesses lose £10 billon per annum from congestion. If this could be reduced, it would have a significant impact on the economy.
- 1 gallon of fuel moves 1 tonne of goods 246 miles by rail, but just 88 miles by road.
- If all freight moved in 2004 had been carried by rail, it would have produced 3.6 million tonnes of CO_2. This is a 90 % reduction against the actual emissions of freight transportation in 2004.

These figures highlight the inefficiencies of transporting freight by road, and demonstrate some of the benefits presented by rail transport. It is clear, therefore, that a modal shift to rail must be considered.

One of the significant drawbacks of freight by rail is its lack of door-to-door capability. It rarely has the ability to transport directly to the customer and must therefore rely on another mode of transport to deliver the last leg of the journey. It is clear, therefore, that the development of road-based transport cannot cease if there is a modal shift to rail. They must continue to develop together, as rail will still need to rely on a separate mode of transport to complete the journey.

2.2 Conventionally Powered Vehicles

HGVs, or 'heavy goods vehicles', are now the most common way of transporting freight throughout the UK and Europe. Figures show [17, 18] that the freight industry has been steadily in decline since 1999, with fewer enterprises and new licenses being issued. However, in recent years, a greater proportion of heavier articulated goods vehicles have been licensed. In 2000, 26 % of all articulated goods vehicles licensed had the capacity to carry over 40 tonnes. By 2010, this figure had risen to 72 % [17, 18].

HGVs have a certain number of advantages. They have the ability to transport goods directly from the source to the customer, with relative ease. This is a source of great convenience, as hauliers do not need to collaborate with other businesses to ensure that the product arrives with the customer. The design of HGVs also makes them well suited for use in either intermodal or multimodal systems. HGVs can be fitted with TEUs (twenty-foot equivalent units), amongst other types of container, and with the right equipment, these can be used interchangeably between rail, road, and sea transport.

However, there are also numerous limitations imposed on HGVs when transporting freight. Regulatory boards limit the weight of the vehicle to 44 tonnes and limit the size of the trailer that can be carried [32]. HGVs can therefore carry significantly less in a single journey than when compared to rail or sea.

HGVs also have speed restrictions; vehicles carrying more than 7.5 tonnes cannot travel faster than 60 mph on dual carriageways [30, 43]. This is slower than the main flow of traffic and limits the speed with which goods can be delivered.

There is evidence to suggest that HGVs can have a negative effect on inner city congestion. Due to their larger length, HGVs can take up the same amount of space that would otherwise be occupied by four passenger cars [47]. Large trucks also struggle to navigate complex road networks such as those often present in cities, as it takes more time for such vehicles to turn right at junctions and travel around roundabouts [47].

Perhaps one of the most potent issues in the twenty-first Century is that of emissions from vehicles. It is widely accepted that HGVs are not 'environmentally friendly', and that they contribute a disproportionate amount of emissions to the atmosphere. A recent study suggested that HGVs contribute 21 % of all CO_2 emissions from surface transport, along with 28 % of NO_X and 16 % of particulate matter emissions. This is despite HGVs only accounting for 1.5 % of all road vehicles [20].

The UK Government has set a target of decreasing the UK's greenhouse gas emissions by 80 %, from a baseline figure of that in 1990, by 2050 [14]. If emissions from transport are not cut, it may be that this target becomes impossible to achieve.

LGVs, or 'Light Goods Vehicles', are also rising in popularity, with an increase of 35 % between 2000 and 2010 [32]. Whilst not all LGVs are used for freight transportation, figures suggest that in 2008, 21 % of all LGVs registered were used for 'collection or delivery of goods' [17, 18].

LGVs are defined as a vehicle whose gross weight does not exceed 3.5 tonnes and have similar advantages to those of the HGV. They have very good door-to-door capabilities, perhaps even more so than their larger counterparts due to their smaller size. They are also readily available on the market, for a much smaller investment than HGVs, allowing small businesses, a doorway into transporting their own freight.

LGVs suffer from many of the same limitations that apply to HGVs. They are restricted to 60 mph on dual carriageways [30], causing many of the same problems regarding speed of transportation and traffic flow as discussed earlier. LGVs are not capable of intermodal transport using a large standardised container, however, can still be involved in a multimodal transport system. The high emissions from LGVs are a cause for concern; a study into CO_2 emissions from LGVs found that whilst emissions from passenger cars decreased by 3 % in a two-year period, emissions of CO_2 from LGVs rose by 22 % in the same period [2]. This is harmful to both the environment, and the Government targets for cutting emissions.

2.3 Intermodal Transport

According to Reis [48], the European Union has realised that the growth of road transport is unsustainable, and therefore, they heavily support the use of intermodal transport. Bontekoning [6] demonstrates that due to the emerging nature of research on the subject, there is no one definition of intermodal transport; indeed each of the authors of the 92 publications reviewed during the report in question used "a definition that reflects the scope of their research", as one would expect. However, most sources tend to agree somewhat with the definition presented by the UNECE in the 2009 glossary for transport statistics: "multimodal transport of goods, in one and the same intermodal transport unit by successive modes of transport without handling of the goods themselves when changing mode" [62]. A proper definition is paramount to the development of a system and allows for greater progression in that field.

Currently, however, there are relatively few systems that are utilizing intermodal transport. Reis [48] tells us that at an EU level, 75 % of freight is transported via road with just 5 % of this transport being classed as intermodal. The paper then goes on to explain the barriers that face intermodal transport; it is complex to produce and relies on integration of disparate transport networks, and the regulatory framework is often inadequate.

Another issue raised by numerous authors is the increased drayage that can be introduced into an intermodal system. If we take an example of using rail to transport freight, the goods must arrive at the rail station and subsequently be transported to their final destination at the end of the rail journey. These initial and final transportation legs are known as drayage movements. Bontekoning [6] shows that "despite the relatively short distance of the truck movement compared to the line haul, drayage accounts for between 25 and 40 % of origin to destination expenses". These high drayage costs can have a significant impact on the profitability of an intermodal system. A paper from Craig [9] determined that "for the overall intermodal shipment to be more efficient than truckload, the length of the linehaul must be long enough to offset the lower efficiency of drayage". Reversed, that is to say the length of the drayage must be short enough for the system to reap the benefits of the higher efficiency of rail. The results from this study concluded that the efficiency and benefits from using intermodal transport must be evaluated on a case-by-case basis.

One way to reduce the inefficiencies added by drayage distances would be to use zero-emission vehicles such as electric vans or bicycles to transport the freight to and from the rail terminals. This would allow for low emissions in the initial and final drayage legs; however, relatively close proximity to the station would be necessary.

2.4 Electric Freight Transportation

Electric vehicles are much more environmentally friendly than their petrol or diesel powered counterparts. Deemed 'ultra-low emission vehicles' (ULEVs) by the government, each must release below a specified value of greenhouse gases during operation.

Methods of urban freight distribution have been centred on the use of conventionally powered HGVs and LGVs for many years, with little in the way of a shift towards a more environmentally conscious mode of transportation. Indeed until recently, this image was mirrored across the transport market as a whole; in 2009, less than 0.1 % of cars registered in the UK were electric [28]. However, more recently there has been a large increase in the demand for such vehicles [51, 52]. In 2014, the sales of ULEVs increased fourfold on the value from the previous year, from 3586 to 14,498 [55]. Such a rise in demand could be attributed to many factors: an increase in the number of models available on the market, a rise in the accessibility of charging points, or the availability of grants to purchase vehicles. However, the underlying motives behind the purchase of an ULEV are the desire for transport that produces less CO_2 that increases air quality, and that leaves less of a long-term impact on the environment [54]. Why then, has this jump in sales of electric vehicles not been mirrored in freight transportation?

A report published by PTEG [45] states that there has been "no significant shift towards low emissions vehicles in the LGV market", with natural gas/electric powered vehicles accounting for less than 0.1 % of vans registered in 2013. Electric HGVs are available on the market; however, they have a range of around 100 miles, which is not suitable for long freight journeys. During initial research, several electric LGVs were sourced, all of which had a similar range to that of the electric HGV, around 100 miles [44]. This significantly lessens the potential for the use of electric vehicles for freight transportation, which often requires long journeys. It is clear, therefore, that it will be some time before electric vehicles are able to compete as a stand-alone method of freight transportation.

This is not to say that electric vehicles do not have their place in goods transportation. When combined with another mode of transport, or used to perform 'last mile' deliveries, small electric vehicles allow for vast reductions in CO_2 emissions per package.

It can also be said that electric vehicles are quieter than conventionally powered transport. Whilst running, they are virtually silent, and even hybrid vehicles will only use the internal combustion engine for a small part of the journey. The introduction of more electric vehicles could significantly reduce noise pollution in city centres.

The government is keen for people to make the transition to Ultra-Low Emission Vehicles, and is therefore offering grants, to make the vehicles more accessible. A grant of up to £5000 is available for this type of vehicle, until 50,000 grants have been issued, or before the beginning of 2017, whichever comes first [41, 42].

2.5 Bicycle Couriers

Bicycle couriers are a mode of freight transportation that has been increasing in popularity in recent years, and many cities have at least one courier service operating in their centre. A bicycle is a zero-emission vehicle that either relies solely on power provided by the rider, or is electrically assisted through the use of a small motor. When used as a method of freight transportation, it has the ability to help reduce congestion and to improve both noise and air pollution in city centres.

A study funded by the Intelligent Energy Europe Programme of the European Union [27] suggests that "51 % of all motorized trips in EU cities that involve the transportation of goods could be shifted to cargo bikes". This sort of modal shift would have a significant impact in city centres, particularly in cities that suffer from high congestion levels.

Bicycles have the ability to use both roadways and cycle paths, which means that in many cases, it can be faster to travel around a city centre using a bicycle than using a motorized vehicle such as a car or HGV. If cycle logistics were to become a large-scale operation, investment in 'cycle highways' would be necessary. Indeed there has been a plan proposed for two cycle superhighways to be built in and around the capital city, London, in an attempt to persuade more of the general public to cycle [58]. These could also be used for the transportation of freight, allowing for fast and carbon neutral delivery directly to the heart of London.

Another major benefit of using cycle couriers is that the bicycles would not be susceptible to congestion charges, which are present in some cities throughout the UK. This charge must be absorbed by the haulage company in some way, and it is therefore likely that the cost will be transferred to the customer.

However, by design, bicycle courier services cannot carry the same volumes of freight that most other forms of transportation can, simply due to the fact that cycling with large bulky goods would become impractical. A cycle courier, 'Outspoken! Delivery', based in Cambridge and Glasgow can transport up to 250 kg of goods using their 'Maximus Bikes', which are used "almost exclusively in the city centre for last mile logistics" [16]. This is clearly not in the same range as the 44 tonnes that a HGV is capable of carrying, however, may be sufficient for many 'last mile' applications.

2.6 Case Study-GNewt Cargo

A London-based courier service, GNewt Cargo evolved from the desire to revolutionise inner city logistics in one of the most congested and polluted cities in the UK. When the business was formed in 2009, cargo cycles were their main form of transportation; however, as the business has grown, more and more electric vehicles have been brought into service. The company is thriving; every year since its formation, it has grown by at least 50 %, and in 2014, it doubled its electric vehicle fleet. Now, up to 17,000 parcels are delivered daily, and as all vehicles are 100 % zero emission at tailpipe, huge savings are made in CO_2 emissions [50].

Using a combination of electrically assisted cargo tricycles and electric vans, along with an urban micro-consolidation centre to increase the loading factor of each vehicle, GNewt Cargo has produced a 54 % fall in total CO_2 emissions per parcel carried. In the centre of London itself, the reduction was 83 % per parcel, demonstrating that such systems carry huge potential [7].

2.7 Urban Consolidation Centres

Rooijen [49] defines an Urban Consolidation Centre (UCC) as "an initiative which uses a facility, in which flows from outside the city are consolidated with the objective to bundle inner-city transportation activities". By loading a greater number of goods onto each vehicle, the total number of trips into a city centre can be vastly reduced, which is one of the defining attributes of the UCC. A major cause of inefficiency in HGV freight distribution is the low volumes carried in the last mile, which the use of an UCC could help to stop. Using an UCC to combine part-loads could also vastly reduce the total road miles travelled, which would have a significant impact on inner city congestion.

Regarding the use of bicycle couriers, Schliwa [50] concludes that "Urban Consolidation Centres were identified to be the major complementary infrastructure to enable cycle logistics operators to keep the length of the journeys, and hence the delivery time, short", and even refers to them as a "necessary precondition" to the success of a bicycle courier service. It is therefore clear that for a system which chooses to use a cycle courier for last mile transportation, the use of an UCC is crucial to whether the scheme is a success or failure.

Allen et al. [3] has identified, however, that in comparison to other EU countries such as Germany and Italy, the UK is poor at implementing trials of UCCs. In the UK, "two-thirds of UCCs identified did not proceed beyond a research project or feasibility study", compared to other EU countries where "the vast majority of UCC studies proceeded to either a trial or a fully operational scheme".

The paper then goes on to specify that the lack of trials in the UK is caused by a lack of funding; there is money available for research projects however funds are not sufficient to invest in numerous trial schemes.

Allen et al. [3] comes to the conclusion that "UCCs have the ability to improve the efficiency of freight transport operations and thereby reduce the congestion and environmental impacts of this activity". The paper also identifies that UCCs allow the opportunity for zero-emission vehicles to carry goods the last mile into city centres, thus further increasing the efficiency of the operation.

2.8 Short Distance Light Rail

Many cities in the UK have an existing light rail network, whether it is a metro-based system, or a network of trams, and there is increasing scope to use these systems for the transportation of freight in city centres. As opposed to the heavy rail freight industry, which would need investment, in many cases transporting goods via light rail would simply utilize the existing infrastructure. Often little alteration would be needed, with current platforms and train carriages sufficing, albeit in a modified form. A recent study of urban freight distribution [53] suggests that the use of easily portable, standardized containers such as the 'Bento Box' would be suitable for use on light rail systems which had simply had the grab poles and seats removed. Trains that were no longer fit for passenger use could be repurposed; indeed there is a plan to upgrade the existing metro fleet in Newcastle, with the new fleet beginning operation in 2025 [37]. The old rolling stock could be modified and used in freight transportation, both reducing the need for investment and 'recycling' the unwanted carriages.

There are many different forms of light rail system in operation; however, this paper looks at the one operated by Nexus throughout Newcastle and the surrounding area. The Tyne and Wear Metro fleet, despite having been in operation since the opening of the network in the 1980s, is in the top third in terms of overall energy efficiency per train-kilometre during operation [37]. There is an initiative in place to generate much of the electricity used by the Metro using renewable sources such as wind turbines and solar panels, with an aim to become a "100 % green energy transport network" in the future. It is estimated that 10–15 % of this energy could be generated at Metro stations [37]. There are no data available in the public domain regarding the actual value of Metro energy efficiency, and for a fair comparison with other modes such data would need to be analysed. The same can be said when looking at emissions of nitrous oxides and other harmful pollutants; the potential benefits cannot be properly evaluated without access to emissions data.

Using light rail would have a significant impact on the congestion in city centres. Section 3.5 looks in detail at the number of vehicles that could be removed from the roads around Newcastle city centre if the Metro network was used to transport just a fraction of goods from local businesses. Reducing the number of HGVs on the road would have a significant effect on the safety of both pedestrians and motorists. A study [45] suggests that not only are HGVs involved in a disproportionate number of both pedestrian and cyclist fatalities, but an accident involving a cyclist and a HGV is over twelve times more likely to be fatal than a cyclist-car collision. Such statistics show that a decrease in HGV numbers would vastly improve the conditions for those regularly travelling through inner city roadways on bicycles and by foot.

Metro systems are, however, not capable of providing 'last mile' transportation. In the case of Newcastle, the goods would need to be brought from an underground station to street level, to then be distributed to the final destination via another mode of transport. This would add complication and potentially expense to the system and would have a significant impact on the transportation of time-sensitive goods.

Concerns have also been raised regarding the potential for disruption to passenger services if light rail was used to carry freight. In many urban transportation systems, trains run at the minimum specified headway, meaning that if freight trains were to be introduced, the existing passenger timetable would have to be altered. It has been suggested that the Tyne and Wear Metro system is operating at very low efficiency in terms of the number of trains running, which suggests that there is scope for the introduction of freight rail in-between passenger services.

2.9 Case Study: Monoprix

As part of the Global Report on Human Settlements 2013, a case study was prepared on the transportation of commercial goods by Dablanc [13]. As a part of this, an in-depth evaluation of an intermodal transport system adopted by Monoprix was performed. Monoprix is a large chain of supermarket stores, with more than 300 urban supermarkets throughout France. In November 2007, the company began running a train service carrying goods for the supermarket from a warehouse in the suburbs of the city to Bercy rail station. From there on, Compressed Natural Gas (CNG) vehicles were used to perform last mile operations. In this manner, 65 stores in Paris are supplied, along with 25 stores in other regions.

This method of goods transportation has achieved significant environmental benefits upon the previous method. According to Dablanc [13], there was a reduction of 700,000 lorry miles travelled, along with a 36 % decrease in particulates emitted, a 56 % decrease of nitrous oxides and a drop of 47 % in CO_2 emissions, with 410 tonnes less of CO_2 emitted per annum.

Such impressive figures speak for themselves; however, as with anything, there are limitations to the success. Dablanc also reports much negative feedback regarding noise emissions. Monoprix opted to use passenger rail facilities at off peak times such as overnight, however residents in the surrounding area complained of increased noise due to the late night running of trains.

The matter of the funding and the cost of the system must also be raised. The municipality invested €11 million to allow the scheme to function; however, there is no charge to Monoprix for using the system, so this money will not be paid back. The cost per pallet was also increased by 26 %, an undesirable side effect that may be transferred directly to customers [13]. A suggestion from Dablanc was to run two trains each night, allowing for "better use of the employees", and "a more optimal use of the facility".

2.10 Lessons Learnt

When reviewing the data regarding making modal shifts and the huge variety of other modes of transport available for movement of freight, it is surprising that more businesses have not opted to make a transfer. Huge savings in carbon dioxide emissions can be achieved by a switch to any mode of transport other than conventionally powered vehicles, which are the least efficient yet the most commonly used.

However, it is clear that some modes are more suited to certain applications than others. It would not be possible to use a cycle courier to transport goods from one end of the country to the other, any more than it would be appropriate to use a freight train to deliver goods the last mile. A combination of these different modes of transport seems to be the best approach, with a multimodal system allowing for the combination of the greater efficiency of freight by rail with the zero emissions of cycle couriers or electric vans.

It is also clear that the optimum system for each city must be determined on a case-by-case basis. Each city is different, and what may work in one place would not in another. A scheme that would function using an underground system would need to be adapted for use with trams. There is not a universal template that can be applied to all cities, and thus, it was decided that the potential scheme in Newcastle should be further analysed using modelling software, to ascertain the benefits that are possible.

3 Analysis of Potential System Using COBALT

3.1 Review of Software

There are many software packages which would have been suitable for the analysis of a modal shift to light rail in Newcastle, which were initially considered for the investigation of the scheme. Xpress MP is a linear programming software, allowing many different scenarios to be mathematically modelled. A basic model was developed, which would have allowed the user to input of the cost per unit volume of various different modes of transport, determining the most cost effective over a set distance. This model could then have been repeated, using emissions data instead of cost, to determine which mode of transport would be the most environmentally friendly over a distance. This is recommended in the future, to gather more information about the potential scheme. It was not used due to issues with operating systems, however, is available for free download for members of a university.

Other modelling systems were also considered, such as Arena and SIMUL8, which are both classified as event-based simulation software packages. These have been successfully used in studies of similar systems in the past, with Motraghi & Marinov [36] using Arena to analyse an inner city freight by rail scheme for Newcastle upon Tyne, and Abbott & Marinov [1] studying a high-speed railway system through use of the same software. This highlights the benefits of using event-based simulation; it can be adapted for many different purposes and used to analyse many different systems.

Specific transport modelling software is also available which allows users to analyse railway systems. OpenTrack is a specific rail modelling system, allowing for analysis of many different modes of transport including high-speed rail, metro systems, heavy haul freight and other track-based modes. This would also be suitable for a study to determine where would be the optimum placement of a dedicated freight train in a passenger timetable. RailSys has been successfully used in several papers to analyse a transport timetable, which would be a useful addition to the work presented in this report. Chen and Han [65] used OpenTrack in an analysis of the Nanjing-Shanghai high-speed railway, to study the carrying capacity of the system. Such a model could be adapted to suit the potential Newcastle scheme.

After a review of the various software packages available to analyse potential modal shift, COBALT was chosen for use in this study. As shown in the Monoprix case study, there is often an investment required to allow such a system to begin functioning; an UCC may need to be developed, or trains may need to be repurposed for the transportation of freight. This cost is often borne by the local authority and short of charging for use of the system; there is little way that these funds can be repaid. COBALT was chosen to demonstrate that there is a distinct monetary benefit to the implementation of such a scheme in Newcastle. Whilst initial funding may be needed, it may be the case that the cost of this investment is reaped back in the savings from accidents and notably the reduction in the use of emergency services to deal with these.

3.2 WebTAG-COBALT

COBALT, or COst and benefit to accidents-light touch, is an Excel-based software developed by the Department for Transport. It is an analysis tool and determines the impact on accidents as part of an economic appraisal for a road scheme [59]. Data regarding each section of road and each junction that would be affected by the scheme are required, such as Annual Average Daily Traffic data, local accident rates, speed limit and link length. The user calculates data regarding the 'With Scheme Flows', containing information about the traffic flows under the proposed scheme. A parameter file is made available by the DfT, containing information pertinent to the financial implications of accidents such as 'casualty rates' and 'cost per casualty'. When the analysis is run, this assigns a monetary value to the number of accidents that would be prevented by the implementation of a scheme. Below is an example input file, provided by DfT, to demonstrate some of the necessary data (Fig. 1).

This is a very useful analysis tool, as not only does it force the user to perform an in-depth study of the area which would be affected; it has the ability to calculate the economic benefits to society should the scheme be implemented.

Once the relevant data are obtained and analysis is performed, COBALT presents an output file, with detailed analysis on a year-by-year basis regarding the financial impact of the system. It will determine if the number of road accidents would increase, remain unchanged or decrease, and the severity of accidents that could be prevented. An example (condensed) output file is shown below in Fig. 2.

3.3 Killingworth

Inspired by the successes of the case studies presented earlier and the desire to reduce noise and carbon emissions in the city centre, this report aims to determine whether a multimodal system utilizing the Tyne and Wear Metro

```
Link Input Section

Link Classification Subsection
Link         Road          Length   Speed Limit
Name         Type          (km)               (mph)
L119         4             0.38               40

Link Flow Subsection
Link         Base Year     Without-Scheme Flows                              With-Scheme Flows
Name         Flows         Year 1  Year 2  Year 3  Year 4  Year 5            Year 1  Year 2  Year 3  Year 4  Year 5
L119         3538          4101    5510                                      4085    5489

Link Local Accident Rate Subsection
Link         Observed Accidents    First    Observed    Local Severity Split
Name                               Accident Year        Ratio         Year
L119         8,3,5                 2007                                      *A locally observed pre-calculated rate
```

Fig. 1 Example COBALT input file [59]

Fig. 2 Example COBALT output file [59]

```
[Section 1]       Summary Statistics

[Section 1.1]     Economic Summary

                  Total Without-Scheme Accident Costs =    661,334.4
                     Total With-Scheme Accident Costs =    654,079.4

              Total Accident Benefits Saved by Scheme =      7,255.0


     Costs and benefits discounted to 2010 in multiples of a thousand pounds.

[Section 1.2]     Accident Summary

                       Total Without-Scheme Accidents =     13,146.3
                          Total With-Scheme Accidents =     12,951.3

                       Total Accidents Saved by Scheme =        195.0


[Section 1.3]     Casualty Summary

        Total Without-Scheme Casualties (Fatal) =              109.5
                                      (Serious) =            1,376.4
                                       (Slight) =           17,970.8
           Total With-Scheme Casualties (Fatal) =              109.5
                                      (Serious) =            1,363.3
                                       (Slight) =           17,692.3

        Total Casualties Saved by Scheme (Fatal) =             -0.1
                                      (Serious) =               13.1
                                       (Slight) =              278.5
```

Fig. 3 Location of Killingworth [29]

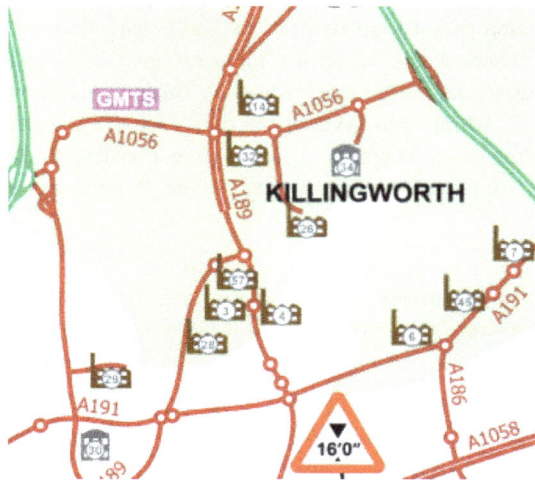

Fig. 4 Businesses around Killingworth [29]

Fig. 5 TWFP map [61]

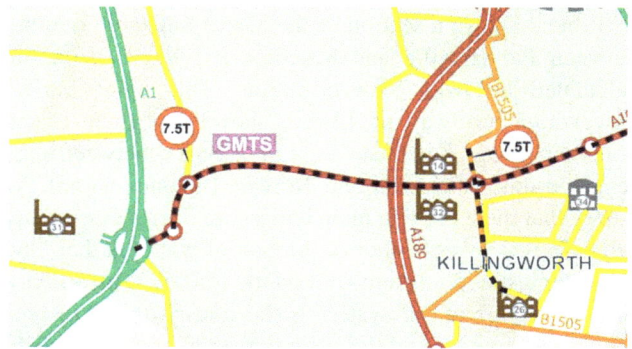

Fig. 6 TWFP map [61]

optimum route to access freight destinations within Tyne and Wear [61]. These maps have therefore been used to find the route that would be taken by a vehicle making a delivery to the city centre. Examples of these maps are shown in Figs. 5 and 6, with the dashed line being the preferred route for freight to travel by.

By determining the route taken by each business, it gives a clear picture of how the surrounding roads would be affected if a number of vehicles were taken from the roads. All businesses were directed to the city centre via either the A1, or the A19, two of the main roads that lead into Newcastle.

The data taken from the Tyne and Wear Freight Maps allowed the 'link length' to be found and provided a starting place from which other data such as speed limit and accident rate could be found. This then helped to build the 'Link Input Subsection' of the COBALT file.

3.4 Metro Links

The closest Metro station, Palmersville, can be seen in Fig. 3, in red. All of the businesses in the study were chosen as they are within a ten-minute road journey of Palmersville. Trains run regularly from this station to

system, combined with electric vehicles and bicycle couriers, would be feasible and beneficial for Newcastle upon Tyne.

The suburb of Killingworth to the north of the city centre was chosen as a study, due to the large numbers of businesses located in a small radius (Fig. 3). Figure 4 is a schematic of the businesses located in and around Killingworth.

The Tyne and Wear Freight Partnership maps were developed to allow logistics companies to determine the

Table 1 Palmersville to monument metro timetable [38–40]

Depart	Arrive	Changes	Time	Headway
20:25	20:41	0	00:16	00:13
20:38	21:14	0	00:36	00:02
20:40	20:56	0	00:16	00:13
20:53	21:29	0	00:36	00:02
20:55	21:11	0	00:16	00:13
21:08	21:44	0	00:36	N/A

Monument in the city centre, beginning operation at 0550 and ceasing at 2353. Journey time varies; the fastest train takes 16 min to arrive at Monument Metro station and journey length varies between 16 and 36 min, depending on which direction the train takes towards the city centre.

Table 1 shows a section of the Metro timetable running between Palmersville and Monument, [38–40], with the calculated headway between trains. This value varies; however, a headway of 13 min shows that there is the potential to run a dedicated freight service between passenger trains. Arvidsson and Browne [4] put forward the theory that there are five main barriers to the success of any urban freight by rail scheme, the first of which is that "the scheme must not interfere with public traffic". The primary role of any urban rail system is the transportation of personnel, not freight, and the system must be treated as such. Clearly passenger services must have priority; however, by studying Metro timetables in depth, we can find the optimum time to run freight services, with little or no disruption to the every-day traveller.

Following discussions with Nexus, it has become apparent that there is a link connecting Manors Metro station and Jesmond Metro station, which is just over 1 km in length. This link can be seen in Fig. 7, in red. These two stations are not connected for every-day use; however, this connection is used to allow a faster return to the depot late at night when services terminate. It is rare that this line is used during the day, and this would therefore allow more time to unload freight from trains, without needing to allow another metro carriage to use the line. In time, if the scheme was further developed, a distribution centre would be recommended, to improve the efficiency of the sorting and dispatch of goods to their final destination. This would also be the destination from which last mile couriers would depart into the city centre. It is less than 2 km from the optimum placement of a distribution centre along this track to Monument, which is a feasible distance for a cycle courier to travel.

3.5 Distance Travelled by Road

Many of the businesses in Killingworth are industrial estates or business parks, containing large numbers of individual business premises. There is also a large shopping centre in the village. Using a conservative estimate that 50 % of businesses located at each site make just one journey each week into Newcastle city centre, along with a third of businesses from the shopping centre, it can be calculated that a total of 4485.48 km is travelled cumulatively each week, if return journeys are included. The breakdown of the journey for each business is shown in Table 2, which also gives a reference for the information regarding each business. Due to time constraints, online information was used; however, if the project was to be

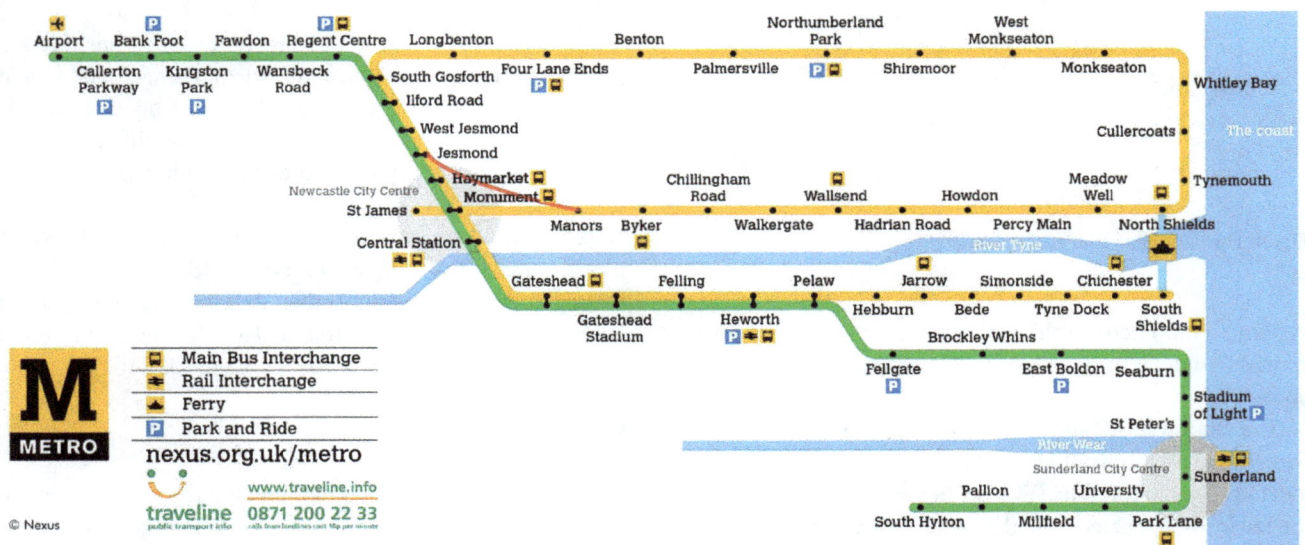

Fig. 7 Metro line linking Manors and Jesmond [38–40]

Table 2 Calculation of road kilometres travelled by businesses surrounding Killingworth

No. on TWFP map	Businesses	Journey length/km	Vehicles removed from road	Distance travelled weekly (one way)/km	Distance travelled weekly (return journey)/km	Referenced from
14	29	13.54	14.5	196.3	392.7	[23–26]
32	10	13.54	5	67.7	135.4	[11, 12]
34	68	13.54	34	460.4	920.7	[34]
26	7	14.54	3.5	50.9	101.8	[64]
7	20	12.32	10	123.2	246.4	[11, 12]
45	75	12.62	37.5	473.3	946.5	[23–26]
6	28	14.43	14	202.0	404.0	[23–26]
28	57	15.19	28.5	432.9	865.8	[23–26]
3	10	13.49	5	67.5	134.9	[23–26]
4	25	13.49	12.5	168.6	337.3	[46]
Total	329	136.7	164.5	2242.74	4485.48	

continued, it would be beneficial to visit each site, to verify that the information found online was correct.

It is probable that this estimate of kilometres travelled is less than the actual value; the number of goods deliveries to the city centre will likely be higher than just half of businesses making one journey weekly; however, if there are still benefits to society with such a small number of journeys being considered, then the potential for expansion is much greater.

The calculation for reduction of road kilometres travelled is not one which was required for the COBALT input file; however, it was found to be useful as a reference number for comparison with other schemes. Case studies such as Monoprix and Tesco claim reductions of "400,000 road kilometres weekly" [57] and "700,000 road miles per annum" [13], and it was felt that this form of analysis should be performed for the potential Newcastle scheme. When the number of road kilometres prevented is calculated on a yearly basis, the value is 233,245 km, showing that such a reduction would be directly comparable with that achieved by Monoprix, albeit initially on a smaller scale. Should more businesses become involved, this number would rise greatly.

3.6 Data Acquisition

For the input file to be of value, accurate data regarding all aspects of each road link and junction were needed. This included values for the traffic flow through both junctions and links, along with the speed limit and length of each section of road. The local observed accident rate was also needed, and the 'with scheme' traffic flows for a number of years were predicted.

There were numerous sources used to find the data needed. All traffic flow data were taken from [19]. Using the interactive map provided (Fig. 8), the traffic data along

a specific road could be determined. This method was also used for junction flows.

To calculate the number of vehicles that would be removed from the roads under the new system, the weekly reduction of vehicles from each business was taken and multiplied by 52 (the number of weeks in a year). The sum of all of the vehicles from the businesses using the same roads was then calculated, giving a total flow reduction for each section of road. This was then subtracted from the flows provided by the DfT to provide the 'With Scheme Flows' section.

The data regarding local accident rates were difficult to come by. The English road comparison safety website was closed on the 27th March 2015, as "equivalent information is now widely available online from other sources" [21]. A link was given to www.crashmap.co.uk, which provided an interactive map, with a marker at every point where an accident had occurred. Figure 9 shows an example of the data; each orange marker is a 'slight' accident, each red marker is 'serious' and black, not shown in the image, is a fatal accident. This was used to study each road link and junction in detail, and the number of accidents each year was manually counted. This means the data are accurate;

Fig. 8 Traffic count points Killingworth [59]

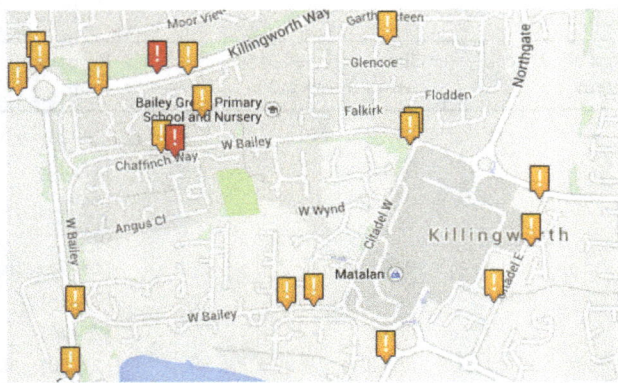

Fig. 9 Local accident data Killingworth [10]

however, for a larger system, this would be impractical. It is recommended that another source be found for local accident data.

Speed limit data were found using a combination of local knowledge and maps provided by Google. Again whilst this is accurate, it would be impractical for a system involving a larger number of roads, and a better method should be found if the analysis was repeated.

COBALT requires the user to define the length of time that is being studied, requiring the current year, a 'base year' with which to compare data, and then up to five specified dates in the future at which all information will be calculated. For this system, the current year is 2015, with a base year of 2012. Information regarding the scheme was calculated at three intervals: in 2020, 2023, and 2035. For this to occur, the file requires a prediction of traffic flow data for the future years. This was calculated using the "Automated Traffic Growth Calculator" [15], using the current traffic flow data provided by DfT. It has also been predicted that the scheme will remain the same for the first few years of operation up until 2020, then would grow by a half by 2023, and then double by 2035.

3.7 Scenarios

To increase the value of the model, it was decided to re-run the input file, under a number of different scenarios. This would allow the problem to be analysed in more depth, as when a greater number of solutions are studied, it is more likely that one of them will be relevant to the 'real-life' application.

Initially, the assumption was made that 50 % of businesses make one weekly journey to the city centre. Through time, the validity of this assumption was questioned; what if the number was actually less, or indeed was much higher? The input files were varied accordingly, to allow for just 25 % of businesses making a journey, all businesses making a journey, or all businesses making two trips into the city centre weekly.

The projected growth was also an assumption. As stated previously, it was predicted that the scheme would have grown by half by 2023, which would involve 15 business parks using this form of freight transportation, and have doubled by 2035, giving a total of 20 business parks participating in the scheme. This cannot be accurately predicted, so it was decided that the analysis should be run again, without the projected growth included. In other words, the number of businesses that use the Metro to transport freight would remain the same between 2017 and 2035. The desired outcome of this was to see how much of an effect the growth would have on accidents prevented.

There are 329 businesses contained in the 10 industrial estates and business parks around Killingworth. The model was adapted to determine the number of businesses that would need to be involved in the system to allow for a reduction in accidents and therefore the generation of significant monetary savings. This form of analysis would prove useful if the scheme was to be taken further; by determining the savings in cost with the introduction of more businesses, it would be possible to see how many companies would need to use the scheme to offset any initial investment cost.

This final form of analysis is not as accurate as the previous model, as there is no way to determine which out of the 329 businesses would be the first to join the scheme; hence, there is no way to know the exact roads affected. With this in mind, this version of the model was adapted just to look at four main road links, the A1056, A1, A167 and A189. Initial data were consulted, and it was found that seven out of ten of the business parks used these four links to reach the city centre. This simplification introduces an element of error into the system; however, it is necessary if the modelling is to occur.

4 Results

4.1 Projected Growth Included

This series of results involves the system evolving and more businesses joining over time. It was predicted that by 2020, the scheme would have grown by half, and by 2035, the scheme would have doubled in size. This growth would be facilitated through the remainder of the business parks in and around Killingworth becoming a part of the scheme, as just a handful was chosen for this study. In reality, some of the roads used by these new businesses would be different to those entered into the COBALT input file; however, as the TWFP maps directed all businesses to use the A19 or A1, the model will still be accurate to a level at which a basic analysis can be made.

It can be seen from Table 3 that even with a relatively small number of businesses making a modal shift, the

Table 3 Data with projected growth

Businesses involved	Money saved by scheme	Accidents saved by scheme	Casualties		
			Fatal	Serious	Slight
25 % make 1 trip	£50,426.00	994.2	10.7	107.7	1398.9
50 % make 1 trip	£94,308.40	1805.7	21.1	205.2	2543.7
100 % make 1 trip	£189,588.00	3549.7	43.8	414.4	4985.0
100 % make 2 trips	£277,238.50	4976.9	68.6	618.3	6983.8

initial savings are high. If just 25 % of businesses made 1 trip weekly, which relates to around 80 separate businesses becoming involved in the scheme, a reduction in flow of 4732 vehicles per annum would be achieved. Due to the predicted growth of the scheme, by 2035, the end of the time period specified in COBALT, this number would have risen to 9464. This reduction in flow contributed to the reduction in road accidents, and therefore, the increase in monetary savings which can be seen between the 'growth' and 'no growth' systems. This demonstrates that there is a potential for such a system to benefit the local authority; a very large reduction in vehicles on the road could be achieved if all of the businesses in Killingworth used the Metro to transport their freight twice a week, rather than driving a conventionally powered vehicle into the city centre. This would not only have an effect on casualties and money saved, but would cut CO_2 emissions, noise pollution and particulate matter emissions.

It should be noted that this calculation of flow reduction is including the return journey, as using the Metro to transport freight would prevent a vehicle from both travelling into the city centre and then returning to the business.

4.2 No Growth of Scheme

These results do not include the growth that is predicted. The reduction of vehicles that would be achieved in the first year of the scheme remains the same through to 2035. This allows for analysis of the system, even if the predicted growth was found to be unrealistic.

Table 4 shows that the projected growth of the system has an effect on the monetary gain from the scheme. This is to be expected; if the scheme does not grow, then the number of vehicles removed from the roads remains the

same, thus the reduction in accidents is less. Savings are still significant, but may not be large enough to match the initial investment, depending on the scale of finance needed. However, the greater the number of businesses involved in the scheme, the smaller the gap between the growth and no growth schemes becomes relative to the overall monetary savings. For example, the difference in money saved if just 25 % of businesses make 1 trip is £16,481, which as a fraction of overall savings without projected growth is just under a half. When all 329 businesses use the metro instead of 2 trips weekly, the difference between the growth and no growth schemes is £36,322, which as a fraction of the no growth scheme is around 0.15. This can clearly be seen in Fig. 10; as more businesses become involved in the scheme, the projected growth makes less difference as the overall monetary savings become larger.

4.3 Savings for Both Schemes

This is a comparison of the savings for both the scheme with projected growth and without. It can be seen that the relationship between the two is not a set value, nor is it a set percentage of total savings. Given that the scheme was initially set to double by 2035, it would be expected that this would have a large effect on the monetary savings of the scheme; however, this is not the case.

There is a difference between the growth and no growth schemes each year; however, it does not seem to represent the fact that double the number of businesses would be using the Metro to transport freight. It would be predicted that if the number of vehicles removed from the road doubled, then the savings in both accidents and money would also roughly double. The graph above shows that this is not the case; in fact, this is not even close to the

Table 4 Data without projected growth

Businesses involved	Money saved by scheme	Accidents saved by scheme	Casualties		
			Fatal	Serious	Slight
25 % make 1 trip	£33,945.20	667.2	7.4	70.8	926.6
50 % make 1 trip	£65,700.10	1237.8	15.2	139.7	1717.5
100 % make 1 trip	£135,965.40	2506.2	33.0	293.4	3465.5
100 % make 2 trips	£240,916.00	4358.2	59.7	528.1	6037.8

Fig. 10 Comparison of growth
inclusive/exclusive scheme

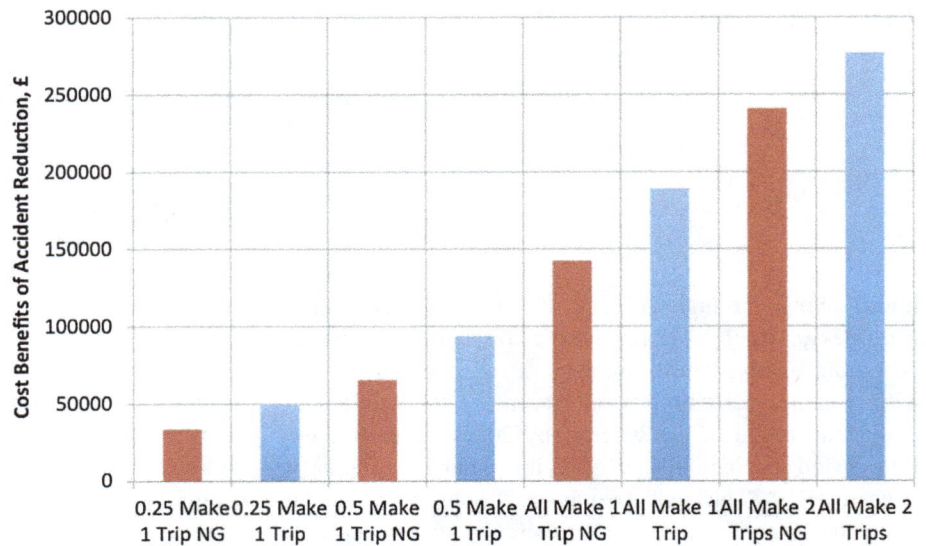

Fig. 11 Money saved versus
number of businesses

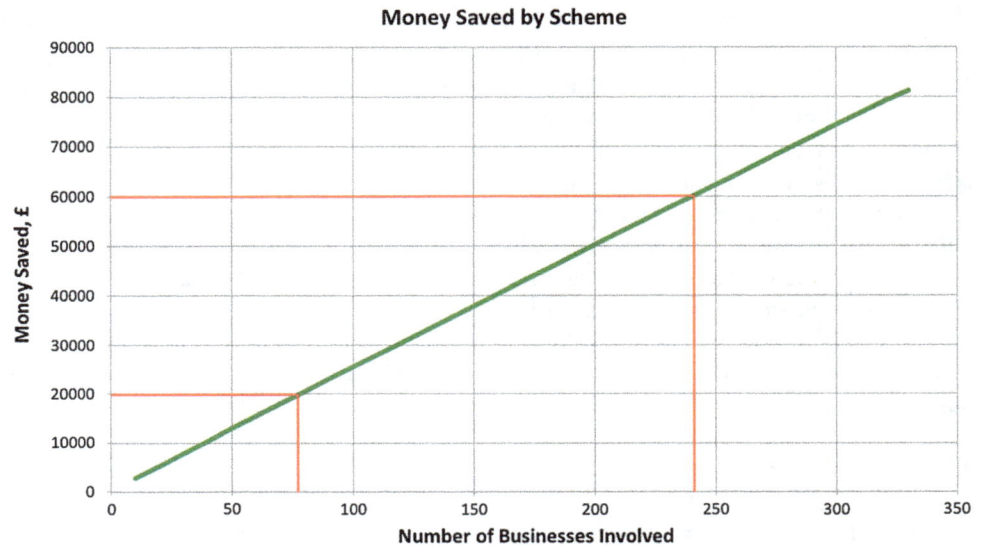

relationship. There is only ever a fractional difference between the schemes.

4.4 Money Saved by Scheme

The number of businesses was increased in increments of ten, from 10 to 330; the relevant data such as reduction of flows were altered and the input file for each number of businesses was run through COBALT. The relationship was found to be highly linear, with only minor fluctuations. This graph would allow a local authority to determine the number of businesses that would need to be involved in the scheme to gain the same savings in accident costs as the initial investment. For example, as shown in Fig. 11, if the amount of investment in the scheme was £20,000, with

around 80 businesses making one trip to the city centre weekly, the money invested would be reaped in accident savings. However, if the initial investment were larger, for example, at around £60,000, a much larger number of businesses would need to be involved to make it financially viable. This is as would be expected; however, a graph such as this would allow the local authority to decide if the initial money needed for the scheme to function would be a good investment, and to determine what the take-up of the scheme would have to be before the money was paid back.

These data do not include projected growth of the scheme. The data do also based on the assumption that half of the businesses involved would make one trip weekly to the city centre. The number of vehicles removed from the road in 2020 is predicted to be the same as the number

Fig. 12 Businesses involved in scheme

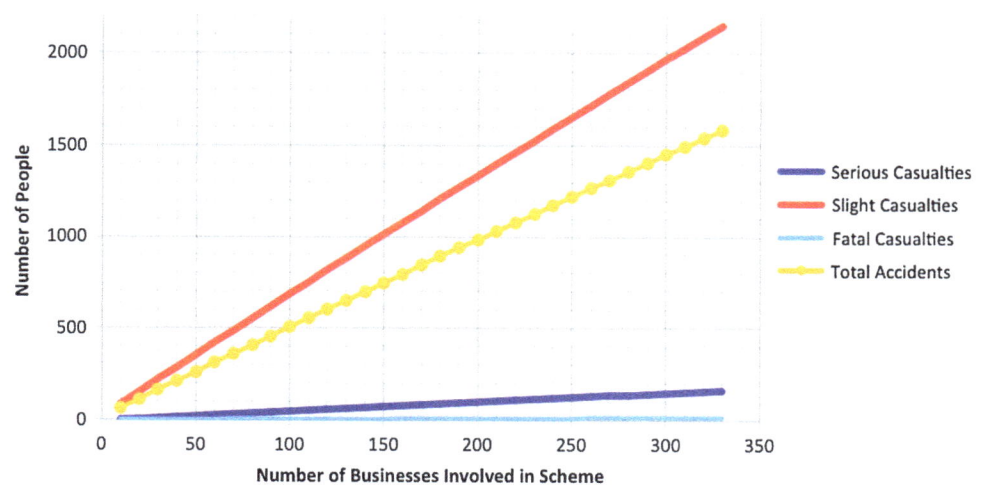

removed in 2035, so if the scheme grew and more businesses were involved, the data would need to be altered to accommodate this. The traffic flow data growth was predicted using the Defra traffic calculator, as before [15].

4.5 Casualty Rates

As data regarding the local severity split of the casualties were not available, the default accident rates from the parameter file were used. As expected, the number of casualties saved increases in a linear fashion with the number of businesses involved in the scheme; however, the gradient of each line varies (Fig. 12). The number of fatal casualties rises most slowly; it cannot be seen from the graph, however, with 10 businesses involved, there are no savings in fatal casualties, and with 330 businesses, there are 17.5 fatalities prevented. This is in contrast to the 'slight casualties' section, which rises from 86.2 to 2146.2. This is clearly because slight casualties are most common in traffic collisions, indeed this can be seen from the fact that the number of slight casualties prevented is greater than the total number of accidents. This is presumably due to the fact that there is often more than one person in a vehicle, and that usually two vehicles are involved in a collision.

5 Observations Presented By Study

5.1 Monetary Savings

It is evident from the COBALT results that even a small reduction in the number of vehicles on the road can have a significant effect on the costs of dealing with collisions. It was predicted before any analysis was performed that the scheme would produce a reduction in the number of crashes on the road; however, it was not thought that such large

monetary savings could be made. Data from Fig. 11 shows that with 40 businesses involved in the scheme, making one trip weekly into the city centre, the monetary savings are just over £10,000, showing that the take-up of the scheme does not necessarily need to be high to have a positive impact on traffic flows in the local area. However, the larger the scheme grows, the more of an effect the scheme will have. As expected, monetary savings increase with inclusion of businesses, so the more companies are encouraged to participate in such a scheme, the greater impact it will have.

5.2 Accident Savings

Regardless of any monetary savings that the scheme may bring, the reduction in casualties that could be achieved would have a large impact on local communities. If all 329 businesses studied reduced their travel by one journey weekly, 2,146 casualties could be prevented, 17 of which would be fatal. Even with just 10 businesses involved, 3 serious and 86 slight injuries could be prevented, demonstrating again that even a small reduction in flow would positively affect the roads surrounding Newcastle. The analysis using COBALT shows that such a scheme can not only positively impact the businesses directly involved in the scheme but could save lives and reduce the strain on hospitals and other emergency services through a reduction in the number of collisions.

Unfortunately COBALT does not have the ability to differentiate between different modes of transport, so there was no way of specifying that it would be HGVs and LGVs which were removed from the roads. This would further increase the safety of the public, particularly that of pedestrians and cyclists, as there is a much higher change of serious or fatal injury in a collision between a cyclist and a HGV. Any number of reductions in casualties is a benefit to society, and thus, the scheme should be investigated further.

5.3 Modal Choice

It is clear from all of the information presented throughout this report that a modal shift will be necessary for the transportation of freight in future years. Many separate sources have come to the conclusion that road-based freight transportation cannot continue to grow in the manner which is has done in recent years if we are to achieve emissions guidelines set out both by our government and the EU.

It is clear that there are many options available to logistics providers to transport freight, each with their own benefits. The most attractive option in terms of long-distance transportation is freight rail; however, this cannot be used to a large extent without being regenerated. Despite the investment needed, the positives offered by long-distance freight rail are numerous, allowing for large reductions in congestion, huge economies of scale and much lower fuel consumption when compared to road-based transportation. Many companies have achieved large emissions savings through a modal shift [56, 60], and where possible other businesses should follow. This is the underlying conclusion in almost all of the literature reviewed, and for such a shift to occur, there needs to be development and innovation in the way in which freight is transported. It is not enough to simply state that rail should be used; ways must be devised in which this is made easier and more convenient for the logistics providers, so that freight by rail becomes a desirable option once more.

For the carbon–neutral transport of freight in cities, it is clear that bicycle couriers and small electric vehicles are showing a surge in popularity for last mile operations. These also require investment, in terms of consolidation and distribution centres; however, the reductions in distance travelled by each package, noise emissions and carbon emissions are vast, and these vehicles should therefore be considered as a viable option for short distance freight transportation.

5.4 Use of Metro Systems

There is also significant evidence presented that short distance rail services such as metros have great potential for the transportation of freight, particularly in areas that have pre-existing infrastructure. This theory has been tested across Europe, with success in many cases, and could be mirrored in many cities across the UK. The majority of UK city centres are historic and therefore not suited for HGV or LGV transportation. A reduction in these types of vehicle could improve the image of the city whilst freeing many pedestrian walkways of parked vehicles unloading freight.

It is also clear that there are often gaps in the timetables of metro systems, which would allow a dedicated freight train to operate. Due to issues with safety and speed of unloading, combining passengers and freight on the same train would be unrealistic; however, with careful planning, it would often be possible to run a dedicated freight train and cause no disruption to passenger transport. This must be the case if such a system is to succeed, and great care must be taken to achieve minimal passenger disruption.

The use of inner city metro systems presents an opportunity to further utilize existing infrastructure, for relatively little investment, to cut emissions, and for faster and more efficient freight transportation.

5.5 Newcastle upon Tyne

It has been suggested that the Tyne and Wear Metro system is not currently running at its full potential. The operation of freight trains would allow for greater usage of the network, and due to the long headway between trains in this specific timetable, it is highly likely that there is a large enough gap in the timetable to allow for another train to run. Indeed if there was a large interest in the scheme, it may be possible for several extra trains to be included in the timetable.

The town of Killingworth shows great potential for the use of the Metro to transport freight. There are a large number of businesses situated in the town, the majority of which are a short distance away from the local metro station. The station is not underground, which allows for easier loading of goods, as they do not have to be transported from street level.

6 Conclusion

The evidence presented throughout this paper is demonstrative of the fact that a multimodal system using light rail is something which should be seriously considered for Newcastle upon Tyne. The Metro system has infrastructure that could be adapted for use in such a system, with many outdoor platforms and a significant amount of unused space at many stations. The timetable for the Tyne and Wear Metro is such that there is plenty of opportunity for the addition of at least one, if not several dedicated freight trains, with little or no disruption to passenger services. It has been shown through the use of software provided by the Department for Transport that there would be distinct savings in the number of casualties on the road, which in turn would present monetary savings through dealing with less collisions. The area surrounding Newcastle has a large number of businesses, and if these were to opt to transport their freight by metro, it could have a significant effect on inner city congestion, road collisions and emissions from transport.

A modal shift of some description will be necessary to achieve targets set out by the Government, and a system with such high potential for emission reductions should be seriously considered.

7 Future Work

There are many industrial estates and business parks situated in the areas surrounding Newcastle, with the majority being in close proximity to a Metro station. Over time, it may be that a trial system is implemented in one of these towns, which would require significant collaboration between separate businesses, Nexus and the local authority. Before a trial system is developed, it would be necessary to contact each of the businesses and find detailed information about their usage of the scheme, for example, their current choice of transport, the volume of goods which they would transport, and the frequency at which they would use the system.

It would also be beneficial to find other methods of data capture, to ensure that the model presented is accurate. This was one of the major difficulties; often the relevant information is not easily available. In the long term, much of the data such as accident rates, speed limits and traffic flow could be obtained through discussion with the local authority, to ensure the relevance and accuracy of the modelling.

Only one small section of the viability of this scheme has thus far been analysed. It would be valuable to model the system using other software to find out more information. Particular areas that need further study include:

- The Tyne and Wear Metro timetable should be studied in depth, to determine the optimum placement of a freight train to minimize disruption to passenger services.
- A thorough look should be taken at the current Metro carriages, to see if these would be suitable for the transportation of freight, or if a new vehicle would need to be used.
- Research should also be undertaken into the use of driverless vehicles to transport freight. This would keep the running costs of the system low, as a driver would not have to be employed. This may not be viable; however, it would be beneficial to at least consider as an option.
- A study into the use, location and development of an Urban Consolidation/Distribution Centre should be completed. As demonstrated, these facilities allow for a much more efficient multimodal system, and the location of such a centre in Newcastle should be considered.
- Market research should be completed for a wide number of businesses, in order to ascertain the potential for take-up of the scheme. If it is found that the desire

to participate in such a scheme is very low, then there is little point implementing it in its current form.

- A model should be developed to look at the carbon efficiency of using the Metro to transport freight. Little data could be found regarding emissions from the Tyne and Wear Metro, and such a study would be necessary to see if it could compete with other modes of transport in this area. A key reason behind the implementation of such a scheme would be to reduce CO_2 emissions, and for that to occur the Metro would have to present significant savings over road-based freight transportation.
- The potential for such a system should be studied in many of the towns around Newcastle that are connected to the city centre via the Metro.

These forms of analysis would allow the scheme to be assessed from all angles, and once further examination has been completed, the feasibility of using the Tyne and Wear Metro to carry freight can be properly evaluated.

References

1. Abbott D, Marinov MV (2015) An event based simulation model to evaluate the design of a rail interchange yard, which provides service to high speed and conventional railways. Simul Model Pract Theory 52:15–39
2. AEAT (2010) Light goods vehicle-CO_2 emissions study: final report. [Online] Available at http://www.lowcvp.org.uk/news,dft-publishes-study-on-light-goods-vehicle-co2-emissions_1421.htm. Accessed 15 Apr 2015
3. Allen J, Browne M, Woodburn A, Leonardi J (2012) The role of urban consolidation centres in sustainable freight transport. Transp Rev 32(4):473–490
4. Arvidsson N, Browne M (2013) A review of the success and failure of tram systems to carry urban freight: the implications for a low emission intermodal solution using electric vehicles on trams. Eur Transp 54:1825–3997
5. Behrends S, Lindholm M, Woxenius J (2008) The impact of urban freight transport: a definition of sustainability from an actor's perspective. Transp Plan Technol 31(6):693–713
6. Bontekoning YM (2004) Is a new applied transportation research field emerging? A review of intermodal rail–truck freight transport literature. Transp Res Part A 38(1):1–34
7. Browne M (2011) Evaluating the use of an urban consolidation centre and electric vehicles in central London. IATSS Res 35(1):1–6
8. European Commission (2011) White paper on transport—roadmap to a single European transport area—towards a competitive and resource-efficient transport system. Publications Office of the European Union European Commission, Luxembourg
9. Craig AJ (2013) Estimating the CO_2 intensity of intermodal freight transportation. Transp Res Part D 22:49–53

10. CrashMap (2015) CrashMap. [Online] Available at www.crash map.co.uk Accessed 26 Apr 2015

11. CYLEX (2015) Map of Benton Square Industrial Estate. [Online] Available at http://www.newcastle-upon-tyne.cylex-uk.co.uk/map/benton%20square%20industrial%20estate.html. Accessed 3 Apr 2015

12. CYLEX (2015) Map of Harvey Combe from Newcastle Upon Tyne. [Online] Available at http://www.newcastle-upon-tyne.cylex-uk.co.uk/map/harvey%20combe.html. Accessed 3 Apr 2015

13. Dablanc L (2013) Commercial goods transport, Paris, France. Paris, IFSTTAR

14. DECC (2013) Greenhouse gas emissions. [Online] Available at https://www.gov.uk/government/policies/reducing-the-uk-s-greenhouse-gas-emissions-by-80-by-2050. Accessed 15 Apr 2015

15. Defra (2010) How do I predict future traffic flows from available counts and what part does TEMPRO play? [Online] Available at http://www.laqm.defra.gov.uk/documents/TEMPRO_guidance.pdf. Accessed 13 Apr 2015

16. Delivery O (2015) Outspoken! Delivery. [Online] Available at http://www.outspokendelivery.co.uk/. Accessed 19 Apr 2015

17. DfT (2011) Domestic activity of GB-registered heavy goods vehicles. [Online] Available at https://www.gov.uk/government/uploads/system/uploads/attachment_data/file/8966/domestic-activity-of-GB-registered-heavy-goods-vehicles.pdf. Accessed 15 Apr 2015

18. DfT (2011) Road freight economic, environmental and safety statistics. [Online] Available at www.gov.uk/government/uploads/system/uploads/attachment_data/file/8968/road-freight-economic-environmental-safety-statistics.pdf. Accessed 15 Apr 2015

19. DfT (2013) Traffic counts. [Online] Available at www.dft.gov.uk/traffic-counts/area.php?region=North+East&Ia=Newcastle+upon+Tyne. Accessed 15 Apr 2015

20. DfT (2014) Low Emission HGV Task Force: Recommendations on the use of methane and biomethane in HGVs. [Online] Available at https://www.gov.uk/government/uploads/system/uploads/attachment_data/file/287528/taskforce-recommendations.pdf. Accessed 15 Apr 2015

21. DfT (2015) English road safety comparison website closes on 27 March 2015. [Online] Available at https://www.gov.uk/government/news/english-road-safety-comparison-website-closes-on-27-march-2015. Accessed 9 Apr 2015

22. DfT (2015). Vehicle Licensing Statistics: Quarter 4 (Oct-Dec) 2014. [Online] Available at https://www.gov.uk/government/uploads/system/uploads/attachment_data/file/421337/vls-2014.pdf. Accessed 15 Apr 2015

23. Endole (2015) Companies in NE12 8ET. [Online] Available at http://www.endole.co.uk/company-by-postcode/ne12-8et. Accessed 3 Apr 2015

24. Endole (2015) Companies in NE12 9SW. [Online] Available at http://www.endole.co.uk/company-by-postcode/ne12-9sw. Accessed 3 Apr 2015

25. Endole (2015) Companies in NE12 9SZ. [Online] Available at www.endole.co.uk/company-by-postcode/ne12-9sz. Accessed 3 Apr 2015

26. Endole (2015) NE12 5UJ. [Online] Available at: www.endole.co.uk/company-by-postcode/ne12-5uj. Accessed 3 Apr 2015

27. FGM-AMOR (2015) Moving Europe forward. [Online] Cyclelogistics Available at: http://cyclelogistics.eu/docs/111/D6_9_FPR_Cyclelogistics_print_single_pages_final.pdf. Accessed 19 Apr 2015

28. Franco J (2009) UK Pledges Electric Car and PHEV Subsidies. Octane Week

29. Google (2015) Google maps. [Online] Available at https://www.google.co.uk/maps/. Accessed 23 Apr 2015

30. Government (2015) Speed limits. [Online] Available at https://www.gov.uk/speed-limits. Accessed 15 Apr 2015

31. Headicar P (2009) Transport policy and planning in Great Britain, 1st edn. Taylor & Francis, New York

32. HMRC (2013) Moving goods by road. [Online] Available at https://www.gov.uk/moving-goods-by-road. Accessed 15 Apr 2015

33. IMechE (2009) Rail freight: getting on the right track. [Online] Available at http://www.imeche.org/docs/default-source/key-themes/Rail_Freight_Report.pdf?sfvrsn=0. Accessed 15 Apr 2015

34. LocalStore (2015)Shops. [Online] Available at www.localstore.co.uk/a/402/killingworth/. Accessed 3 Apr 2015

35. Marinov M (2012) Urban freight movement by rail. J Transp Lit 7(3):87–116

36. Motraghi A, Marinov MV (2012) Analysis of urban freight by rail using event based simulation. Simul Model Pract Theory 25:73–89

37. Nexus (2014) Metro strategy 2030. [Online] Available at http://www.nexus.org.uk/sites/default/files/Metro%20Strategy%20Background%20document.pdf. Accessed 17 Apr 2015

38. Nexus (2015) Journey planner. [Online] Available at: http://jplanner.travelinenortheast.info/nexus. Accessed 23 Apr 2015

39. Nexus (2015) Journey planner. [Online] Available at http://jplanner.travelinenortheast.info/nexus. Accessed 23 Apr 2015

40. Nexus (2015) Metro map. [Online] Available at http://www.nexus.org.uk/sites/default/files/Zone%20Map%20TVM%20Map_2.pdf. Accessed 4 May 2015

41. OLEV (2015) Take-up of plug-in car grant continues to rise. [Online] Available at https://www.gov.uk/government/news/take-up-of-plug-in-car-grant-continues-to-rise. Accessed 15 Apr 2015

42. OLEV (2015) Ultra low emissions vehicles. [Online] Available at www.gov.uk/government/organisations/office-for-low-emission-vehicles. Accessed 15 Apr 2015

43. Perry C (2014) HGV speed limits on dual carriageways. [Online] Available at https://www.gov.uk/government/speeches/hgv-speed-limits-on-dual-carriageways. Accessed 15 Apr 2015

44. Peugeot (2013) Peugeot electric partner van. [Online] Available at www.peugeot.co.uk/news/new-peugeot-partner-electric-van/. Accessed 15 Apr 2015

45. PTEG (2015). Delivering the future-new approaches to urban freight

46. Quorum (2015) Occupiers. [Online] Available at www.quorumbusiness.co.uk/space/occupiers/. Accessed 3 Apr 2015

47. Rail FO (2011) Road congestion implications of bigger HGVs. [Online] Available at www.freightonrail.org.uk/NoMegaTrucksRoadCongestionImpactsOfBiggerHGVs.htm. Accessed 15 Apr 2015

48. Reis V (2014) Analysis of mode choice variables in short-distance intermodal freight transport using an agent-based model. Transp Res Part A 61:100–120

49. Rooijen TV (2012). Local impacts of a new urban consolidation centre: the case of Binnenstadservice.nl. In: Procedia social and behavioural sciences. Mallorca

50. Schliwa (2015) Sustainable city logistics: making cargo cycles viable for urban freight transport. Rese Transp Bus Manag

51. SE (2013) Go electric. [Online] Available at http://www.smithelectric.com/go-electric/less-expensive-to-incorporate-and-maintain/. Accessed 15 Apr 2015

52. SE (2013). Models and Configurations. [Online] Available at http://www.smithelectric.com/smith-vehicles/models-and-configurations/. Accessed 15 Apr 2015

53. Singhania V (2014) Urban freight distribution: council warehouses and freight by rail. Transp Probl 9:29–43

54. Smith WJ (2010) Can EV (electric vehicles) address Ireland's CO_2 emissions from transport? Energy 35(12):4514–4521

55. SMMT (2014) UK new car registration. [Online] Available at http://www.smmt.co.uk/2015/01/uk-new-car-registrations-december-2014/. Accessed 21 Apr 2015

56. Tesco (2014) Our carbon footprint. [Online] Available at www. tescoplc.com/assets/files/cms/Resources/Environment/Our_Carbon_Footprint.pdf. Accessed 15 Feb 2015

57. Tesco (2015) Reducing our impact on the environment. [Online] Available at http://www.tescoplc.com/index.asp?pageid=632#tabnav. Accessed 22 Apr 2015

58. TfL (2014) Cycle superhighways. [Online] Available at http://www.tfl.gov.uk/campaign/cycle-superhighway-consultations. Accessed 19 Apr 2015

59. Transport DF (2013) COBALT user guids. [Online] Available at: https://www.gov.uk/government/uploads/system/uploads/attachment_data/file/262973/cobalt-user-manuel.pdf. Accessed 26 Apr 2015

60. TWFP (2010) Tesco sets the pace on low carbon and efficiency. Queens Printer and Controller of HMSO

61. TWFP (2015) Freight maps. [Online] Available at http://www.tyneandwearfreight.info/maps/freight/tyneandwear/freight_map.aspx. Accessed 23 Apr 2015

62. UNECE (2009) Glossary for transport statistics. [Online] Available at http://www.unece.org/trans/main/wp6/transstatglossmain.html. Accessed 22 Apr 2015

63. Woodburn AG (2003) A logistical perspective on the potential for modal shift of freight from road to rail in Great Britain. Int J Transp Manag 1(4):237–245

64. Yell (2015) Companies in Stephenson Industrial Estate. [Online] Available at www.yell.com/s/companies-stephenson+industrial+estate-newcastle+upon+tyne.html. Accessed 3 Apr 2015

65. Chen Z, Han BM (2014) Simulation study based on opentrack on carrying capacity in district of Beijing-Shanghai high-speed railway. Appl Mech Mater 505–506:567–570

15

A Newly Designed Baggage Transfer System Implemented Using Event-Based Simulations

Daniel Brice[1] · **Marin Marinov**[2] · **Bernhard Rüger**[3]

Abstract This paper proposes a newly designed system for baggage transfer, which utilises the Nexus Metro system in Newcastle-Upon-Tyne by running a pendulum freight train system between the Haymarket and Newcastle Airport to carry travellers' baggage. This system is capable of serving all passengers departing from Newcastle Airport in a day, with a capacity of 9750 bags across 26 freight train journeys. Following the initial solution two more solutions were designed with the aim of maximising the utilisation of the metro tracks by saturating the system with freight trains on a 24-h system. All solutions have been replicated using models designed and validated by event-based simulation using SIMUL8, a simulation modelling software package.

Keywords Systems design · Baggage transfer · Simulations

1 Introduction

1.1 Motivation

With a number of travellers using cars possibly due to baggage transportation complications, a new system could encourage more users of sustainable transport, possibly providing environmental benefits. In the absence of baggage on passenger rail, the current passenger users could benefit from a more comfortable baggage-less journey. People may therefore benefit from a social point of view, it is also more likely that people will travel if the journeys can be embarked upon with fewer complications and more comfort. It would appear that there is a need for a new system which can improve the utilisation of sustainable transport such as rail and facilitate the passenger.

1.2 Objectives

The objective of this study is to propose a new baggage transfer system to facilitate the passenger in travelling with baggage and improve utilisation of sustainable transport.

1.3 Methodology

Case studies of current baggage transfer systems are evaluated to identify strengths and weaknesses. These can be investigated through online research. Following the case studies, work is started to design a new system incorporating this newfound knowledge as well as using innovative ideas to further benefit the system.

A proposed design is evaluated using simulation-based software, in the form of SIMUL8. The utilisations of the tracks on this design are analysed. Following this, two more designs are created which focused on saturating the system with freight trains in an attempt to maximise the track utilisations.

All of the designs proposed focus on a pendulum system between a collection hub at Haymarket and Newcastle Airport. Finally, an alternative design to a currently non-functional system which should facilitate the passenger to

✉ Marin Marinov
  marin.marinov@ncl.ac.uk

[1] School of Mechanical and Systems Engineering, Newcastle University, Newcastle upon Tyne, UK

[2] NewRail, School of Mechanical and Systems Engineering, Newcastle University, Newcastle upon Tyne, UK

[3] TU-Wien, Wien, Austria

Editor: Prof. Guoquan Li

the degree whereby sustainable transport will become a more viable option is designed and evaluated.

1.4 Contribution to Knowledge

This system design study contributes to the development of a new concept for handling passenger's baggage. Specifically it proposes a new way of transporting heavy bags by introducing a new system for baggage transfer between a city centre and an airport. It is one of the very first studies to propose that check-in of a heavy bag can occur in a city centre and then transported to an airport separately from the passenger who owns it. This is to eliminate the hassle of carrying heavy bags when we travel.

For the new designs, collection of information about policies and practices of freight forwarding companies was needed. Therefore, a market research was undertaken to understand if such a system has potential for market uptake. The design work required the identification of technical parameters of the new system followed by evaluations of its performance using simulation modelling.

2 State of Practice

2.1 Company Policies

Before beginning the process of designing concepts, it was important to analyse current systems in use. In the case of designing a baggage transfer system, these cases would be companies that transfer luggage on a domestic level. Three companies were looked at as having examples of a system that transports baggage across Europe: DHL, FedEx and SendMyBag.

Company websites were studied to confirm that the companies were functional in the field of non-passenger transport. The company websites were limiting in the technical details they would provide about the system. This led to the second step of the case studies: contacting the companies directly to ask questions about their systems.

The information that was asked of the companies was the following:

- The mode of transport that is used to transfer the baggage, i.e. car, plane or train.
- Length of time the baggage spends in storage.
- What form the bags are transported in, i.e. crates, loose or plastic film.
- The time patterns of the transfer of the baggage, do the bags travel overnight, during daytime or both.
- The method of tracking that is used for the baggage.
- The amount of automation involved in the process.

- The amount of times baggage is separated and reconsolidated.
- The standards that are followed by the company with regards to baggage transfer.

This work did not lead to any gain in knowledge of current systems as none of the companies provided a response with answers.

2.2 European Baggage Transfer

It was decided that a cost analysis of delivery services may be of useful information. Five companies were analysed using their online tariff lists to understand the cost of sending different mass bags across different distances. The companies were: DHL, ParcelForce, DPD, FedEx and Yodel. Graphs were plotted to identify the trends in their pricing.

The masses of bag used in the study were 5–30 kg in intervals of 5 kg to represent the mass of bags most likely to be transported by a person. Five different locations were chosen to deliver to from the UK. These locations were selected based on being in different tariff zones for DPD. These locations were: France, Austria, Italy, Finland and Greece [1].

Figure 1 shows the impact that the mass of a bag and distance to travel has on the price of a delivery using DPD.

This is an example of service charges being as expected, where the cost of the delivery increases linearly with the mass of the bag as well as increases with the increase in distance.

Some companies such as Yodel would contrast to this with a pricing structure that would fluctuate inexplicably, Fig. 2. This suggests there could be complicated variables that will influence the costs of transferring baggage. Therefore, clearly it may not always be the more straightforward linear costing as shown in Fig. 1.

A graph, Fig. 3, was also made to find the cost that companies would charge to deliver bags of different masses across the UK. In this case, the same distance was used each time and only the bag mass was changed.

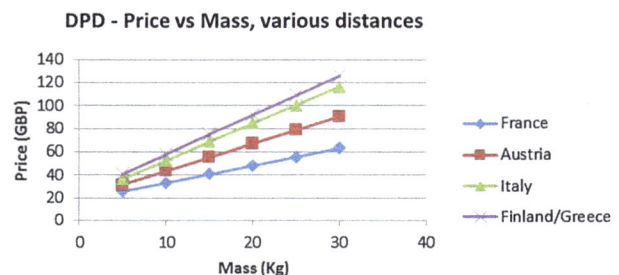

DPD - Price vs Mass, various distances

Fig. 1 DPD, price versus mass chart, data from DPD [1]

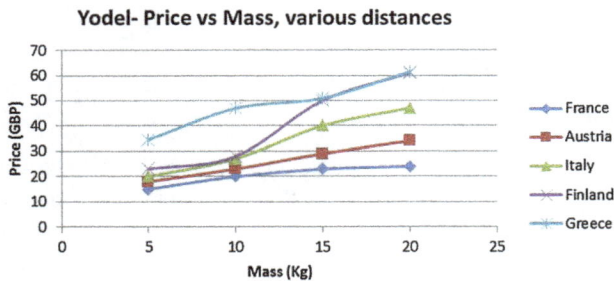

Fig. 2 Yodel, price versus mass chart, data from Yodel [13]

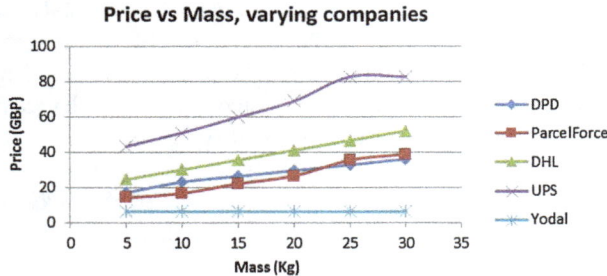

Fig. 3 All companies price versus mass chart, data from UPS [14], DPD [1], Yodel [13], Parcel Force [15] and DHL [16]

Most of the companies have a similar pricing system where the price increases as the bag mass does. Yodel, however, oddly does not change the cost of the delivery at all for any bag mass changes, this was even confirmed by Yodel staff using the online Q&A service.

This is evidence that there are relatively no more charges incurred in transporting heavier bags using land-based transport. This is contradictory to the understanding that the heavier bags would cause an increase in fuel consumption. It is possible that the price for this increase in fuel consumption is negligible in comparison to all the other costs induced.

2.3 Travelling by Car

A number of travellers use a car to travel as handling personal baggage is too impractical using other methods. This is clearly not an ideal solution to the problem as vehicles are not only an inconvenience to transport long distance, due to possibly needing to pay expensive tolls on roads, ferry costs and inevitably incurring large petrol costs.

In a world working towards a more sustainable future, where the accumulations of greenhouse gases are not such a cause for concern, it would be ideal for travellers to use public transport and not rely completely on personal vehicles to travel. There are also many who cannot afford the luxury of their own automobile and this option is therefore completely unavailable to them.

2.4 Luggage in Advance

Similar to the options that were available in 1960, 'Passengers luggage in advance' explained by Peter Kenyon, there are current systems that will deliver a suitcase to a different country for a small fee. Companies like 'send-mybag.com' have stated that they use courier services such as UPS to provide such services. One flaw in using this service is that there are usually tight restrictions on the times for bags to be picked up from the customer. This can make the service unusable in cases where a person cannot be at home during the pickup times, a very real problem for those who work 9–5, the same time window where the bags would need to be picked up.

2.5 Public Transport

This is probably the most common method after using a car. This system is the most sustainable, however, many are discouraged from this service due to the need to handle their bag the entire duration of the journey. Some have been quoted saying, "I once travelled with a bag and I was so paranoid the whole journey, I wanted to get off at every stop and check it was still there". [2]. How are people expected to want to use public transport with fears such as this? There are also more practical problems such as people not being physically able to handle heavy bags on services where they are required to load and unload the bag themselves.

After looking at all the available options it appears that baggage still creates a problem for those travelling. A new design for a system which can resolve this problem by handling the baggage that people travel with would therefore be highly desirable.

2.6 CAT: City Airport Train Vienna

The City Airport Train connects the Vienna city centre with the airport in 16 min every one and a half hours and is a good example of baggage handling system in operation. The system is operated parallel to commuter trains and Inter-City-trains. The big benefit of the CAT service with the others is that passengers are able to check-in and drop off their baggage at the Vienna City Air Terminal directly in the city centre up to 75 min before take-off. The procedure is the same as at the airport. The baggage must be delivered at drop counters like at the airport and will be transported on a conveyor belt to a storage room. There it will be loaded on a trolley which is delivered to a separate and securely locked storage compartment in the train. At the Vienna Airport the trolley will be unloaded from the CAT-train and the baggage will be put on a conveyor belt which brings in the baggage into the airport baggage system.

From the moment the customer delivers the baggage to the CAT-counter, the baggage is separated from the passengers, also when loading and unloading the train, so the total system meets the requirements of the air transport legacy.

Every person with a valid ticked is allowed to use the train, tickets are also available in the train. But only people who have a CAT-ticket are allowed to use the check in and baggage drop off. Additionally only passengers who take airlines which are in co-operation with CAT can use the baggage drop off. Currently, 20 carriers have got a contract with CAT so their passengers can check in at Vienna City Air Terminal.

CAT also had the idea of transporting baggage the other way around and to check it through from the plane into the train. The Vienna City Air Terminal also has a separate IATA-code but two reasons preclude this idea till now. Firstly, at the Vienna City Air Terminal an additional customs office must be installed and secondly, it is very difficult to guarantee that the baggage is in the same train as the passenger. If the situation happens that the passenger reaches the train but the baggage misses it the passenger has to wait half an hour at the City Air Terminal for the next train delivering the baggage.

3 Literature Review and State of the Art

Simulation modelling allows proposed designs to be analysed; they can be validated and verify hand calculations. Operating efficiencies can be measured and utilisations maximised. In this particular design, simulation modelling allows the utilisation of each section of the track to be analysed individually.

Simulation modelling has already been used effectively to model different scenarios such as rail yards [3], freight implementations to a passenger rail [4] and also complete rail networks. Specifically simulation modelling has been used to explore aspects of the metro system [20], utilisations have been meassured for stations and railway tracks and transit times recorded for trains to move between stations and their time stationary at each station. The work discussed in this paper builds on this work where utilisations of railway sections are analysed similarly using simulation modelling software [4].

There is software designed with the focus on simulating the detailed movement of rail vehicles, such as Villon. This software allows details such as exact infrastructure modelling to be modelled. There is also an increased amount of flexibility and detail in the operation modelling of a network using Villon.

Easier to use and more broad ranged, softwares such as 'SIMUL8' and 'Arena' have also be used for less complex modelling of rails. Some have used non-specific software to model tracks and achieve optimal track layouts. Simulation modelling software which utilised SLAM II language for their analyses' was used. Using analytical models to study delays, capacity of tracks between Downtown Los Angeles, Long Beach and Los Angeles ports was decided against compound delays and ripple effects at areas such as complex junctions causing problems [5].

Simulation modelling can be more favourable than analytical modelling as real life systems can be of great complexity, which is an area where analytical modelling may struggle to achieve clear analysis. Simulation modelling however is able to consistently achieve these accurate results.

After looking at the work of others using simulation modelling, it was a clear choice to model the system using such a software; SIMUL8 in this case. Schedules of passenger and freight trains were implemented to the system and adjusted with ease. SIMUL8 also allowed evaluation of the line between Haymarket and the airport to not only ensure that each design was validated but also allow their track utilisations to be compared to one another.

4 Technical Considerations of Design

4.1 Containerisation

Different containerisation options for baggage to be transported via aircraft have been investigated [6]. There are points of interest relevant to designing a system. The paper illustrates the importance in sorting the baggage as soon as possible and small a number as possible, i.e. it would be better to sort 10 sets of 5 bags than one set of 50. The reason for this is that there is an increased probability that baggage will be lost due to errors and delays when sorted in a larger quantity. The paper concludes that decentralising the sorting method reduces handling costs and transfer times. Though this is mainly applicable to the sorting of containers for airplanes it can be considered relevant to railway vehicles too.

4.2 Multi-modal Transport

Advantages and disadvantages of combining modes of transport for freight transfer have been investigated, with railway and road vehicle being one of these more relevant cases [7]. Figures 4 and 5 show a comparison of trucks versus combined transport and the position of optimal multi-modal transport point. Advantages of road transport are raised. Trucks are more flexible and can reach almost

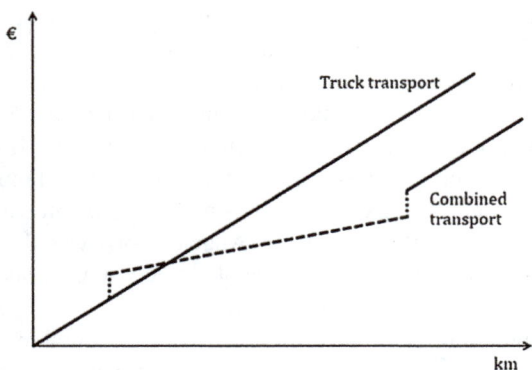

Fig. 4 Trucks versus combined (Reis et al. 7, pp 22)

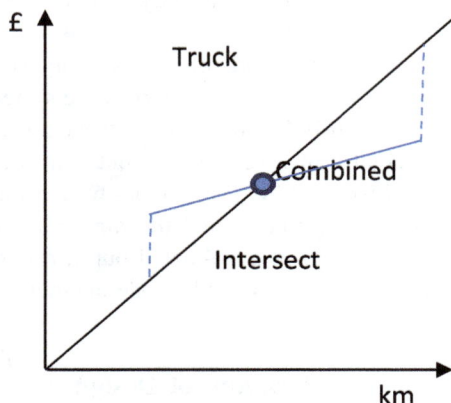

Fig. 5 Optimal multi-modal transport point

any point, not limited to stops along a single line. They are compatible to travel through any country, as all countries contain roads, most of which are passable by truck. There is also no faster way to deliver goods for short to medium distances. Finally, delivery by road incurs the least costs as there are only fixed costs for truck such as maintenance and fuel consumption costs. There are however problems associated with delivery by vehicle such as the fact that this method does not run on a tight schedule as a railway solution does. Road vehicles are also large contributors of CO_2 emissions and are becoming increasingly undesirable within society due to this. The costs of running a vehicle on the road are constantly increasing and will continue to increase with the exhaustion of fossil fuels; this makes road-based transport a temporary solution and not so viable for the future.

After a certain distance, it becomes more suitable to use combined transport as a solution to baggage transfer.

To move this intersection point towards the left on, making combined transport cheaper for shorter distances, a couple of changes can be made. Reducing the distance between baggage hubs and the train stations reduces the pre-carriage travel costs. There can also be improvements

made to the transfer units used on both train and road vehicle, making the crossover of cargo between the two more simple and efficient.

Figure 6 illustrates just how much more CO_2 emissions are contributed by road than other transport methods. This will of course become outdated very fast by the injection of electric and hybrid vehicles into society.

As this project is based on the transfer of baggage between two fixed points with a readily available metro line it would be most suitable to use railway only and not use two modes of transport together, which would complicate the system and scheduling.

4.3 The Client

The average age of a leisure traveller is 47.5 years old and the average age of a business traveller is 45.6 years old [8].

More importantly in 2012, 33 % of domestic business flights included air travel as opposed to just 11 % of leisure trips. 79 % of leisure trips used a car compared to the 48 % of business trips which used a car. These facts further demonstrate that cars are the regular choice of leisure travellers, most likely due to the handling of their baggage.

Though the information here is focused on travellers within the US, where there is obviously a different infrastructure to that of Europe, it is still a fair indicator of the target audience of the system design.

The passengers who the system is set to facilitate are those who wish to travel without needing to have a baggage constantly on their person. In the case of leisurely travellers this could be most effective as leisurely travellers can have the tendency to buy many goods such as souvenirs whilst on holiday without thinking about the logistics of their newly prized possessions. The system will allow them to send additional bags back home with time constraints of their selection. Many leisure travellers will also stay in hotels whilst abroad, this can cause concerns for those who have a flight home late in the day but are required to check

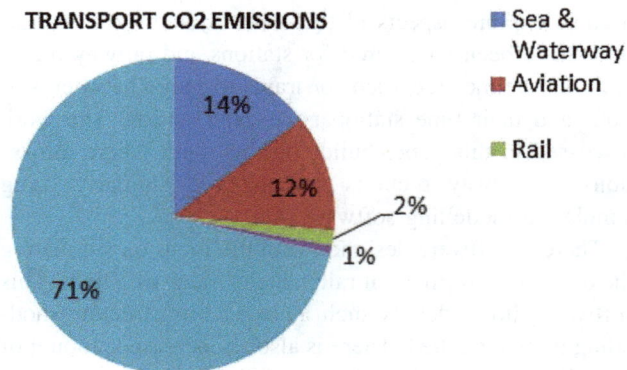

Fig. 6 CO_2 emissions [7]

out of their hotel earlier in the morning. The system will prevent this large group of people from being stuck with their bags for the duration between these times. Instead these travellers can drop their bags off and be free of the constant worry attached to the baggage.

There are also those who travel for business trips, these can often be people who will have a flight booked at a very short notice and will require a bag to be packed up and sent ahead of them to a destination. It is common for these trips to go to heavily populated areas such as busy cities where the use of cars are highly problematic, causing the handling of baggage to result in limitations of freedom when they arrive at their destination.

The final group of users would be those who have too much baggage to travel with at once. For example students relocating to university may wish to take their many bags to a hub over the course of a week or so for a later flight. This is a facility which is currently not so readily available.

5 Physical Design Elements

There are a number of physical aspects of the baggage transfer system which have been looked at. The real world system from the customer dropping a bag off at the collection hub to bags being unloaded from containers at the airport has been designed.

5.1 Baggage Traffic

The physical elements of the design are based on the volume of traffic that flows through the system. This volume of traffic is estimated based on a variety of assumptions.

To estimate the number of travellers from Newcastle Airport, timetables for departures were looked at. The timetable on Monday the 6th of May was studied; this timetable showed 60 flights departing from Newcastle Airport across the duration of the day. Similarly Sunday the 5th of May contained 61 flights and Tuesday the 7th May contained 60. These flights all had a fairly similar distribution of flights across the day, i.e. eight flights between 6:00 and 7:00 am, earliest flight between 5:00 and 6:00 am and last flight between 22:00 and 23:00.

An assumption that every plane that leaves the airport is full of bags that have come through the system was made. To find the number of bags, now having the number of flights from the airport in a day, it was necessary to find the number of passengers on each flight and the amount of cargo that each plane would travel with. To gain these statistics the fleet of a popular airline company was studied. EasyJet has a fleet of 198 airplanes, 138 or 70 % of this fleet being the Airbus

A319 airplane [9]. The A319, which has cargo capacity of 5 LD3 containers and a passenger capacity of 124 [10], is therefore a suitable model to treat each flight from Newcastle Airport as. Therefore, multiply the number of passengers per flight by the number of planes that leave a day.

$$60 \text{ flights} \times 124 \text{ passengers} = 7440 \text{ bags per day}, \quad (1)$$

$$\frac{7440 \text{ bags per day}}{24 \text{ hours}} = 310 \text{ bags processed an hour}. \quad (2)$$

This will be the traffic of bags under the stated assumptions if every person on every flight had a bag that they choose to send through the proposed system.

5.2 The Collection Hub

5.2.1 Location of the Collection Point

This is the origin of the bags' journey within the system. The location of the collection hub for this system has been set at Newcastle's Haymarket. The location of the collection hub is critical to how well it will facilitate the passenger. Haymarket is situated at the end of the busy high street Northumberland Street, centre of the city. It is also a busy changeover station for the Newcastle Metro service. This means that a large volume of people will travel past the Haymarket, it is also reasonably accessible with nearby roads and multi-storey car parks. These factors make Haymarket the ideal location for the bag collection in the system. The only complications that may arise would be infrastructure developments in such a busy area.

5.2.2 Baggage Collection

Passengers are required to drop each bag off to the Haymarket hub to have them sent through the system. There will therefore be a need for a means of collecting these bags. This is done using one large area with different service points to accommodate to the large flow of customers. An assumption is made that it will take approximately 5 min to process a customer's bag. This results in the following calculation to calculate the number of service points required.

$$\frac{310 \text{ bags an hour}}{60 \text{ min an hour}} \times 5 \text{ min per bag} \approx 26 \text{ service points}. \quad (3)$$

The 26 service points required is split up into 10 service points with a human point of contact and 16 points which are automated. These provide a slightly faster option for those who do not need assistance with the process.

5.3 Freight Trains, Containers and Cradles

5.3.1 LD3 Containers

The LD3 containers that are carried by airplanes have also been used in the transit of bags from the collection point to the airport. Using these containers allows consolidation of the bags at after bag sorting at the Haymarket and allows cargo to be moved about with greater ease than would be possible with loose bags.

These containers are of the following overall dimensions height 1.63 m, width 1.54 m and length 2.01 m.

Using the containers to transport consolidations of 25 bags allows containers of approximately 500 kg to be handled by machinery before the track. This means there is less labour intensive lifting of bags on and off trains, which results in faster loading and unloading times (Fig. 7).

5.3.2 Loading and Unloading Freight Trains

The system was designed with the aim to transport 15 containers per train. This means loading and unloading 15 containers at both Haymarket and the airport. It is therefore ideal to have these 15 containers placed into one freight cradle structure. This cradle structure needs to support the 7.5 tons that the 15 containers produce together. Using one cradle on each freight train allows cradles to be prepared before a freight train arrives at the loading/unloading station, again reducing time spent loading or unloading.

The cradles of large mass are required to be lifted off a freight train whilst empty and replaced with a loaded cradle. To allow this loading and unloading of such heavy goods, automation is necessary. An option that would be very suitable is the use of a gantry crane. A gantry crane allows an empty cradle to be removed and replaced with a loaded cradle with ease.

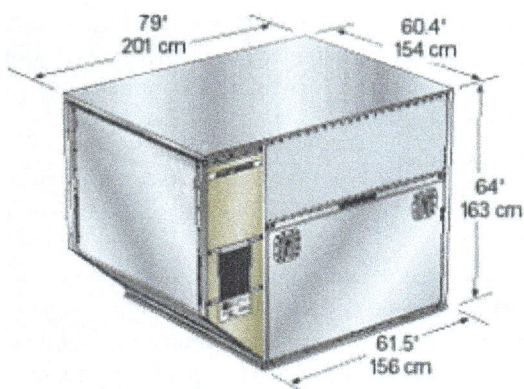

Fig. 7 LD3 container [17]

5.3.3 Freight Trains

The number of freight trains that the system required originates from a few assumptions. It was assumed that for every flight there would be 5 containers to be transported. The type of freight train that has been looked at to be used is the British Class 66 due to its overwhelming popularity within the UK rail industry.

5.3.4 The Service

As previously discussed, Sect. 5.3 the client, there are two main groups of customers for the system, these are the business passengers and those travelling for leisure. The design is less likely to fail if the bags are not always being dropped off just before they need to be processed through the entire system.

To avoid this situation the service options to the customer are split into immediate processing of the bag, processing within 5 days or within 14 days. Using these three service options helps reduce the amount of bags needing immediate processing. Incentivising the 5- and 14-day options for with reduced rates ensures that more people drop their bags many days in advance, keeping the system from suffering a large spike in baggage traffic. With reduced rates for later deliveries the system also appeals to groups of people who can afford to drop their bag off earlier, taking another concern off their minds on the day of their travels.

5.3.5 Storage and Logistics

There are three different storage areas, the 14-day storage, the 5-day storage area and the 24-h storage facility. It can be assumed that there is a maximum capacity of 7440 bags in the 24-h facility, as this is the most the system can process. To store such volumes of baggage, a level of automation is beneficial. The bags are stored using racking and crane systems. Vanderlande offer a system called 'BAGSTORE', this system stores the bags individually on large shelf structures that span across a warehouse. These are 10 rows of shelves which are 10 high and 75 long. This therefore accommodates the storage of 7500 bags, more than are required to be stored a day. The bags are retrieved by automated cranes which pull the bags out of their shelves to be transported to the appropriate LD3 container. This storage is also used in the 5- and 14-day storage rooms, of much smaller volumes than the 24-h facility.

Transporting the bags to, around and from these storage rooms is another important aspect of the design. It is most desirable to use automated options again here due to the volume of bags to be transferred. There are a couple of options for this transfer of baggage; there is a standard

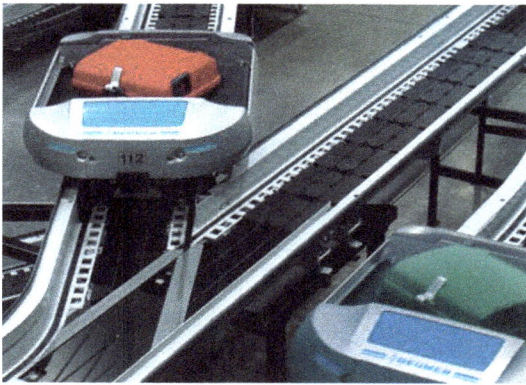

Fig. 8 BEUMER AUTOVER [18]

Fig. 9 Vanderlande BAGSTORE [19]

option of using conveyor belts and there is also the more expensive option of using a system such as BEUMER AUTOVER. An independent carrier system (ICS) such as the BEUMER AUTOVER offers benefits in control and tracking of the individual bags (Figs. 8, 9).

The collection hub is best split into two areas, the unloading/loading area and the storage areas (Fig. 10). The bags originate from the collection point. From here they move to the 14-day storage, the 5-day storage or the 24-h storage. The 14- and 5-day storage options both have means of moving bags through to the next duration storage, i.e. the 14-day storage moves to 5 day storage and the 5-day storage to 24-h storage at the appropriate times. The majority of bags from the collection point move directly to the 24-h storage, bypassing the 5-day and 14-day options. This system means that the bags are always moving towards their final destination within the hub, the track where they are placed onto the train.

Aside from the storage facilities there is the loading/unloading area. This is an area where there is a constant circulation of LD3 containers. These containers start being filled with bags moved out from the 24-h storage and are

placed onto the freight cradles when full. This loaded cradle will then wait on the cradle track. When a freight train enters the hub, the cradle of empty containers is removed by the gantry crane and placed on the cradle track, the loaded cradle is then moved up the cradle track and loaded onto the train by the same gantry crane. Once the train leaves, the empty containers are removed from the cradle, which then returns to the original position and is circulated back to the LD3 loading bay to await more bags for the next train.

6 SIMUL8, Event-Based Simulation Modelling

6.1 Simulation Modelling

To analyse the performance of the designs they were created using SIMUL8. The simulation models are created to mimic the most realistic behaviour possible. The designs are run in their optimal state. Therefore, there are no delays of trains or other possible real-time problems for the model to simulate. These models are used not only as a measure of their performance, with regard to track utilisation, but also as a validation that the system can be created and can function. The freight trains placed onto the system are done so without interfering with any of the passenger trains; this is one of the key rules of design employed (Table 1).

There are four scenarios modelled using SIMUL8. These consist of one passenger only system, a scenario where freight trains are placed on the system only where required based on flight times from Newcastle Airport and two other scenarios where the aims are to increase track utilisation by saturating the system with freight trains.

6.2 Start Points

Each model has multiple start points which are the locations where the trains enter the system. There must be a unique start point used for each group of trains, i.e. South Gosforth, Regent Centre and regular passenger trains.

The start points control the input of work items into the system i.e. the origin of trains in the model. These can be input with a frequency, mathematical functions or as in this case using a schedule sheet (Figs. 11, 12).

6.3 Work Items and Labels

Each simulation model requires work items, the trains in these cases. Though there is only one work type used in the proposed designs, this work item is shared across four different trains: regular passenger trains, freight trains, trains that stop at Regent Centre and trains that stop at South Gosforth. To split the trains into these four different

Fig. 10 A Sketch of the collection hub

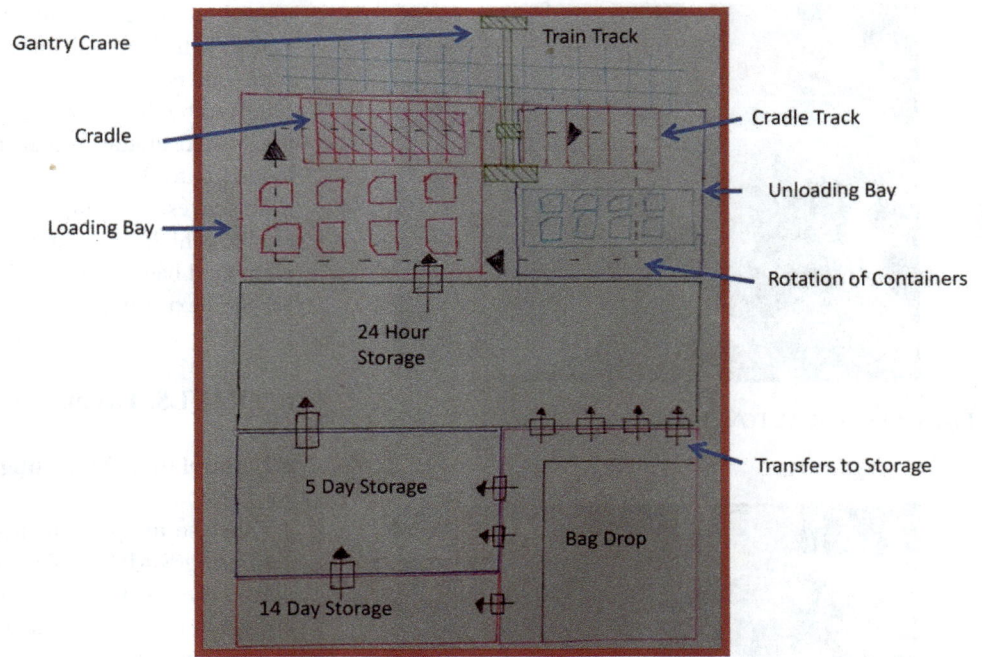

Table 1 Scenarios

Scenario 0	Passenger service only
Scenario 1	Passenger + freight trains when required
Scenario 2	Passenger + freight trains to increase track utilisation 1
Scenario 3	Passenger + freight trains to increase track utilisation 2

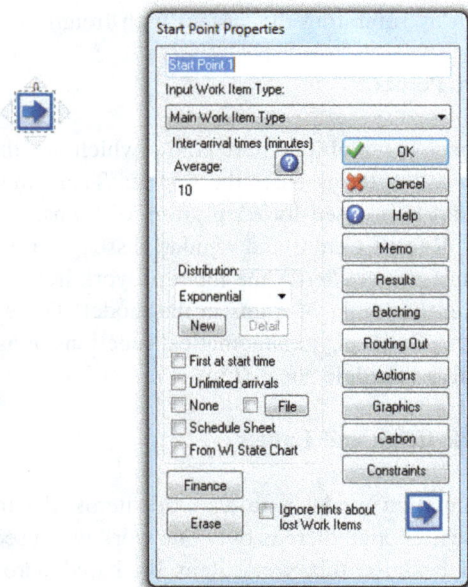

Fig. 11 Start point

groups a different label value is assigned to each of the trains from their start points. These labels allow the trains to be routed differently later in the model by activities (Fig. 13).

6.4 Schedule Sheets

Each start point has a unique schedule. These schedules can be created externally on a software such as excel and imported via the paste option to SIMUL8. The requirements of the schedule sheet are that a time and quantity (batch size) is stated. There are rules to how the trains are ordered in the schedule sheet, such as the trains must be entered into the sheet chronologically (Fig. 14).

6.5 Activities

The activities are the most used function in the simulation models. These activities are used as both the stations as well as the tracks between stations. Activities apply an action to the work items that travel through them, for instance making them wait for a fixed amount of time (Fig. 15).

6.6 Routing Out

Each activity within a model has a routing out option, used when there is more than one exit from an activity that work items can travel down. These routing out options can use many forms such as mathematical functions or circulate,

Fig. 12 SIMUL8 labelling

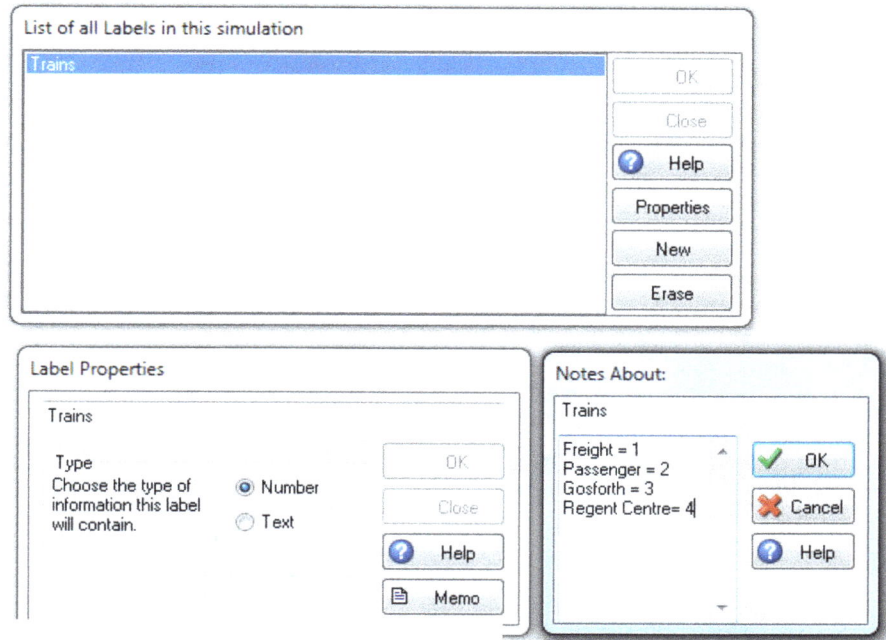

Fig. 13 SIMUL8 work items

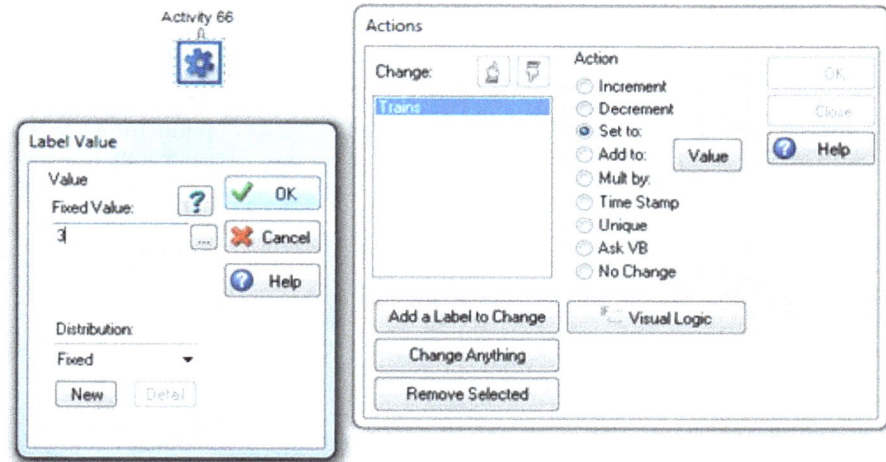

which releases work items in the same manner that cards are dealt with (Fig. 16).

The option used to control the journeys of the different trains is Label, allowing the user to control each train by the value that they were assigned at their start points.

6.7 Activity Bypasses

Freight trains and passenger trains will use the same metro line from the Haymarket to the airport, however the freight trains need not stop at stations as the passenger trains do. As an option to treat the two different groups differently by an activity was not found, bypasses were created which

exist as activities which move the freight trains to the next activity (track) whilst making passenger trains move to the activity before the track (the station).

As it can be seen from Fig. 17 the freight trains, which are set to value 1, will bypass Callerton Parkway and move straight to the track towards Bank Foot. This is not for the case for passenger trains, value 2, which will be sent to the Callerton Parkway station to wait their duration. It is important to note that these activities, which are attached to every station in the model, have a processing time of 0 and therefore do not interfere with the model in anyway. These activities only act as a controller for the trains moving through the model.

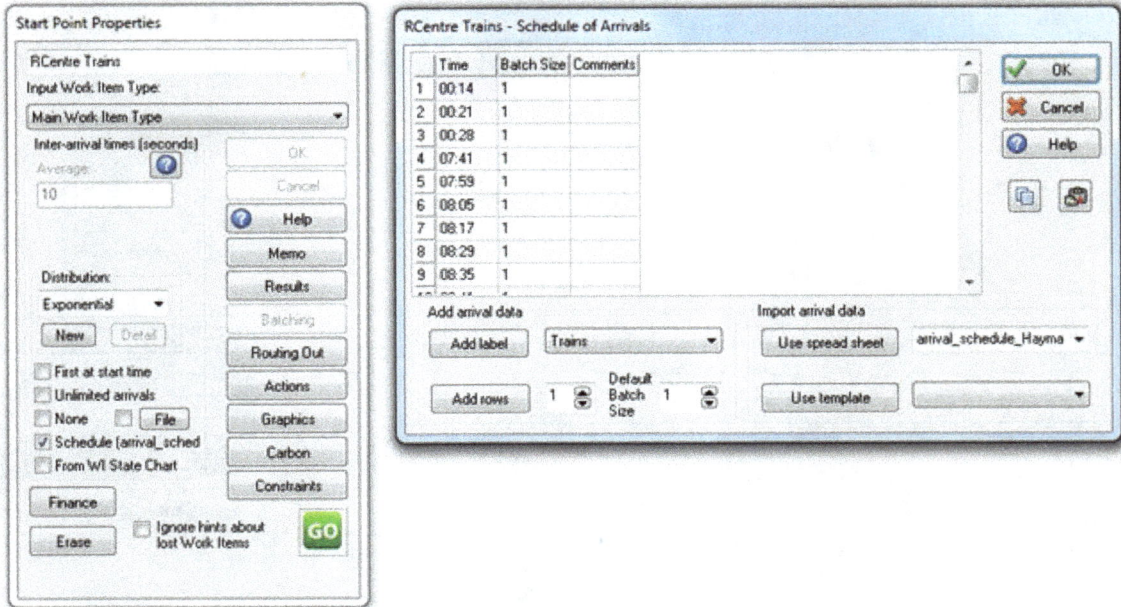

Fig. 14 SIMUL8 schedule sheets

Fig. 15 SIMUL8 activities

6.8 Clock Settings

The clock settings are an important part of the simulation model. A unit of time for the model is chosen here, seconds being the most suitable for the design models as the transit times and station times are recorded in seconds. A warm-up period is also chosen here. This is a period of time which can be used to move the system into the middle of its duration before starting to take measurements.

The clock settings also involve choosing a start time and duration for each day. To measure the utilisations of tracks on the metro line it is essential that the duration be 24 h as freight trains may be implemented onto the system over 24 h, the start time of each day is not as important so midnight was chosen. The final important option from clock settings is the duration that the simulation will run for (Fig. 18). To analyse the system it is best to run it over the course of a week, as there will be many more trains run over the system this way than if just a day were used.

6.9 Utilisations and Blockages

Each activity will produce a result for the percentage of time that the activity was in use and not in use. This is the percentage of time that there is a train on the track as opposed to the track being empty. This is how data have been collected for the utilisation of each track.

The system can also be validated through the results produced by all of the activities. The results show a percentage of time that the activity was blocked and a maximum number of work items (Fig. 19). If the maximum is 2 it means that two trains have been on the track together at the same time, this would mean the system has not worked as there should never be two trains on the same track. Similarly if the blockage value is not 0 it means there has been a reason why the work item could not leave the activity, normally caused by the next item having a capacity limit preventing work items entering. This would also be an indicator of the system failing.

Fig. 16 SIMUL8 routing out

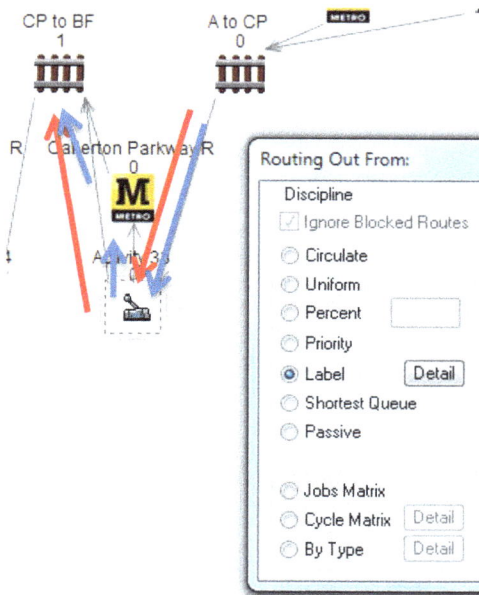

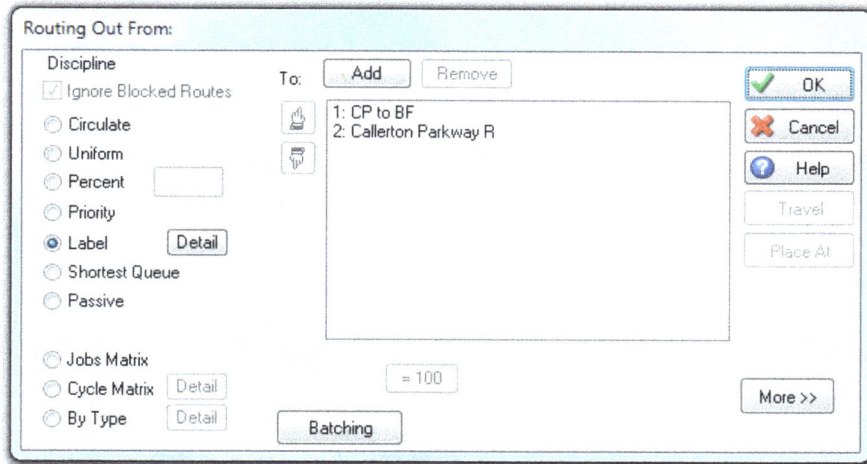

Fig. 17 Model bypasses, *blue arrow* shows passenger route and *red arrow* shows freight route

6.10 End Points

End points record max, min and average transit times that work items take to reach the end of the system. These points also count the number of work items that made it through the system to the end (Fig. 20).

7 Applications of Modelling

7.1 Scenario 0

The first step in modelling the new designs was to create a simulation model for the regular metro system between Haymarket and the airport. This would then be validated

and have its utilisations of tracks recorded. There are no freight trains on this model.

To make the simulation model as close to the reality as possible it was essential to implement the correct times for transit of the trains between stations, as well as the correct time that a train would spend waiting at a station. The times used in the model originate from readings taken by Marinov and Motraghi [4] (Tables 2, 3).

7.1.1 Schedule

The schedule for the trains implemented onto the system is identical to the online timetable that metro provides for trains going towards the airport from the Haymarket [11].

Fig. 18 SIMUL8 clock properties

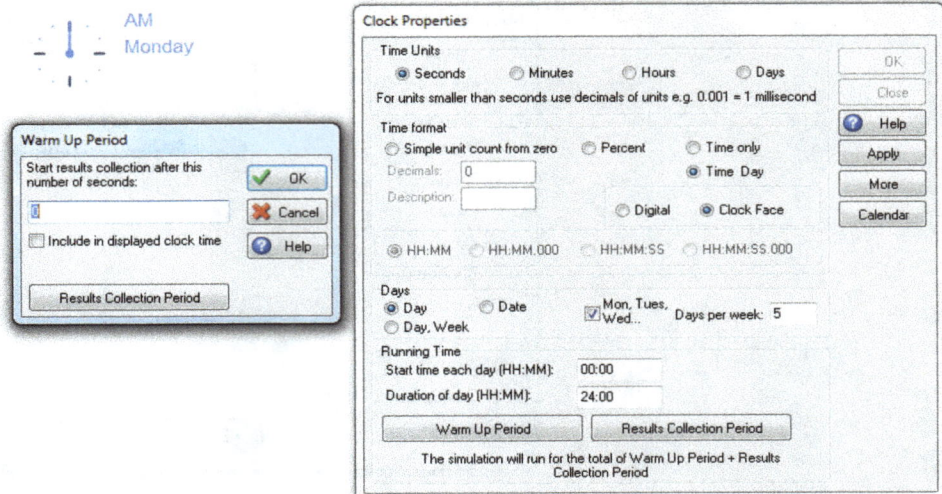

Fig. 19 Activity results

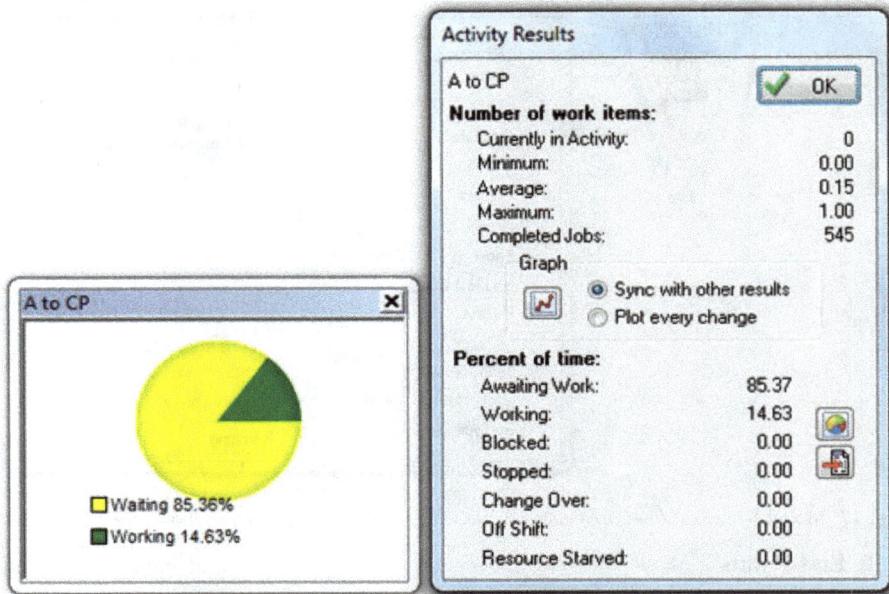

This is split up into three different start points in the model: the trains going to the airport, trains ending at Regent Centre and trains ending at South Gosforth. Each start point has its own timetable for the different groups of trains (Table 4).

The addition of the freight trains to the system has been done without interfering with the passenger system. The passenger trains therefore control where freight trains can be added to the system during the day.

7.1.2 The Model

The simulation model as shown in Fig. 21 emulates the behaviour of the metro train system. There are 3 start points active for South Gosforth, Regent Centre and Newcastle Airport trains. Once the airport passenger trains get to the airport they wait and return down the bottom line which is a mirror image of the top line (Haymarket to Callerton Parkway) (Tables 5, 6).

End Results

Haymarket R	✔ OK
Work Completed:	543
Time in system:	
	All
Minimum:	2447.00
Average:	2563.37
Maximum:	2600.00
Standard Deviation:	65.35
Time in system within limit:	
Time limit: 10 seconds	
Percentage within limit:	0%

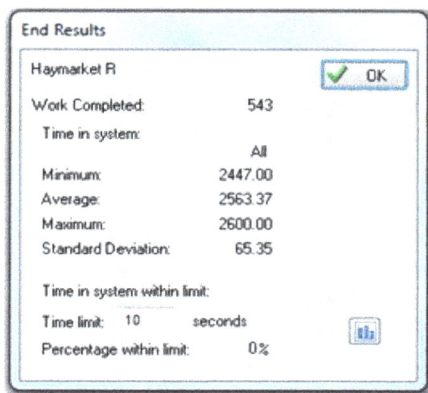

Fig. 20 End results

7.1.3 Validation

As this is the most basic model (no freight trains) to be presented it is crucial that it is validated to check that the model does indeed function as the real world network does. There are two means of validation performed on the model. Firstly the results of the end line are analysed. The total time that a passenger train should take is the total waiting time in all the stations up to the airport, multiplied by two for the return journey, plus the time spent at airport and all the times between stations.

As it can be seen from the end of line results, Fig. 22, the minimum, maximum and average time are all 2600. This then validates the model as it shows the exact amount of time each train took that hand calculations have shown it should take.

Table 2 Station wait times

Station	Recorded times spent at station (seconds)			Average time (to the nearest second)
Airport	232	259	356	282
Callerton Parkway	14.5	15.1	16.8	15
Bank Foot	11.4	12.7	16.2	13
Kingston Park.	11.5	17.1	17.8	15
Fawdon	10.8	15.3	16.5	14
Wansbeck Road	13.4	14.2	15.7	14
Regent Centre	12.7	13.6	18.4	15
South Gosforth	12	17.4	25.4	18
Ilford Road	11	15.1	16.6	14
West Jesmond	12.9	18.5	19.8	17
Jesmond	17.1	17.4	18.2	18
Total				435

Table 3 Train travel times

Stations		Recorded travel times (seconds)			Average travel time (to the nearest second)
From	To				
Airport	Callerton Parkway	109	115	123	116
Callerton Parkway	Bank Foot	156	163	171	163
Bank Foot	Kingston Park	66	68	77	70
Kingston Park	Fawdon	103	113	116	111
Fawdon	Wansbeck Road	56	65	67	63
wans beck Road	Regent Centra	30	47	69	49
Regent Centre	South Gosforth	101	109	117	109
South. Gosforth	Ilford Road	61	68	98	76
Ilford Road	West Jesmond	41	42	46	43
WestJesmond	Jesmond	89	93	103	95
Jesmond	Hay market	99	116	117	111
Total				1006	

Table 4 Airport timetable

Timetable to a airport										
06:02	07:59	08:53	10:08	12:08	14:08	16:08	17:17	18:19	20:18	22:48
06:25	08:05	08:56	10:20	12:20	14:20	16:20	17:20	18:28	20:33	23:03
06:47	08:08	09:08	10:32	12:32	14:32	16:32	17:23	18:31	20:48	23:18
06:58	08:17	09:20	10:44	12:44	14:44	16:35	17:32	18:42	21:03	23:33
07:08	08:20	09:32	10:56	12:56	14:56	16:44	17:35	18:53	21:18	23:48
07:20	08:29	09:35	11:08	13:08	15:08	16:47	17:44	19:05	21:33	23:59
07:32	08:32	09:44	11:20	13:20	15:20	16:56	17:47	19:18	21:48	REPEAT
07:41	08:35	09:47	11:32	13:32	15:32	16:59	17:56	19:33	22:03	
07:44	08:41	09:53	11:44	13:44	15:44	17:05	18:07	19:48	22:18	
07:56	08:44	09:56	11:56	13:56	15:56	17:08	18:16	20:03	22:33	

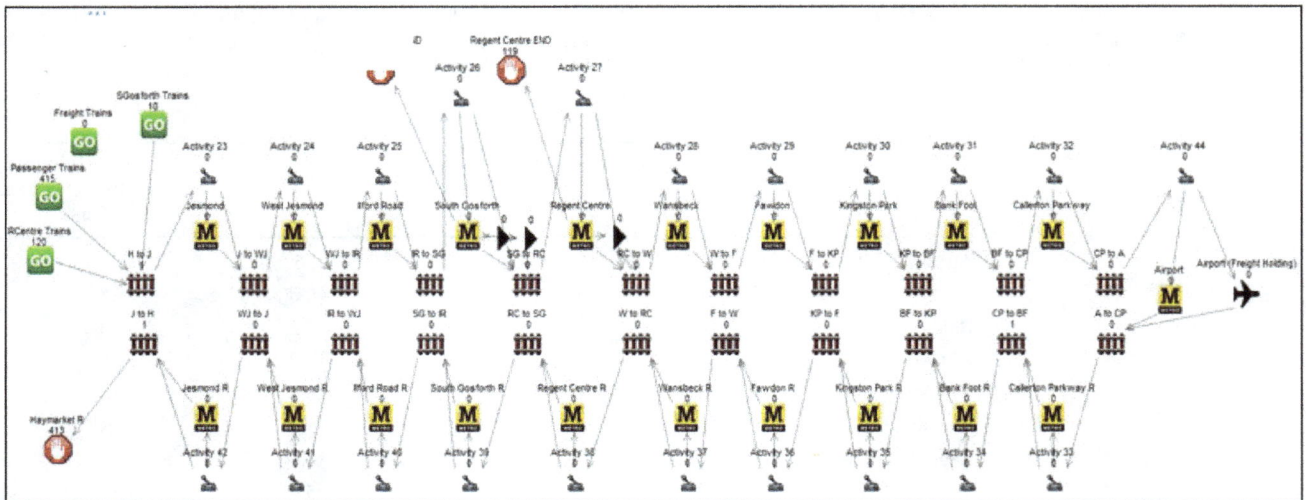

Fig. 21 Simulation model: screen-shot, Scenario 0

Table 5 South Gosforth timetable

Timetable to South Gosforth
09:53
18:28
REPEAT

Table 6 Regent Centre timetable

Timetable to Regent Centre		
00:14	08:41	17:23
00:21	08:53	17:35
00:28	09:35	17:47
07:41	09:47	18:16
07:59	16:35	23:48
08:05	16:47	23:59
08:17	16:59	REPEAT
08:29	17:05	
08:35	17:17	

The results of each activity can also be analysed. If at any point there is a value other than 0 for percentage of time blocked, then the trains are clashing with one another on the track and the model needs to be adjusted.

Time spent at stations + time spent on tracks

+ time spent at airport = transit time

$$((435 - 282) \times 2) + (1006 \times 2) + 282 = 2600 \text{ s}$$

(4)

7.2 Scenario 1

This is the first baggage transfer system proposed. The intentions of this system are to facilitate the airport users. The quantity of trains is driven by the number of flights from Newcastle Airport [12], as shown in Table 7.

A business traveller may have a short-notice flight booked and would be required to drop his baggage off shortly before embarking upon his journey. This system

Fig. 22 Scenario 0 end and activity results

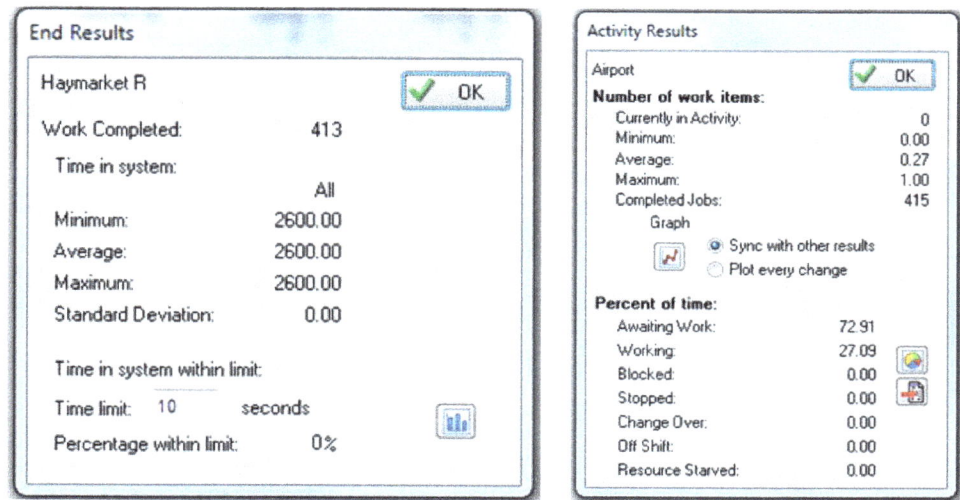

Table 7 Freight trains for flights [12]

Times		Departures	
From	To	Airplanes	Freight trains
03:00	04:00	0	1
04:00	05:00	0	3
05:00	06:00	1	3
06:00	07:00	8	1
07:00	08:00	9	2
08:00	09:00	3	1
09:00	10:00	5	2
10:00	11:00	1	1
11:00	12:00	4	1
12:00	13:00	1	1
13:00	14:00	3	i
14:00	15:00	3	2
15:00	16:00	2	2
16:00	17:00	4	1
17:00	18:00	4	1
18:00	19:00	2	2
19:00	20:00	2	0
20:00	21:00	4	1
21:00	22:00	0	0
22:00	23:00	1	0
		Total	26

accommodates the short notice with fast delivery of a bag to the airport to catch the same flight as the passenger. To get these bags onto the same flight as the passengers it would be essential that they drop their bags off at the collection point in time for a freight train that leaves two hours before the airplane it is intended for. For example if an airplane leaves within the hours of 5:00 and 6:00 a freight train will be required to leave between 03:00 and 04:00. As every airplane carries 5 LD3 containers and each train carries 15 it is clear that for every 3 flights in an hour there will be a freight train required 2 h before.

7.2.1 Schedule

The freight trains enter the line at any point within their designated hour that they do not interrupt the passenger trains. Table 8 shows the proposed schedule of freight trains, the schedule of passenger trains remains unchanged. The total of 26 freight trips a day equates to the maximum transfer of 390 containers a day or 9750 bags.

7.2.2 The Model: Freight Trains Airport Holding

The only changes to the model shown in Fig. 23 from the passenger only model is the addition of freight trains to the system and the use of the 'Airport (Freight Holding)' which in the previous model was not routed to from Activity 44.

The timing that a freight train spends in the airport holding is what controls the separation of freight trains and passenger trains from the airport to Haymarket. The time has been chosen to re-sync the difference between freight trains and passenger trains that existed when they originally departed Haymarket.

Difference between freight and passenger after airport

$$= \text{station waiting times} + \text{passenger airport waiting time}$$
$$153\,\text{s} + 282\,\text{s} = 435\,\text{s} \qquad (5)$$

It is therefore essential that each freight train waits for 435 s, 7 min and 15 s, at the airport to ensure that the difference due to transit times between the passenger trains

Table 8 Scenario 1 timetable

Scenario 1 (departure times)	
03:00	11:02
04:00	12:02
04:15	13:02
04:30	14:02
05:00	14:15
05:15	15:02
05:30	15:15
06:10	16:02
07:05	17:14
07:26	18:04
08:26	18:25
09:02	20:00
09:15	REPEAT
10:02	

and freight trains is removed from the equation, and the schedules are re-synced.

There are rules to the timings that must exist between freight trains and passenger trains. These are explained in Fig. 24.

This means that freight train to freight train separation must be 8 min or more, passenger trains must leave at least 5 min in advance of freight trains and freight trains must leave at least 5 min in advance of passenger trains.

7.2.3 Validation

The same validation techniques can be used as shown for Scenario 0. The transit times are checked and the activities are checked for blockages.

Train Separations			
Type	Cause	Calculation (Seconds)	Spacing rounded up (min)
Freight - Freight	Airport holding	435	8
Freight - Passenger	Callerton Parkway - Airport	(1159 - 1006) +116 = 269	5
Passenger - Freight	Callerton Parkway - Bank Foot	163	3

Fig. 24 Train separations

Each freight train takes 2447 s to complete its transit from Haymarket to airport and back again.

$$\text{Freight transit time} = (1006 \times 2) + 435 = 2447 \quad (6)$$

Transit times calculated over the course of a day are as follows:

$$\text{Average transit time}$$
$$= \frac{(26\ \text{freight} \times 2447\ \text{s}) + (81\ \text{passenger} \times 2600\ \text{s})}{(81 + 26)\ \text{trains}} \quad (7)$$
$$= 2562.82\ \text{s}$$

Results obtained from simulation show the same figure for the average (Fig. 25).

7.3 Scenario 2

Scenario 2 is created to fulfil the same purposes as Scenario 1 with an additional function. Scenario 2 attempts to saturate the system using the same infrastructure. The reason for saturating the system is to see what the maximum capacity of the system is. This is done by adding freight

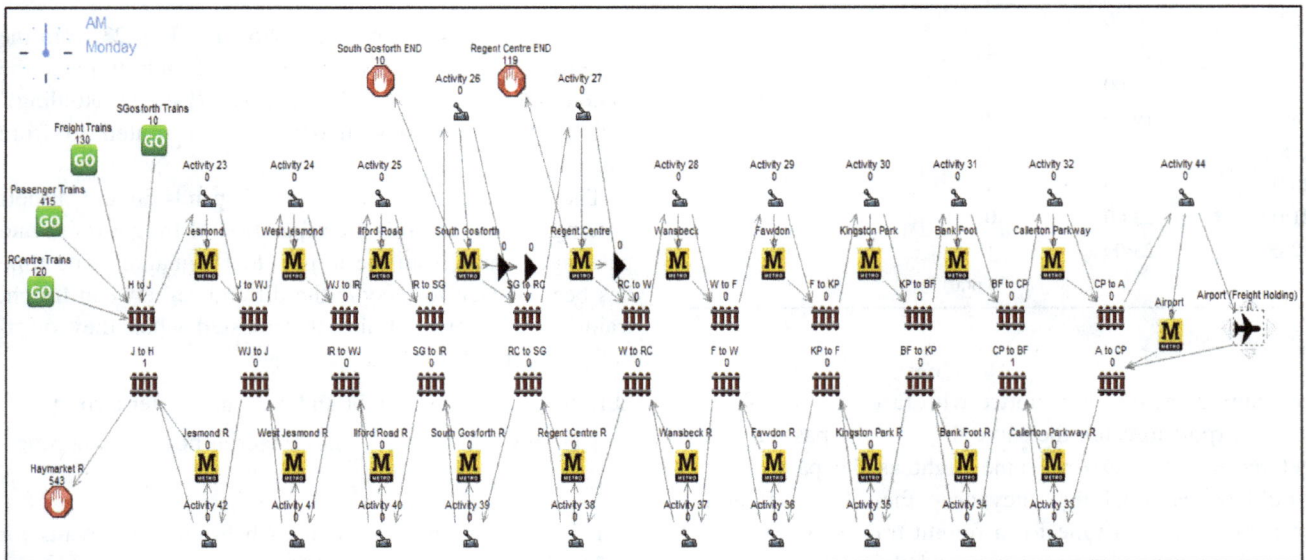

Fig. 23 Scenario 1 model

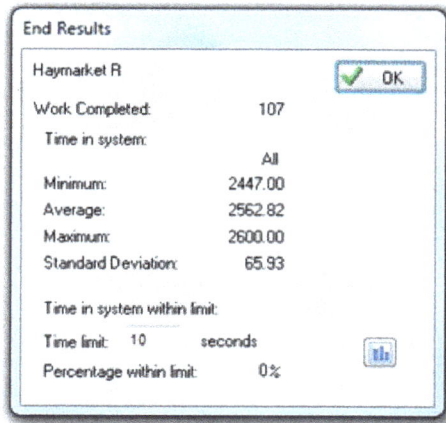

Fig. 25 Scenario 1 end results

trains to the system overnight and occasionally in gaps during the day.

7.3.1 Schedule

The number of freight trips that have been added to the system since Scenario 1 is 91, resulting in a total of 117 freight trips. For Scenario 2 the timetable is shown in Table 9. Overnight when the track is not in use the system can run 7 freight trains an hour. The factor that limits the number of freight trains that can be added is the 8 min separation between each train.

The 117 freight trips equate to the maximum transfer of 1755 containers a day or 43875 bags.

Table 9 Scenario 2 timetable

Scenario 2 (departure times)										
00:05	02:00	03:32	05:08	07:03	09:40	12:13	14:37	17:13	19:40	22:25
00:32	02:08	03:40	05:16	07:13	10:01	12:25	14:49	17:29	19:54	22:40
00:40	02:16	03:48	05:24	07:25	10:13	12:37	15:01	17:40	20:09	22:55
00:48	02:24	04:00	05:32	07:37	10:25	12:49	15:13	17:53	20:24	23:10
01:00	02:32	04:08	05:40	07:49	10:37	13:01	15:25	18:03	20:39	23:25
01:08	02:40	04:16	05:48	08:13	10:49	13:25	15:37	18:13	20:55	23:40
01:16	02:48	04:24	06:07	08:25	11:01	13:37	15:49	18:25	21:10	23:54
01:24	03:00	04:32	06:15	08:47	11:13	13:49	16:01	18:37	21:25	REPEAT
01:32	03:08	04:40	06:30	09:01	11:37	14:01	16:25	18:48	21:40	
01:40	03:16	04:48	06:38	09:13	11:49	14:13	16:40	19:00	21:55	
01:48	03:24	05:00	06:52	09:25	12:01	14:25	16:52	19:11	22:10	

Table 10 Scenario 3 timetable

Scenario 3 (departure times)										
00:04	01:24	02:24	03:24	04:24	05:24	07:13	10:49	14:13	17:29	20:55
00:08	01:28	02:28	03:28	04:28	05:28	07:25	11:01	14:25	17:40	21:10
00:32	01:32	02:32	03:32	04:32	05:32	07:37	11:13	14:37	17:53	21:25
00:36	01:36	02:36	03:36	04:36	05:36	07:49	11:37	14:49	18:03	21:40
00:40	01:40	02:40	03:40	04:40	05:40	08:13	11:49	15:01	18:13	21:55
00:44	01:44	02:44	03:44	04:44	05:44	08:25	12:01	15:13	18:25	22:10
00:48	01:48	02:48	03:48	04:48	05:48	08:50	12:13	15:25	18:37	22:25
00:52	01:52	02:52	03:52	04:52	05:52	09:01	12:25	15:37	18:48	22:40
00:56	01:56	02:56	03:56	04:56	05:56	09:13	12:37	15:49	19:00	22:55
01:00	02:00	03:00	04:00	05:00	06:07	09:25	12:49	16:01	19:11	23:10
01:04	02:04	03:04	04:04	05:04	06:15	09:40	13:01	16:25	19:40	23:25
01:08	02:08	03:08	04:08	05:08	06:30	10:01	13:25	16:39	19:54	23:40
01:12	02:12	03:12	04:12	05:12	06:38	10:13	13:37	16:52	20:09	23:54
01:16	02:16	03:16	04:16	05:16	06:52	10:25	13:49	17:03	20:24	REPEAT
01:20	02:20	03:20	04:20	05:20	07:03	10:37	14:01	17:13	20:39	

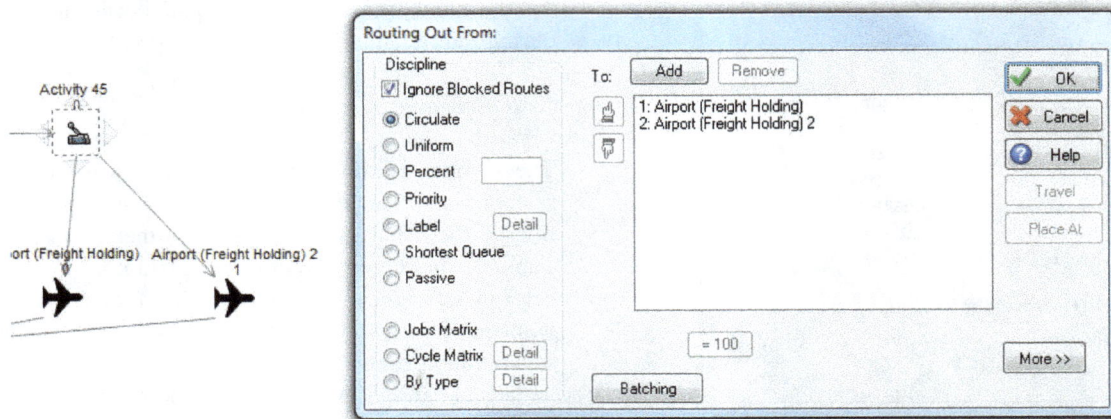

Fig. 26 Scenario 3 model

Table 11 All utilisations

Section	Scenario 0	Scenario 1	Scenario 3	Scenario 3
H–J	13.99	17.33	29.02	34.93
J–WJ	11.96	14.82	24.83	29.89
WJ-1	5.41	6.71	11.24	13.53
1-SG	9.57	11.86	19.86	23.91
SG–RC	13.47	16.75	28.22	34.02
RC–W	4.71	6.18	11.33	13.94
W–F	6.05	7.95	14.57	17.92
F–KP	10.66	14.00	25.67	31.58
KP–BF	6.72	8.83	16.19	19.91
BF–CP	15.66	20.56	37.69	46.37
CP–A	11.14	14.63	26.83	33.00
A–CP	11.14	14.63	26.80	32.97
CP–BF	15.63	20.54	37.63	46.31
BF–KP	6.71	8.81	16.16	19.88
KP–F	10.64	13.98	25.62	31.53
F–W	6.04	7.93	14.54	17.89
W–RC	4.70	6.17	11.31	13.92
RC–SG	10.45	13.73	25.15	30.95
SG–I	7.28	9.57	17.52	21.57
1-WJ	4.12	5.41	9.91	12.20
WJ–J	9.10	11.96	21.90	26.96
J–H	10.62	13.96	25.57	31.48
Max	15.66	20.56	37.69	46.37
Min	4.12	5.41	9.91	12.20
Average	9.35	12.11	21.71	26.58

7.3.2 Validation

The model is the same model used for Scenario 1, validation has only been performed to see that the new freight trains schedule does not cause any build-ups on the system. This was done by looking for any blockages within any of the activities, of which there were none.

7.4 Scenario 3

This final scenario has been designed with the same targets in mind as Scenario 2. The key change is that this time there is a second Airport holding area on the model. This would mean that the airport holding area can now hold two trains at a time without them interfering with one another.

7.4.1 Schedule

The difference between Scenario 2 and Scenario 3 with regards to freight trips is the number that can be placed onto the system overnight. As there is a second holding area now available at the airport for freight trains, the original requirement that each train be separated by 8 min can be halved down to 4 min. This means that 15 trains can now be placed on the track every hour overnight as opposed to the seven trains that could be used in Scenario 2.

There are now a total of 163 freight trips every day to the airport; timetable is shown in Table 10. The 163 freight trips equate to the maximum transfer of 2445 containers a day or 61,125 bags.

7.4.2 The Model

The key changes to the simulation model from Scenario 2 are that there is a new schedule sheet placing many more freight trains onto the system and there is a second 'Airport (Freight Holding)' activity.

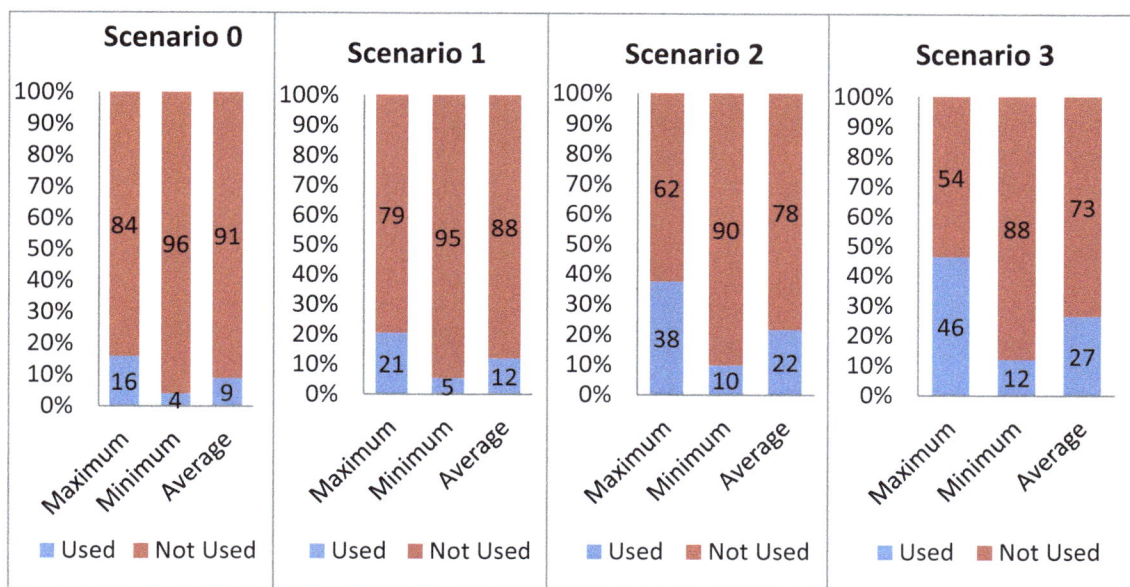

Fig. 27 Min, max and average utilisations

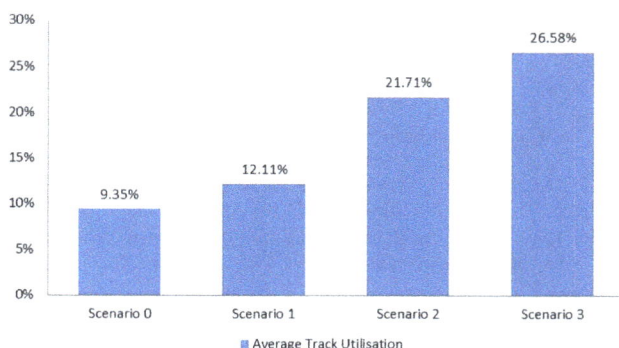

Fig. 28 Overall utilisations

For previous scenarios freight trains would move to the airport holding from Activity 44, now these trains move to a new splitting activity, Activity 45. At Activity 45, these freight trains are split using the circulate option, similar to a deck of cards being dealt, this means that the Activity exit will swap each time a train passes (Fig. 26).

Validation techniques are the same as used in Scenario 2.

8 Track Utilisations of Proposed Designs

The utilisations for each section of the line using each scenario can be seen in Table 11, Figs. 27 and 28.

There is an average difference of approximately 3 % between Scenario 0 and Scenario 1. Scenario 2 however almost doubles the utilisation of Scenario 1 and Scenario 3 further improves the work done by scenario 2 with approximately an additional 5 % utilisation. Overall, the utilisation of Scenario 3 is nearly three times as much as that of the passenger only system.

9 Conclusions

This study presents a new design for an innovative baggage handling system which transfers baggage between a collection point, Haymarket, situated in the city centre of Newcastle-Upon-Tyne and Newcastle Airport. There are also two additional solutions which saturate the system with freight trains, resulting in greater utilisations of the metro tracks.

Validations have shown that freight trains can indeed be added to the current metro system to enable freight to be transferred between Haymarket and the airport. It was also found that this system could have a capacity large enough to accommodate every single passenger on every plane leaving from Newcastle Airport on an average day.

Exploring the opportunity to use the metro line overnight for the transit of freight trains, it was found that restrictions in scheduling freight trains in parallel to passenger trains could be avoided, leading to shorter times between freight trips and ultimately more trains per hour. A proposed system boasting a large capacity of 61,125 bags a day was validated and analysed.

10 Further Work

The proposed system could possibly be improved upon by looking at changing passenger train schedules to find an optimal schedule where additional freight trains could be run during the day. The simulation models were all run under optimal conditions, it would also be recommended that these models be run with imperfections incorporated such as random delays and inefficiencies to see how the systems behave. Finally, it might be of interest to perform cost analyses on the three different system proposals to see if the additional costs of Scenarios 1 and 2 can be justified by their greatly increased capacities.

References

1. DPD (2014) International tariff guide. http://www.dpd.co.uk/content/product-services/dpd_international_tariff_guide-opt.pdf. Accessed 13 Mar 2014
2. Passenger Focus (2011) Coach passenger needs and experiences
3. Marinov M, Viegas J (2011) A mesoscopic simulation modelling methodology for analyzing and evaluating freight train operations in a rail network. Simul Model Pract Theory 19(1):516–539
4. Marinov M, Motraghi A (2012) Analysis of urban freight by rail using event based simulation. Simul Model Pract Theory 25:73–89
5. Leachman R, Dessouky M (1995) A simulation modelling methodology for analyzing large complex rail network. Simulation 65(2):131–142
6. Robuste F (1992) Analysis of baggage sorting schemes for containerized aircraft. Transp Res Part A 26(1):75–92
7. Reis V, Meier F, Pace G, Palacin R (2013) Rail and multi-modal transport. Res Transp Econ 41(1):17–30
8. U.S Travel Association (2014) U.S travel answer sheet. http://www.ustravel.org/sites/default/files/page/2009/09/US_Travel_AnswerSheet_March_2014.pdf. Accessed 10 Apr 2014
9. AirFleets (2013) EasyJet. http://www.airfleets.net/flottecie/EasyJet.htm. Accessed 26 Mar 2014
10. Airbus (2014) A319 specifications. http://www.airbus.com/aircraftfamilies/passengeraircraft/a320family/a319/specifications/. Accessed 26 Mar 2014
11. Nexus Metro (2014) Metro timetable Haymarket. http://www.nexus.org.uk/sites/default/files/metro/stations/MT1312.A4.HAY_.pdf. Accessed 20 Mar 2014
12. Newcastle Airport (2014) Departures and arrivals. http://www.newcastleairport.com/arrivals-departures. Accessed 02 Apr 2014
13. Yodel (2014) Quick quote. http://www.yodeldirect.co.uk/. Accessed 13 Mar 2014
14. UPS (2014) Shipment rates. http://www.upscontentcentre.com/html/uk. Accessed 13 Apr 2014
15. Parcel Force (2014) UK and international prices. http://www.parcelforce.com/sites/default/files/UK%20and%20International%20Retail%20Tariff%20Prices%20FINAL.pdf. Accessed 13 Mar 2014
16. DHL (2014) Pricing guide 2014. http://www.dhl.co.uk/content/dam/downloads/gb/express/shipping/rate_guides/dhl_express_pricing_guide_2014.pdf. Accessed 13 Mar 2014
17. Triple Eagle (n.d.) Air container specs. http://www.triple-eagle.com/files/Air-container-specs_Pdf.pdf. Accessed 10 May 2014
18. BEUMER GROUP (2014) Airport baggage handling systems. http://www.beumergroup.com/products/airport-baggage-handling-systems/high-speed-transportation-systems/beumer-autoverr/. Accessed 18 Apr 2014
19. Vanderlande (2014) Products and solutions. http://www.vanderlande.com/en/Baggage-Handling/Products-and-Solutions/Early-Bag-Storage/BAGSTORE.htm. Accessed 19 Apr 2014
20. Grube P, Nunez F, Cipriani A (2011) An event-driven simulator for multi-line metro systems and its application to Santiago de Chile metropolitan rail network. Simul Model Pract Theory 19(1):393–405

16

Development of a Behavior-Based Passenger Flow Assignment Model for Urban Rail Transit in Section Interruption Circumstance

Jing Teng[1] · Wang-Rui Liu[1]

Abstract At present, the urban rail transit (URT) system has achieved network operation in many major cities of China. But, little attention has been given to the vulnerability of the URT system. The purpose of this study is to assign the passenger flow under the condition of section interruption in URT system. Two surveys (a passenger behavior survey and a stated preference survey) were conducted and a multinomial logit model was developed. The results show that although the first choice of passengers in emergency situation is to stay in URT system by a circuitous way, more than half of the respondents express interest in the temporary shuttle bus. For the temporary shuttle bus, the sensitivity analysis show that the relative speed is more important than crowding degree for passengers. The significant variables mostly fall in the personal attributes such as income, gender, age, etc. The impacts of trip feature factors are similar to the previous research in normal situations. These results provide basic support for passenger flow assignment at the shuttle bus level and reducing the risk of crowding at some special stations. Moreover, it is also good for reducing passenger delay and recovering the trip.

Keywords Urban rail transit · Section interruption · Passenger flow assignment · SP survey · Multinomial logit model

✉ Wang-Rui Liu
  liuwangrui@sina.com

[1] Key Laboratory of Road and Traffic Engineering, Ministry Education, Tongji University, 4800 Cao'An Road, JiaDing District, Shanghai, China

Editor: Marin Marinov

1 Introduction

URT, with its large capacity and high reliability, is gradually developing to be the favorite traffic mode in metropolis. However, for the concentrated passenger flow, limited space, closed running, and high capacity characteristics, once an emergent event happens in URT system and operation is interrupted, will come out in a huge amount of delay and spread rapidly in network.

According to the process of emergent measures in URT system, the organization of passenger flow can be divided into two steps—evacuation and recovery. The former is at the beginning of emergent events, trying to evacuate passengers to a safe place as fast as possible; the latter is active after the evacuation step and all potential risks will be cleared. During the recovery phase, the common practice is to maintain the integrated service of OD trips, using other traffic modes, usually bus transit, to replace the interrupting section, which is the research question in this paper.

The basic need in emergency situation is to maintain the accessibility of rail lines and the integrated service of OD trips; that is to say, some indispensable measures need to be proposed on transport organization level. Passenger flow assignment under emergent condition provides essential support for emergency decision, such as adjusting train routings and departure intervals, scheduling plans of shuttle buses (i.e., the temporary shuttle buses), etc.

Quite a few studies [5–8] have focused on passenger flow assignment in URT network. User-equilibrium model and discrete choice model are the most commonly used methods in passenger flow assignment. The factors in the generalized cost function consist of rail-ride travel time and passengers' stay time at a transfer or passing station. An amplification factor is usually introduced to describe transfer time. Though some different disposal ways were adopted for these factors in previous papers, they were not considered from passengers' standpoint. For example, in-vehicle crowding degree is hardly considered in previous papers for it is hard to be measured in practice. In addition, comparing with the variable of the number of stations, most papers prefer travel time. Although travel time seems to be more accurate and objective, the variable of the number of stations has its own advantages, such as, the number of stations between the interrupting section and the origin station, the number of stations between the interrupting section and the destination station. These variables are probably more intuitive and convenient for passengers to choose the best route in emergency situations.

In recent years, some researches [9–11] pay attention to passenger flow assignment under emergent condition in URT network. Delay on passengers and the affected area are considered. The number of affected passengers is also calculated for every affected station. However, these researches are still in the theoretical stage, and many practical factors are out of consideration, which will be hard to provide enough support for the emergent decision makers.

This paper selects factors, in the utility function, from both the passenger behavior survey and previous researches. These factors are brought in the state preference (SP) survey, including the crowding degree, relative speed, the number of stations in the shortest route, etc. In the SP survey, a shuttle bus passageway replacing the interrupting section is assumed to establish an integrated URT network. Then, a discrete choice model is built and analyzed. The objective of this paper is to assign passenger flow under emergent condition and to provide support for the emergent decision makers.

The paper is organized as follows: first, previous researches on passenger flow assignment in normal and emergent situations in URT system are reviewed, followed by a detailed explanation of the methodology and data preparation, including passenger behavior survey, SP survey, and route choice model. Then, with the data collected from a passenger behavior survey and a SP survey in Shanghai URT system, the route choice model is calibrated and analyzed.

2 Literature Review

Several existing reports [1, 2] presented the importance of using the bus transit as the connecting mode when railway or metro emergent events happen. Between the URT and bus systems, the synergy of timetables was the key to realize the intermodal transportation in emergency. Shanghai subway system stipulated [3] the shuttle buses need to be deployed when the delay of the system is more than 30 min. Beijing subway system also stipulated [4] traffic

control need to be implemented and the shuttle bus need to be deployed in emergency. These researches show that the bus is a crucial traffic mode in emergent event of URT system, and the passenger flow assignment is the basis of all emergent decisions.

Previous researches on passenger flow assignment in URT system can be divided into two branches: equilibrium assignment model and utility theory based non-equilibrium assignment model. The papers used travel time, mileage or travel fare in generalized cost function to analyze the passenger's route choice.

Zhu [5] used the method of successive averages to solve the stochastic user-equilibrium problem. He/she described the impact of congestion on passengers' route choices. A generalized cost function with in-vehicle congestion and an amplification factor for transfer time was set up. With the K-th shortest path algorithm to generate the choice set, a route choice model was introduced to perform the stochastic network. Comparing with those practical methods used in China, this model computed more precisely. However, the passenger behavior was not considered. In addition, the effective paths were restricted by a constant or a parameter relating to the shortest path. As different OD pairs have their own transfer times and mileage, it is difficult to practically analyze the effective paths using this criterion. This model was inspired by the flow assignment theory for road traffic, but several features needed to be noticed when this model was applied into URT network: (i) The generalized travel cost is affected considerably by in-vehicle congestion rather than vehicle-to-vehicle congestion; (ii) The transfer time at the transfer station influences the route choices of rail passengers considerably.

Si [6] proposed a modified logit-based passenger flow assignment model, using automatic fare collection (AFC) data as basic data. This paper also adopted an amplification factor for transfer time. In addition, transfer time was considered separately. Compared with all-or-nothing assignment method, this model was more practical and reasonable, according to the Beijing AFC data. Si [7] made a survey about the passenger travel behavior in Beijing URT system. Some parameters were modified and the general framework of passenger flow assignment was presented in this paper. However, the generalized cost functions in these two researches were nearly the same to equilibrium assignment model which considered travel time, transfer time, and transfer times only. Some factors, like crowding level which can reasonably reflect travelers' behaviors, had not been included. Meanwhile, the comparison with all-or-nothing assignment method seemed unpersuasive.

In recent years, a new thought was proposed with URT system's own features. Zhou [8] narrowed down the feasible paths chosen by one passenger based on the train timetable and AFC record data. Assuming no extra activities in destination station, the paper deduced the unique train chosen in the OD pair by the exit time. Then, the route was deduced until the origin station. The result of this model was more accurate and realistic, confirming the train number and stations passed in each OD pair. But, the algorithm was too complicated even in computer technology. In addition, the assumption of no extra activities needed to be improved.

In emergent circumstance, the accurate result and simple algorithm of the model were both important. Too complicated algorithm was not suitable for the quick response of emergent decision. Several researches were focused on emergent condition while most of them dedicated to politics and emergent strategies. And few papers were concerned with passenger flow assignment.

Hong [9] divided emergent passenger flow into three parts: delay passenger flow, detour passenger flow, and loss passenger flow. With graph theory, the paper built evaluation models for all kinds of influenced passenger flow in every impacted station using historical OD matrix in URT network. In addition, with those models, influenced, detour, congested, and loss passenger flow volume with lost time of passenger flow could be calculated out in different stations as time went by. However, the shortest path was only the considered route in route choice set of incidence matrix, which meant that no additional choice routes were provided for passengers and the passenger's behaviors were not considered. In addition, the impacted passenger flow volume calculated in this paper was separated by every station and the flow had no direction. Furthermore, a detailed shuttle bus plan was hardly made in recovery phase by the results.

Pan [10] established a dynamic assignment model of the passenger flow in emergency. The paper analyzed the structure and characteristics of emergent passenger flow and URT network. Although passengers' features were discussed, they were not used in the assignment model. In addition, the basic data, the OD matrix, was not analyzed. Therefore, some key links about passengers' behavior and choice were not sufficiently studied. Liu [11] made the plans of shuttle bus in two conditions, sufficient resource and insufficient resource, which were based on the number of shuttle buses. And, a modified logit model was developed for passenger flow assignment. However, this paper failed to show the results of SP survey and the support to modified logit model was insufficient.

In summary, passenger flow assignment on normality in URT network was studied extensively, and several researches explored the emergent circumstance. But they were still on preliminary stages, and more attention needs to be paid to the passenger behavior and preference.

3 Methodology

3.1 Stated Preference Survey

SP survey is to achieve the subjective preference of the respondents in different assuming conditions. SP survey originates from economics, as a market research tool to understand the consumers' acceptance to different products or service [12]. In the late 1970s, SP survey was introduced to analyze the traffic issues in UK. Until 1983, SP survey was first used in traffic mode choice of citizens by Louviere and Hensher [13]. Nowadays, SP survey has been applied in the study of travel mode choice, parking choice, route choice, etc. [14].

Overall design is used in this scene design, and to combine all the levels of every variable. For example, a study has m factors and every factor has n levels. The total number of scenes is n^m. This approach is able to acquire the comprehensive information and the conclusion is relative accuracy.

3.2 Multinomial Logit Model

Generally, if passenger's perceived travel cost is represented as a random variable consisting of a deterministic component C_k^{rs} and an additive random error. Accordingly, passenger's path choice is actually a probability. The probability of a given path chosen by a passenger can be defined by calculating the probability that the perceived travel cost on such path is lower than that of all other alternatives. Obviously, the choice probability is determined by both of the distribution of random error term and the expected travel cost C_k^{rs} [6]. The multinomial logit model is as the following:

$$P_k^{rs} = \frac{\exp(-\theta C_k^{rs})}{\sum_m \exp(-\theta C_m^{rs})}, k \in K_{rs}, \tag{1}$$

where P_k^{rs} is the choice probability of effective path k ($k \in K_{rs}$) between the OD pair r–s; θ is the dispersion parameter, which is inversely proportional to the standard error of the distribution of the perceived path travel cost [15].

4 Data Preparation

Two surveys were made for this paper: a passenger behavior survey and a SP survey. The passenger behavior survey, as a pre-survey, provided some important basis to the SP survey such as the expected speed and crowding degree of the shuttle bus. The estimation result of models was determined by the SP survey.

4.1 Passenger Behavior Survey

The target of this survey was to understand the demand characteristics of the passenger flow in URT emergency.

The survey sample includes 266 men, accounting for 53.2 % of the total (500), and 234 women, accounting for 46.8 %. In addition, 92.4 % of people surveyed are young (under 30, 43.8 %) or middle-aged (30–60, 48.6 %), respectively, whereas 7.6 % are seniors (above 60). As long as occupation is concerned, there are officials, businessmen, students, temporary employments, etc. These samples are representative and can be used in the study of the demand features of the passenger flow in URT emergency.

Figure 1a shows that 58 % of respondents use URT every day and 88 % of subjects use it at least three times a week. It indicates URT plays a non-substitutable role in citizens' daily life. Figure 1b shows the main purpose is commute (32 %) which is the rigid demand. In addition, most of business (15 %) and journey (11 %) are hardly to be adjusted to other modes as well as some of family visit (10 %), entertainment (11 %), and shopping (17 %). That means as URT emergent events happen, it will have a huge impact.

Figure 2a shows 87 % respondents consider the bus, either the temporary shuttle buses (54 %) or the existed bus lines nearby (33 %), should give support to URT to carry passengers before URT restores to normal service. The ratio of acceptable travel time by shuttle bus over the travel time by URT is shown in Fig. 2b. The percentages of less than two times and 1.5 times are 88 and 64 %. The average travel speed of URT is 3–4 times higher than the bus [16]. Therefore, it is quite challenging to satisfy it.

Figure 2c–d focuses on other two key factors: waiting time and crowding degree. 57 % respondents expect that

the acceptable waiting time of the shuttle bus is 10–20 min. And 20 % expect less than 10 min. Although the average waiting time of the key bus line is 5 min in peak hour [16], it is hard to reach the expectancy under emergent circumstance. Figure 2d describes that some people surveyed (13 %) pay attention to available seats, but most respondents (75 %) just care for the available standing and activity space.

The survey results present in Fig. 2 provide empirical evidence for understanding which is the best alternative traffic mode of passengers' expectation. Moreover, the crowding degree is considered along with the speed of shuttle bus and URT. It provides important guidance on how to set crowding degree and relative speed in SP survey. In addition, the acceptable waiting time of the shuttle bus offers support to shuttle bus organization.

4.2 Stated Preference Survey

A SP survey was prepared for model calibration. This survey was divided into three parts: travel characteristics, scene selection, and personal attributes.

Travel characteristics included travel purpose, type of the luggage, origin station, and destination station.

Scene selection was retrieved by the crowding degree in the shuttle bus and relative speed. This survey provided two levels of in-vehicle congestion and three levels of relative speed. According to passenger behavior survey, the paper assumed the crowding degree in URT was crowded and two levels of crowding in the shuttle bus were set: abundant standing space and crowed. In addition, as the average travel speed of URT is 3–4 times higher than the bus and 88 % of the respondents in passenger behavior survey expect the relative speed is 1/2 or more of the URT,

Fig. 1 Frequency and purpose of using URT system

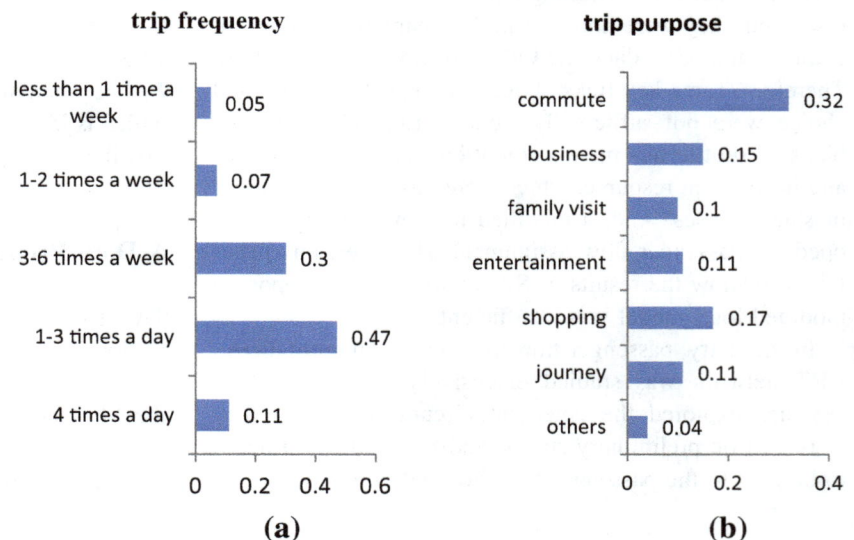

Fig. 2 Traffic option in URT emergent events

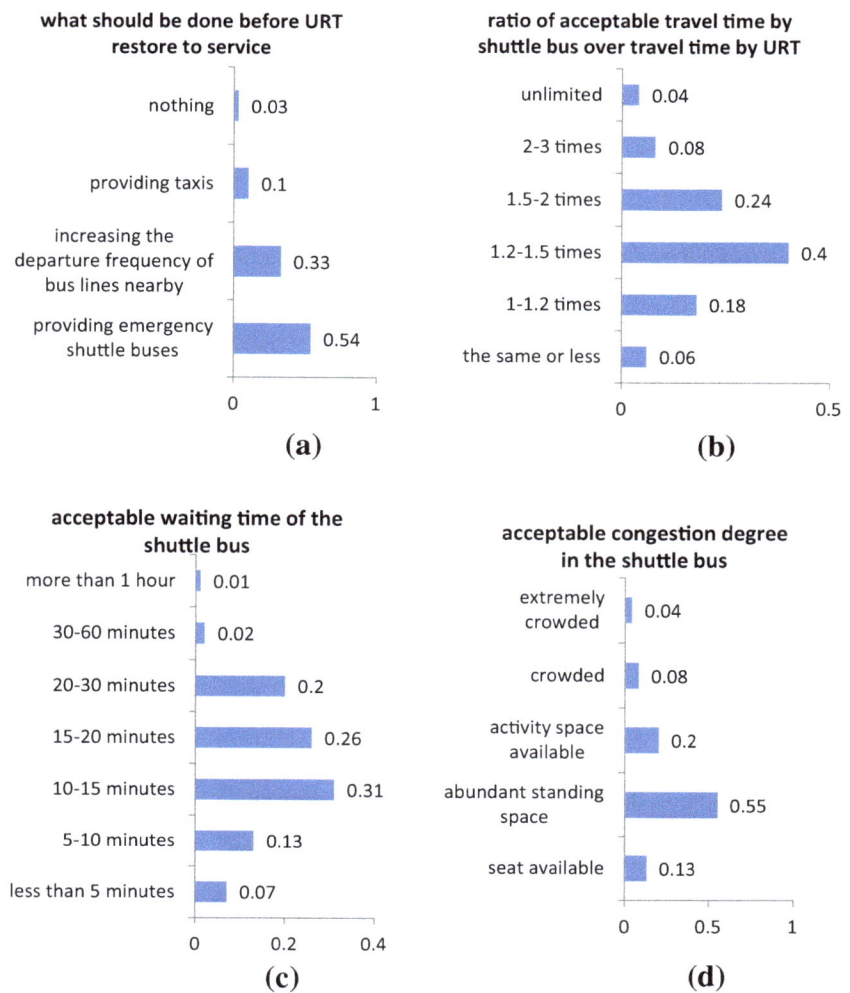

what should be done before URT restore to service

- nothing — 0.03
- providing taxis — 0.1
- increasing the departure frequency of bus lines nearby — 0.33
- providing emergency shuttle buses — 0.54

(a)

ratio of acceptable travel time by shuttle bus over travel time by URT

- unlimited — 0.04
- 2-3 times — 0.08
- 1.5-2 times — 0.24
- 1.2-1.5 times — 0.4
- 1-1.2 times — 0.18
- the same or less — 0.06

(b)

acceptable waiting time of the shuttle bus

- more than 1 hour — 0.01
- 30-60 minutes — 0.02
- 20-30 minutes — 0.2
- 15-20 minutes — 0.26
- 10-15 minutes — 0.31
- 5-10 minutes — 0.13
- less than 5 minutes — 0.07

(c)

acceptable congestion degree in the shuttle bus

- extremely crowded — 0.04
- crowded — 0.08
- activity space available — 0.2
- abundant standing space — 0.55
- seat available — 0.13

(d)

three levels of relative speed were set: 1/2, 1/4, 1/6. The combination of crowding degree and relative speed differs among scenes. Totally, six scenes are provided.

Personal attributes included passenger's gender, age, and income per month. Figure 3 shows the part of Shanghai Metro system.

Meanwhile, the paper supposed the section from Shanghai Railway station to People Square Station of Metro Line 1 was interrupted, as shown in Fig. 3. Then the URT would run on two part routes [17]: (1) departing from Fujin Road to Shanghai Railway Station and then back to Fujin Road; (2) departing from Xinzhuang to People Square and then back to Xinzhuang. The survey stations were selected from North Zhongshan Road to Gongkang Road, including six stations. Moreover, the only direction concerned in this survey was from Fujin Road to Xinzhuang.

Totally, 300 passengers were surveyed, 50 in each survey station, and 545 effective records were collected. The core of this survey is about the preference of passengers'

route choice. The participants were asked to make a selection in a set of four pre-defined traffic modes which decided the travel path. These alternative traffic modes and the corresponding ODs are used to study passengers' trade-off between relative speed and crowding degree. The information provided includes the location of interrupting section, crowding degree in the shuttle bus, and relative speed along these traffic modes. Four choices were prepared and described as

- A circuitous way to the destination station by URT; (mode 1: metro)
- Traveling from Shanghai Railway Station to a certain station by the shuttle bus, then transferring to URT and finishing the trip; (mode 2: metrobus1)
- Traveling from Shanghai Railway Station to the destination station by the shuttle bus; (mode 3: metrobus2)
- Using other traffic modes, such as taxi. (mode 4: Other)

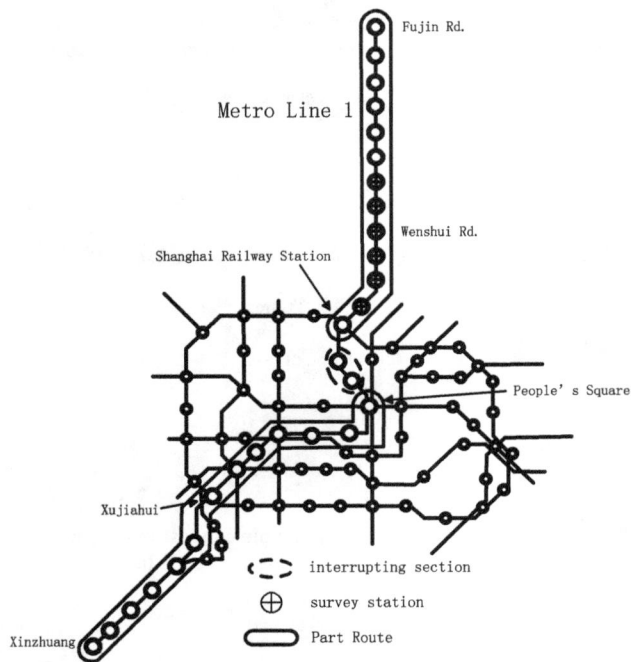

Fig. 3 Location of interrupting section and the organization of part routes

Figure 4 shows the proportion of the different passengers (classified by age, gender, and income) and the different trip purposes among these four alternative modes. It can be seen that, the proportions of passengers choosing mode 1 are consistently the highest, followed by mode 4, mode 3, and mode 2 across all passengers and trip purposes. In addition, in most cases, more passengers prefer mode 3 than mode 2, indicating that they would like to trade the one mode way with less transfer between different traffic modes. The choice proportions vary at a high level in Fig. 4a and c. The passengers over 60 years old show high interests in mode 1 for the following two possible reasons: (1) senior passengers travel free of charge by public transit; (2) a more complicated way and an unfamiliar transfer place may be a big challenge to the seniors. In Fig. 4c, it is reasonable to understand high-income groups prefer mode 4, such as taxi, than the URT and the shuttle bus.

Meanwhile, Fig. 4e shows the crowding degree getting heavy in the shuttle bus, the passengers will abandon mode 2 and mode 3, choosing mode 1 and mode 4. However, the percentage of each choice varies slightly. That is the passenger choice is affected by the crowding degree in the shuttle bus but it is not the key one, in the emergent circumstance. Figure 4f shows that as the shuttle bus speed slows down, the percentage of mode 1 rises. Moreover, the percentages of mode 2 and 3 drop rapidly when the relative speed changes from 1/4 to 1/6. As said in scene selection, 1/4 is the relative speed in normal state. It means that the passengers cannot bear that the shuttle bus speed is slower than the normal speed.

Another important point needs to be focused. Almost 8.4 % of the total effective records choose mode 2. All of the "certain stations" are transfer stations with People Square selected by 75.6 %, followed by East Nanjing Rd., Xujiahui, Changshu Rd., Jiangsu Rd., etc. So, the destination of mode 2 can be narrowed down in the transfer stations and the station of maximum demand is at the first transfer station after the interrupting section.

The surveys' results in Figs. 1, 2, and 4 provide empirical evidence for understanding which factors are included and how they are traded off among each other, when passengers calculate a "utility" for each potential traffic mode, rank them, and make their final choices. Such results present important guidance on how the utility functions and corresponding logit models for traffic mode choices should be constructed. These models are introduced in the following section.

5 Estimation Results

5.1 Variable

According to the conducted survey, the passengers' route choice is influenced by crowding degree in the shuttle bus and relative speed. Other travel characteristics and personal attributes are also considered in the choice model. They are gender, age, income, luggage, purpose, SDB (relative speed), and YJD (the crowding degree in the shuttle bus). Furthermore, several additional travel characteristics are prepared for the choice model based on the OD information. They are

- MRSQ:

$$MRSQ = \frac{\text{The number of stations from destination station to People Square}}{\text{The number of stations from origin station to Shanghai Railway Station}}$$

- SRZB:

$$SRZB = \frac{\text{The number of stations in shortest route after interruption occurence}}{\text{The number of stations in shortest route before interruption occurence}}$$

- HCSC:

HCSC = Transfer times in shortest route after interruption occurrence − transfer times in shortest route before interruption occurrence

- COST: the meaning of this factor varies in every alternative choice. (1) The pricing method of Shanghai URT system is restricted by the origin and destination stations information, without regarding which stations the route passed. That is, the cost in emergency is the same as in normal state. (2) As the shuttle bus replaces the interrupting section, passengers must leave and reenter the URT system. The cost consists of four parts: (a) origin station to Shanghai Railway Station; (b) Shanghai Railway Station to a certain subway station by the shuttle bus. This service is free based on the previous emergent situations. (c) A certain station to destination station. (d) ¥1 privilege is given at reentering the URT system. (3) The only cost passengers need to pay is from origin station to Shanghai Railway Station. (4) Taxi fare is used as the cost of mode 4.

5.2 Utility Function

5.2.1 Initial Utility Function

The paper sets mode 4 as the base case. And the utility functions are described as follows:

$$U(\text{metro}) = a_1 + b_{11} \times \text{gender} + b_{12} \times \text{age} \\ + b_{13} \times \text{income} + b_{14} \times \text{luggage} + b_{151} \times \text{purpose1} \\ + b_{152} \times \text{purpose2} + b_{153} \times \text{purpose3} \\ + b_{154} \times \text{purpose4} + b_{155} \times \text{purpose5} + b_{16} \times \text{mrsq} \\ + b_{17} \times \text{srzb} + b_{18} \times \text{hcsc} + b_{19} \times \text{sdb} + b_1 \times \text{cost} + \varepsilon_1;$$

$$U(\text{metrobus1}) = a_2 + b_{21} \times \text{gender} + b_{22} \times \text{age} \\ + b_{23} \times \text{income} + b_{24} \times \text{luggage} + b_{251} \times \text{purpose1} \\ + b_{252} \times \text{purpose2} + b_{253} \times \text{purpose3} \\ + b_{254} \times \text{purpose4} + b_{255} \times \text{purpose5} + b_{26} \times \text{sdb} \\ + b_{27} \times \text{yjd} + b_1 \times \text{cost} + \varepsilon_2;$$

$$U(\text{metrobus2}) = a_3 + b_{31} \times \text{gender} + b_{32} \times \text{age} \\ + b_{33} \times \text{income} + b_{34} \times \text{luggage} + b_{351} \times \text{purpose1} \\ + b_{352} \times \text{purpose2} + b_{353} \times \text{purpose3} \\ + b_{354} \times \text{purpose4} + b_{355} \times \text{purpose5} \\ + b_{36} \times \text{sdb} + b_{37} \times \text{yjd} + b_1 \times \text{cost} + \varepsilon_3;$$

$$U(\text{other}) = b_1 \times \text{cost} + \varepsilon_4,$$

where gender—0, male; 1, female;

Income—0, below ¥2500/month; 1, ¥2500–¥4000/month; 2, ¥4001–¥7000/month; 3, ¥7001–¥10,000/month; 4, beyond ¥10,000/month;

Luggage—0, no luggage; 1, a briefcase; 2, a trunk; 3, two suitcases or more;

Purpose 1 to 5—commute, business, visit, shopping, other;

Sdb—relative speed;

Yjd—the crowding degree in the shuttle bus.

5.2.2 Significance Testing and Modified Utility Function

Significance tests were conducted for all factors and a factor is considered significant P value < 0.1. Some results of significance tests are shown in Table 1.

There are two points that need to be emphasized. (1) The factors, purpose2 (business), purpose4 (shopping), and purpose5 (other), are all insignificant in utility functions of U(metro), U(metrobus1), and U(metrobus2). These factors will be abandoned from the choice model. (2) SDB and YJD are also insignificant in utility functions. However the six scenes, the organization of the shuttle bus and the basic support in emergency are all decided by these two factors. So, SDB and YJD will be reserved.

The modified utility functions are shown as follows:

$$U(\text{metro}) = a_1 + b_{11} \times \text{gender} + b_{12} \times \text{age} \\ + b_{13} \times \text{income} + b_{14} \times \text{luggage} + b_{151} \times \text{purpose1} \\ + b_{153} \times \text{purpose3} + b_{16} \times \text{mrsq} + b_{17} \times \text{srzb} \\ + b_{18} \times \text{hcsc} + b_{19} \times \text{sdb} + b_1 \times \text{cost} + \varepsilon_1;$$

$$U(\text{metrobus1}) = a_2 + b_{21} \times \text{gender} + b_{22} \times \text{age} \\ + b_{23} \times \text{income} + b_{24} \times \text{luggage} + b_{251} \times \text{purpose1} \\ + b_{253} \times \text{purpose3} + b_{26} \times \text{sdb} + b_{27} \times \text{yjd} \\ + b_1 \times \text{cost} + \varepsilon_2;$$

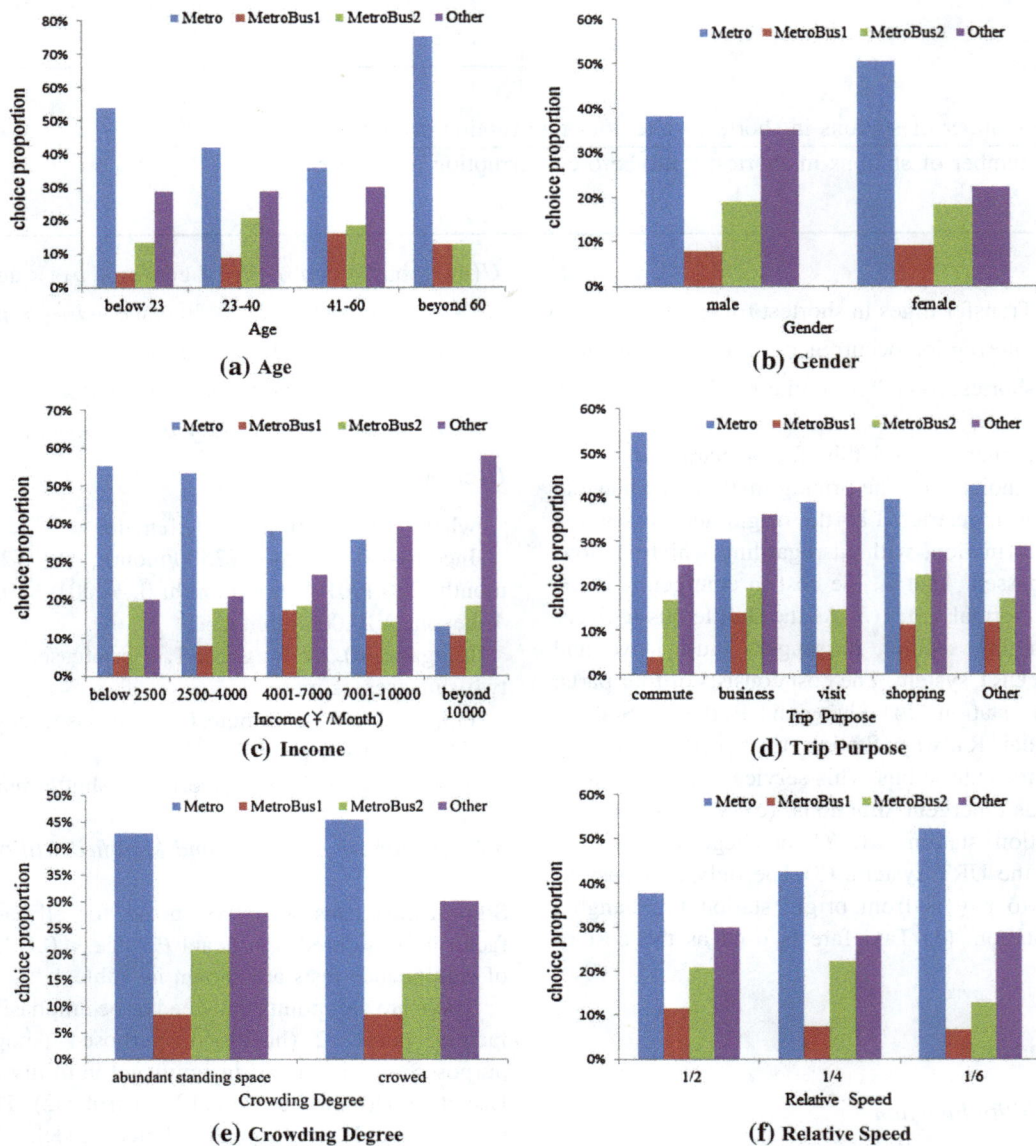

Fig. 4 Analysis of traffic mode choice

$U(\text{metrobus2}) = a_3 + b_{31} \times \text{gender} + b_{32} \times \text{age}$
$\quad + b_{33} \times \text{income} + b_{34} \times \text{luggage} + b_{351} \times \text{purpose1}$
$\quad + b_{353} \times \text{purpose3} + b_{36} \times \text{sdb} + b_{37} \times \text{yjd}$
$\quad + b_1 \times \text{cost} + \varepsilon_3;$

$U(\text{other}) = b_1 \times \text{cost} + \varepsilon_4.$

5.2.3 Fitting Result

The fitting results for the choice model are shown in Table 2.

The fitted utility functions are shown as follows:

$U(\text{metro}) = 2.1716 + 0.5972 \times \text{gender} + 0.0238 \times \text{age}$
$\quad - 0.4897 \times \text{income} - 0.1862 \times \text{luggage}$
$\quad + 0.3271 \times \text{purpose1} - 0.6257 \times \text{purpose3}$
$\quad + 0.0161 \times \text{mrsq} - 1.0162 \times \text{srzb} - 0.0542 \times \text{hcsc}$
$\quad - 1.2603 \times \text{sdb} - 0.0073 \times \text{cost} + \varepsilon_1;$

$U(\text{metrobus1}) = -4.6176 + 0.5136 \times \text{gender}$
$\quad + 0.0434 \times \text{age} - 0.1450 \times \text{income}$
$\quad - 1.0391 \times \text{luggage} - 0.9707 \times \text{purpose1}$
$\quad - 1.5144 \times \text{purpose3} + 1.5529 \times \text{sdb} - 0.1201 \times \text{yjd}$
$\quad - 0.0073 \times \text{cost} + \varepsilon_2;$

Table 1 Some results of significance testing

Factor		P value	Factor		P value
Purpose2	b_{152}	0.1460	Purpose 4	b_{154}	0.6668
	b_{252}	0.5795		b_{254}	0.1179
	b_{352}	0.7032		b_{354}	0.1565
Purpose5	b_{155}	0.9416	SDB	b_{19}	0.1736
	b_{255}	0.2977		b_{26}	0.1701
	b_{355}	0.7925		b_{36}	0.3836
YJD	b_{27}	0.8204			
	b_{37}	0.1420			

$$U(\text{metrobus2}) = -0.2216 + 0.3566 \times \text{gender}$$
$$+ 0.0171 \times \text{age} - 0.3206 \times \text{income}$$
$$- 0.1156 \times \text{luggage} - 0.0976 \times \text{purpose1}$$
$$- 0.7086 \times \text{purpose3} + 0.7228 \times \text{sdb} - 0.3724 \times \text{yjd}$$
$$- 0.0037 \times \text{cost} + \varepsilon_3;$$

$$U(\text{other}) = -0.0073 \times \text{cost} + \varepsilon_4.$$

Some signs of coefficients in the estimated multinomial logit model have important meaning and can give basic support to decision makers. The coefficients of gender in the three utility functions are 0.5972, 0.5136, and 0.3566. Compared with female, male prefers changing the malfunctioning URT system to a new traffic mode. Especially, male shows less interest in mode 1. Well, female seems more conservative and would like to continue in the former mode. The signs of income in utility functions are negative. As income increases, people are partial to abandon the URT to the other traffic modes, such as taxi. The coefficients of luggage in three utility functions are −0.1862, −1.0391, and −0.1156. Public transit possibly leaves a crowded and unsafe impression to passengers, so other traffic modes are more attractive in the URT emergency to the passengers with luggage. Moreover, the passengers with luggage mostly give up $U(\text{metrobus1})$ first because the transfer between the URT and the shuttle bus is tough.

Based on the sign of MRSQ (+), the higher the percentage of distance left in the trip, the more interest passengers show to the mode of metro. As for SRZB (−), the longer the detouring distance is, the low percentage of mode 1 passengers will choose. The HCSC (−) means the transfer times in URT show a negative effect to $U(\text{metro})$.

The factors of SDB and YJD both state passengers would show more interests to the shuttle bus by higher relative speed and lower crowding. Meanwhile, the SDB is more important in metrobus1 than metrobus2 for $b_{26}(1.5529) > b_{36}(0.7228)$. The demand of YJD in mode 3 is much more vital than in mode 2 because the distance in bus in mode 3 is mostly longer than mode 2.

Table 2 Estimation results of the choice model

Variables	Coef.	S.E.	P value
Cost	−0.0073	0.0038	0.0539
U(metro)			
Constant	2.1716	0.8285	0.0088
Gender	0.5972	0.2349	0.0110
Age	0.0238	0.0115	0.0380
Income	−0.4897	0.1011	0.0000
Luggage	−0.1862	0.2221	0.4017
Purpose1	0.3271	0.2530	0.1961
Purpose3	−0.6257	0.3651	0.0865
MRSQ	0.0161	0.0558	0.7729
SRZB	−1.0162	0.5365	0.0582
HCSC	−0.0542	0.1392	0.6969
SDB	−1.2603	0.8063	0.1180
U(metrobus1)			
Constant	−4.6176	1.0925	0.0000
Gender	0.5136	0.3636	0.1577
Age	0.0434	0.0159	0.0063
Income	−0.1450	0.1542	0.3472
Luggage	−1.0391	0.3511	0.0031
Purpose1	−0.9707	0.4680	0.0381
Purpose3	−1.5144	0.6723	0.0243
SDB	1.5529	1.2109	0.1997
YJD	−0.1201	0.3256	0.7122
U(metrobus2)			
Constant	−0.2216	0.7430	0.7655
Gender	0.3566	0.2793	0.2017
Age	0.0171	0.0136	0.2093
Income	−0.3206	0.1175	0.0064
Luggage	−0.1156	0.2615	0.6585
Purpose1	−0.0976	0.3071	0.7507
Purpose3	−0.7086	0.4513	0.1164
SDB	0.7228	0.9211	0.4326
YJD	−0.3724	0.2299	0.1053
Summary statistics			
Number of observations			546
Log likelihood function			−608.2048
R^2			0.0782

Most variables are significant (P value < 0.1) enough. They are gender, age, income, purpose3, SRZB, cost, luggage, and purpose1. The most significant (P value < 0.01) variables are income (P value $= 0.0000$ in mode 1 and 0.0064 in mode 3), age (P value $= 0.0063$ in mode 2), and luggage (P value $= 0.0031$ in mode 2). Some necessary measures need to be paid attention to are (a) Abundant subway staffs need to be prepared in emergency to help the seniors and the temporary passages must be easy to use. (b) A special aisle and some big volume of security check machines need to be provided for

Table 3 Values of variables

Variable	Gender	Age	Income	Luggage	Purpose1	Purpose3
Value	0	30	2	0	1	0
Variable	MRSQ	SRZB	HCSC	Cost in mode 1		
Value	1.25	0.9167	2	4		
Variable	Cost in mode 2		Cost in mode 3		Cost in mode 4	
Value	6		3		47	

Fig. 5 Impacts of SDB values on computed results

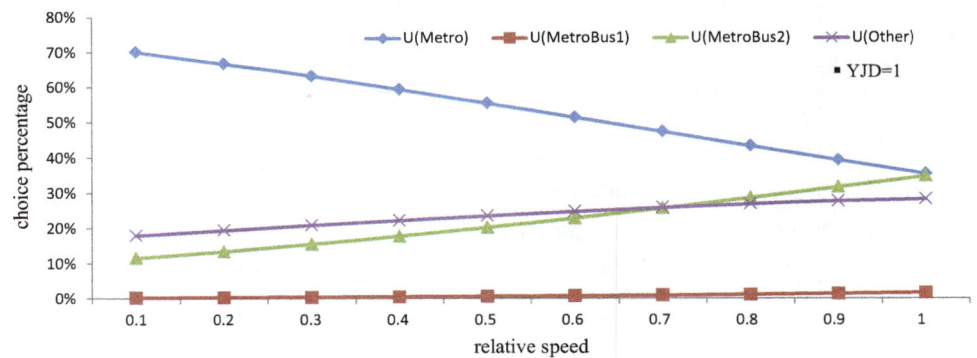

Fig. 6 Impacts of YJD values on computed results

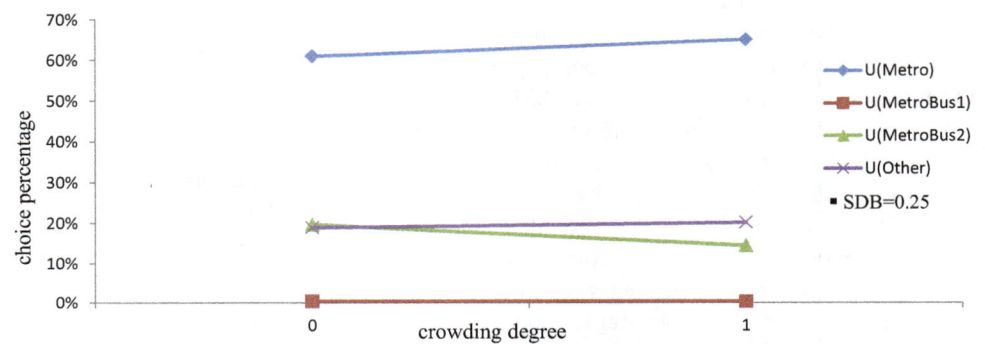

passengers with luggage. Another key variable for the model is cost (Coef. $= -0.0073$; P value $= 0.0539$). Obviously, cost makes a negative effect in the choice model. As the cost of mode 2 is not cheaper than mode 3 in every OD pair, it directly leads that the choice percentage of mode 2 is lower than mode 3. To decrease the pressure on the shuttle, the decision makers can cut down the cost in mode 2 or increase the cost in mode 3. For example, free service can be provided after reentering the URT system or a whole journey ticket can be offered in mode 2. And in mode 3, it is suitable to increase the cost of the long distance journey by the shuttle bus.

5.2.4 Sensitivity Analysis

A further analysis is prepared for SDB and YJD in the proposed model. An OD pair from Wenshui Road to Xujiahui is selected, shown in Fig. 3. The given values of other variables are shown in Table 3.

Sensitivity of Computational Results with Relative Speed assuming the shuttle bus is crowded (YJD = 1), SDB values change from 1/10 to 1 and assignment results are calculated. Figure 5 shows the assignment results when SDB takes different values. It can be seen that the changes of SDB within a certain scope influence mode 1 and mode 3 strongly. But, with SBD increasing, the choice percentage changes regularly. If SDB is very large, passengers will tend to select the path in which the shuttle bus is used to the maximum (mode 3). In addition, the specific passenger group results in the low percentage of mode 2 which is the same with next part (2)).

Sensitivity of Computational Results with Crowding Degree assuming the relative speed is 0.25 (SDB = 0.25), the YJD values are limited to 0 and 1. Figure 6 compares the choice percentage with different YJD values used.

It can be seen that the changes of YJD only have very minor impacts on the assignment results. This can be explained as the follows: as the value of YJD increases, the assignment results of traffic modes, including the shuttle bus, decrease but slightly (mode 2, 0.39–0.37 %; mode 3, 19.65–14.44 %). That means the crowding degree is not the key factor influencing the passengers' choice in emergent circumstance. Passengers may show more interests in the accessibility of the traffic modes. Decision makers can pay less attention to the comfort in the shuttle bus.

6 Conclusion

The assignment of passenger flow is the basic support to emergent organization in the section interruption of the URT system. The literature summaries two types of methods for passenger flow assignment: equilibrium assignment model and utility theory based non-equilibrium assignment model. As the choice result is high correlation to passenger behavior in emergent situation, the latter one is chosen in this paper because more passengers' behavior features can be included in it. Two surveys (a passenger behavior survey and a SP survey) were made and a multinomial logit model was developed in this research.

The passenger behavior survey contains trip features and passenger preference in emergency. More than half of the respondents show interests in the shuttle bus. The relative speed ranging from 1/2 to 2/3 is suitable to the passengers' demand. But, it is quite challenging to satisfy it. Moreover, most respondents (75 %) just care for the available standing and activity space in the shuttle bus and the crowding degree shows slight impact on the passenger choice in the model. Some support in organization is provided by passengers' other preference.

The SP survey presents that the first choice of passengers in emergency is to stay in the URT system by circuitous ways if they exist. A further analysis of variables was made in the choice model. Comparing relative speed and crowding degree, the former is more important to the passenger choice judging by the sensitivity analysis. This result is the same with the passenger behavior survey. The impacts of trip feature factors are similar compared to the previous research in normal state (6) (7), such as HCSC and SRZB. The significant variables mostly fall in the personal attribute. They are income, gender, age, etc. This indicates that the equilibrium assignment model, seldom including the personal attribute and behavior, is not suitable for this situation. Meanwhile, the results of the choice model can assist urban traffic management department to build the scheduling scheme of the shuttle bus. It also can satisfy passengers' travel demand in section interruption circumstance and reduce the risk of crowding at some special stations.

Future research will focus on two aspects. One is to test the model in actual environment. But since section interruption seldom happens in Shanghai URT system, it is hard to complete the empirical work. So, the best practicable way is to expand the sample size. The other one is to build the scheduling scheme of the shuttle bus because the passenger choice will be influenced by the scheduling scheme.

Acknowledgments This research was supported by the Construction Project of Transportation Science and Technology, Ministry of Transport of the People's Republic of China, (Research on the technologies for initiatively preventing and controlling the safety risk of urban passenger network, Grant No. 2015318221020), and the Key Laboratory of Road and Traffic Engineering of the Ministry of Education, Tongji University.

References

1. Transportation Research Board (2007) Transit Cooperative Research Program (TCRP) Report 86. Public Transport Security
2. European Commission Directorate-General for Energy and Transport (2004) Towards passenger intermodality in the EU
3. Shanghai Municipal Transportation Commission (2011) Disposal of Shanghai urban rail transit in emergency (in Chinese)
4. Beijing Municipal Transportation Commission (2007) Disposal of Beijing urban rail transit in emergency (in Chinese)
5. Zhu W, Hu H (2013) Modified stochastic user-equilibrium assignment algorithm for urban rail transit under network operation. J Center South Univ 20:2897–2904
6. Si, BF, Zhong, M (2013) Development of a transfer-cost based passenger flow assignment model for Beijing rail transit network using automated fare collection data. Presented at 92nd annual meeting of the Transportation Research Board, Washington, D.C.
7. Si BF, Zhong M (2013) Development of a transfer-cost-based logit assignment model for the Beijing rail transit network using automated fare collection data. J Adv Transp 3(47):297–318
8. Zhou F, Xu RH (2012) Passenger flow assignment model for urban rail transit based on entry and exit time constraints. Transp Res Rec 2284:57–61
9. Hong L, Gao J (2011) Calculation method of emergency passenger flow in urban rail network. J Tongji Univ (Nat Sci) 10(39):1485–1489 in Chinese
10. Pan HC, Sun YS (2011) Dynamic assignment of emergency passenger flows after metro accidents. In: Proceedings of the 3rd international conference on transportation engineering. Chengdu, China
11. Liu WR, Teng J (2014) Organization of a the shuttle bus under the condition of operation interruption to urban rail transit. In: Proceedings of the 14th COTA international conference of transportation professionals. Changsha, China
12. Zhao SZ, Zhao B (2009) Choice model of trip mode and policy of public transport priority based on SP survey. J Jilin Univ (Eng Technol Ed) 39(2):187–190 (in Chinese)

13. Wang F, Chen JC (2005) Uniformity design method for SP survey in transportation. Urban Transp China 13(3):69–72 (in Chinese)
14. Hu YC, Xu JM (2001) Application of SP survey method in car ownership study. J Highw Transp Res Dev 18(2):86–89 (in Chinese)
15. Sheffi Y (1985) Urban transportation networks: equilibrium analysis with mathematical programming methods. Prentice-Hall, Englewood Cliffs
16. Ma CQ, Wang YP (2007) Competition model between urban rail and bus transit. J Transp Syst Eng Inf Technol 7(3):140–143
17. Shanghai Shentong Metro Group Corporation (2005) The pre-arranged plan in emergent events (in Chinese)

Suitability of Tilting Technology to the Tyne and Wear Metro System

Agajere Ovuezirie Darlton[2] · Marin Marinov[1]

Abstract This paper attempts to determine the suitability of tilting technology as applied to metro systems, taking the Tyne and Wear Metro as its base case study. This is done through designing and implementing of several tests which show the current metro situation and reveals possible impacts on ride comfort and speed, in case tilting technology has been implemented. The paper provides brief background literature review on tilting technology, its different designs and types, control systems, customer satisfaction and history on the Tyne and Wear metro system. Ride comfort evaluation methods, testing of the Metro fleet comfort levels and simulation modelling through the use of OpenTrack simulator software are also introduced. Results and findings include test accuracy and validations and suggest that although tilting technology could be beneficial with respect to speed (minimal improvements) and comfort, implementing it to the Tyne and Wear metro would be an unwise decision owing to the immense amount of upgrades that would be needed on both the network and the metro car fleet. Therefore, recommendations are subsequently made on alternative systems which could achieve or surpass the levels of comfort achievable by tilting technology without the need for an outright overhaul of lines and trains.

✉ Marin Marinov
marin.marinov@ncl.ac.uk

1 NewRail, Mechanical and Systems Engineering School, Newcastle University, 2nd Floor, Stephenson Building, Newcastle upon Tyne NE1 7RU, UK

2 Mechanical and Systems Engineering School, Newcastle University, Stephenson Building, Newcastle upon Tyne NE1 7RU, UK

Editor: Baoming Han

Keywords Tilting technology · Tyne and wear metro · OpenTrack · Simulation · Lateral acceleration · Ride comfort

1 Introduction

Tilting trains, currently gaining popularity in the railway market has been a concept considered ever since the late 1930s. Individuals have pondered on ways of improving rail speeds without compromising passenger ride comfort and possible derailment of the train vehicles. Previously, trains remained limited in speed by the curvature of rail tracks. Trains generally need to slow down to an acceptable speed in other to successfully negotiate curves. A fitting example of disasters caused by over-speeding in curves is the 2013 Spanish high-speed rail disaster [1] where the train derailed due to over-speeding. Despite these incidents, there are still paramount demands on railways to deliver high-speed transportation for their passengers in order to compete effectively with other modes of transportation. A tilting train refers to a train capable of negotiating a bend or curve at speeds greater than that limited by the curve through the use of active or passive tilting mechanisms. When a bicycle or motorcycle negotiates a curve at high speeds, the driver or rider tilts the vehicle at an angle that is a function of horizontal speed and radius of curvature of the curve thus reducing the magnitude of lateral acceleration on the vehicle. This behaviour is also seen in speeding bodies from animals, track athletes, ice skaters and many more. Automobile and rail vehicles cannot mimic the tilts achievable by two wheel vehicles most specifically because they (automobiles and rail vehicles) are four or more wheel vehicles. In the automotive industry for example, attempts have been made to increase speeds around curves through the use of canted roads. Canted roads can most commonly be seen in the racing industry and most commonly in a NASCAR High-speed track. The railway industry can also use cants, but the angles achievable are relatively small and do not necessarily meet with growing high-speed demands. In the automobile industry, road cants or super-elevations can reach angles of 30° or more whilst in railway, cant levels only reach about 3°. Tilting trains use their mechanisms in increasing the angle of tilt hence increasing the amount of speed the train can use in negotiating the curve. Tilting trains are seen as a cost-effective way to meet passenger demands without spending valuable funds on reconstruction of high-speed straight line tracks. The design of a tilting train is shown below. A component called the tilting bolster is mounted on the bogie and the car body is in turn mounted on the bolster. The bolster is connected to swing links which are in turn connected to an actuator. Actuators may be in the form of a hydraulic cylinder using hydraulic oil as its working fluid, an air cylinder using compressed air as its working fluid or an electromechanical actuator. Controls, sensors and programmes are put in place to ensure smooth transitions of the train as it approaches, negotiates and exits the curve. Apart from the benefits of high speed, improved ride comfort is also achieved as the extra cylinders and air springs act as secondary dampers reducing the amount of vibration transmitted from the track, through the bogies to the passengers.

1.1 Motivation

High-speed rail history in the UK suggests that when considering the costs involved in building new rail tracks for the purpose of increased rail speed, tilting technology would be a worthwhile solution. Tilting trains enable increased speed on regular tracks however there are not many tilting trains in service today, especially in the UK. As it turns out, increased speed is not the only benefit of tilting vehicles. Ride comfort and quality can also be significantly increased. Over the next 15 years, the Tyne and Wear metro intends to replace current Metro vehicles with new rolling stock. Would tilting technology be beneficial? Would customer satisfaction be increased through increase in ride quality and/or increased speed (reduced commuting time)? Could tilting technology be applied to intercity trains not necessarily because of increased speed, but increased comfort? These questions pose significant motivation for possible study so as to conclusively understand and determine sufficient and satisfactory answers.

1.2 Aims and Objectives

The objective of this paper is to determine and analyse the suitability of tilting technology to the Tyne and Wear metro. Previous works on tilting technology have mainly been implemented on high-speed rail systems and hardly any research has been done to implement it on intra-city metro systems despite a growing demand for higher comfort and speed. This project attempts to determine the possible impact of tilting on metro systems in terms of speed, ride comfort, motion sickness and also determine if tilting can and should only be useful in the High-speed rail sector. It is worth to note that metro systems hardly travel at the required high speeds which would necessitate the use of tilting technology, however this project will attempt to identify the current levels of comfort in the current metro and also determine how beneficial tilting technology could be to ride comfort and subsequently overall customer satisfaction.

1.3 Research Methodology

This project first determines the current level of customer comfort in the metro by showing the correlation between commuting speed and vibration levels whilst making comparisons with scales determined according to ISO 2631:1997. Track speed and vibration measurements were measured using accelerometer sensors located in a Samsung galaxy grand duos smart phone where smart phone apps, vibration monitoring and speedview 2.31 are used to record these data and return values in graphical format. Comparisons of collected data are then made with speed simulation data from OpenTrack to determine the differences in performance of the train vehicle and correlations made with vibration data thus showing how much of an improvement could be made through the implementation of better ride comfort and speed.

2 Literature Study

A tilting train refers to a train that has a tilting mechanism which enables it to travel at higher speeds compared to non-tilting or conventional trains. Nam-Po Kim and Tae-Won define tilting trains as trains capable of increasing curving speeds without reducing the safety or comfort of passengers [2]. With growing competition from other modes of transport, railway industries have recognised and acted on the need of increasing travel speeds thus reducing travel time on passengers. Another relevant reason for tilting trains stems from the need for better comfort for passengers. Tilting trains achieve these demands excellently without the need of reconstruction of regular lines in order to lay new straight line tracks. Over the years, over 5000 tilting vehicles have been produced worldwide by different suppliers [3].

Despite the many benefits of tilting technology, it still has its drawbacks. The most relevant being motion sickness. Passengers have reported cases of nausea, dizziness and vomiting in severe cases. Motion sickness is not necessarily uncommon in the transportation industries. Non-tilting trains, although not as much as tilting trains, have reported passengers experiencing motion sickness [4]. There is currently a general understanding that having tilting trains which compensate for 100 % of lateral acceleration poses high risks of inducing motion sickness [4]. The Swedish preferred the tilt compensation method to compensate for 65–70 % of lateral acceleration on curves [4] as is currently used in the Swedish X2000 tilting train. Several methods have been employed to find out the underlying cause of motion sickness, and research has shown that passengers complain the most when the train uses a passively tilted system. Actively tilting train does record motion sickness cases however the numbers coincide closely with non-tilting trains [4].

There are several types of tilting technologies currently being used or developed in the railway industry. The Italian pendolino developed by FIAT [5] stands out as the most popular and is being used today in the United Kingdom West Coast mainline. The Swedish X2000 train series developed by ASEA Brown Boveri [5]. Other developed or currently being developed technologies include the long air spring system developed in Japan [6] and the hybrid system also developed in Japan [7].

2.1 Tilting Technology and Mechanisms

The concept of tilting technology lies in the physics of centrifugal and centripetal forces. When a vehicle travels around a curve, bodies in the vehicle experience forces which tend to push them in a direction towards the outside of the curve. This force is called centrifugal force. A more elegant definition states that "Centrifugal force is an apparent force which draws a rotating body away from the centre of rotation and it is caused by the inertia of the body as the body's path is continually redirected". It is important to note that centrifugal force is an apparent force. That is a force introduced into a system in order for Newton's laws of physics to be satisfied. The picture below shows a schematic example of how this force is experienced. In an automobile vehicle, bodies in the vehicle experience this force. If the vehicle goes too fast around the curve, there is a high potential of the vehicle turning over and potentially causing fatal injuries. It is for this reason that vehicles slow down before the beginning of a curve so as to limit the magnitude of this force. This same concept also applies to train vehicles.

Tilting technology enables the trains to tilt at an angle towards the centre of the curve which reduces the magnitude of the centrifugal force experienced by passengers enabling the train to travel at a higher speed whilst improving or maintaining the amount of comfort for the passengers. A schematic diagram of the forces involved is given below. The chosen train vehicle is the Italian pendolino train vehicle with its pendulum system. In actuality, it does not matter which sort of system is chosen as the experienced forces do not change

If θ_v = Car body tilt angle

θ_0 = Track cant angle

and $\ddot{Y}$ = Lateral acceleration (centrifugal acceleration)

$$\ddot{Y} = \frac{V^2}{R}\cos(\varphi_c + \varphi_t) - g \times (\varphi_c + \varphi_t),$$

where V is the velocity, R is the radius and g is the constant of gravity.

As shown in the Fig. 1 above, one can see that the amount of lateral acceleration or centrifugal force experienced by the passengers depends on the summation of both centrifugal and centripetal forces. Tilting technology increases the amount of centripetal force on the vehicle by shifting the centre of gravity of the car body inwards of the curve and this is given by the formula $g\sin(\varphi_c + \varphi_t)$. Centrifugal force given as $\frac{V^2}{R}\cos(\varphi_c + \varphi_t)$ is a function of the speed of the car body, radius of the curve and total angle of tilt. As speed increases, centrifugal force increases significantly. A reduction in the resultant centrifugal force results in better ride comfort for the passengers and higher possible speeds for the train vehicle.

2.1.1 Tilting Technology Types

There are two types of tilting mechanisms

1. Passively tilted trains
2. Actively tilted trains

Passively Tilted Trains Also called naturally tilted trains, these are trains which rely on the physical laws of physics without any actuators, control systems or power sources initiating tilt. The tilt centre of the train is purposely located above the centre of gravity and this enables the train to tilt when negotiating curves due to its weight [9]. Should the tilt centre of the train be located at the same point or below the vehicle centre of gravity, the system would become highly unstable with passengers experiencing high levels of discomfort. According to [3], damping of the tilt motion is required to control the otherwise relatively undamped car body roll motions. This is because if left undamped, the car body would end up reacting in an unconventional manner to the slightest of curves which can be very uncomfortable for the passengers.

Several examples of systems with passive or natural tilting includes

Fig. 1 Pendulum tilting system. **a** Tilting train with tilting bolster, **b** basic concept of tilting [8]

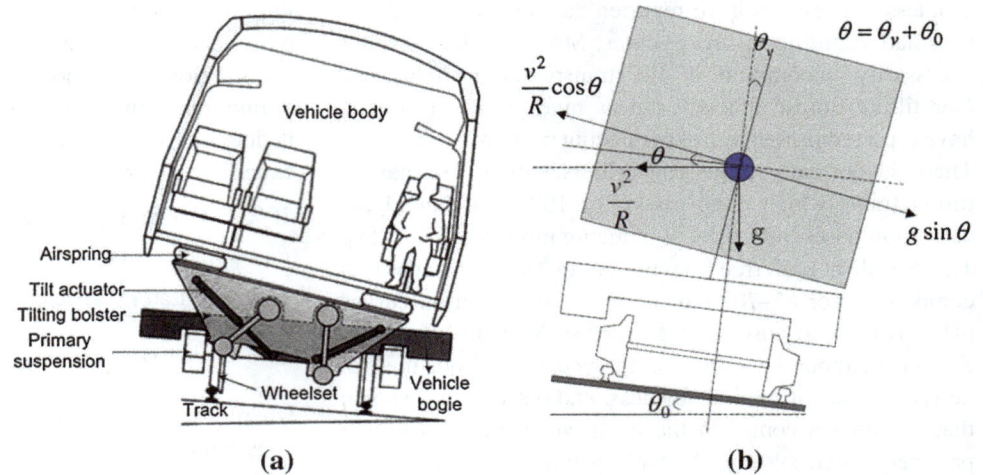

(a)   (b)

1. Swing bolster with circular arc guide
2. High-positioned air spring
3. Inclined anti-roll bar links

Swing Bolster with Circular Arc Guide This is a natural tilting solution used in early Japanese tilting trains. In this system, the car body sits on air springs, which is mounted on a swing bolster (tilting bolster). The tilting bolster in turn sits on rollers or bearings which allow motion along a circular path. According to Persson [3], the car body's centre of gravity is located about 600–900 mm lower than the location of the tilt centre. There are two major tilting components in this system. The first is the tilting bolster and the second is the roller bearings which the bolster sits on. This system allows a 5°–6° tilt capability of the car body. The system uses roll dampers which are installed between the bogie frame and tilting bolster to limit overshooting [3].

High-Positioned Air Spring In this system, centrifugal acceleration forces the car body to tilt around the centre of a pair of air springs which are installed on a pair of pillars which sit on the train bogie. The air springs also acts as a secondary suspension system giving more stability to the car body and this arrangement is mostly used in Talgo systems [3]. Roll stiffness is controlled by an electro-pneumatic valve which connects both air springs. This system has a simple structure and it realises natural tilting without the implementation of complex devices or mechanisms.

Inclined Anti-roll Bar Links In the high-positioned air spring system, the air springs are located closer to the top of the car body. In this system, the air springs are installed below the car body. This system has natural tilting capabilities but can be better considered as a normal train system with better compensation for suspension flexibility. Generally it can achieve a tilt angle of about 2°.

Advantages of passively tilted trains

1. The system is relatively simple to implement.
2. System simplicity enables low initial and maintenance costs.
3. Hardly any control system required.

Disadvantages of passively tilted trains

1. System has higher potential to induce motion sickness in passengers.
2. Safety issues like overturning arise as a result of lateral shift of the car body's centre of gravity.

This system is being phased out as it poses safety concerns due to the lateral shift of the centre of gravity of the car body.

Actively Tilted Trains Actively tilted trains rely on active technology controlled by sensors and programmed electronics. Tilt is executed by actuators which may be hydraulic or electromechanical in nature. Also the centre of gravity does not change its position thus it is considered a much safer alternative to the passive tilt system [10]. Control systems are installed in order to initiate tilt in a timely fashion so as to efficiently utilise the vehicle capabilities. Most especially in natural tilting systems where there are no controls, a train vehicle may begin to tilt too late and this can be highly uncomfortable for passengers and potentially catastrophic to infrastructure. Control systems will be covered in detail in the next section.

Identified examples of systems with active tilting mechanisms include

1. The Japanese pneumatic system
2. Hybrid tilting system
3. Italian pendulum tilting system

The Japanese Pneumatic Mechanism This system developed in Japan comprises a simple air cylinder used as an actuator to initiate tilt. The car body sits on a pair of air

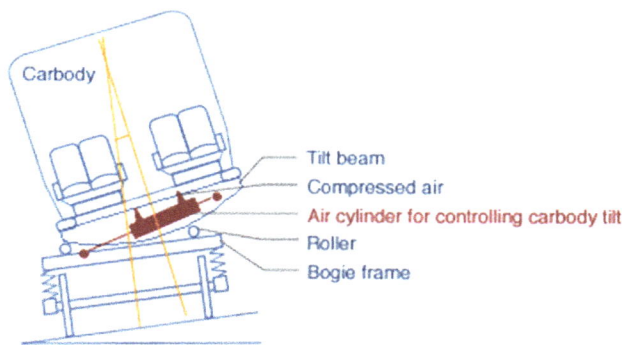

Fig. 2 Schematic illustration of Japanese tilting mechanism [11]

springs which act as dampers and they in turn sit on the tilting bolster. The bolster is mounted on rollers which allow the body to roll into its desired tilting position as shown in Fig. 2. The system shares the same power source with the pneumatic brake system. It is a compact (space saving) system, which is preferable for limited under-floor mounting space of the rolling stocks used on narrow gauged tracks. Without using hydraulic power, the system can be easily cleaned, maintained and generally have low cost of implementation. Also the simplicity of the system contributes to the element for speed up and assures less pressure or damage to the tracks. This system has its merits but scores low in terms of vibration and passenger comfort. A critical look at the schematic diagram would show that the tilting bolster sits on bearings for smooth tilting. However, this design would mean that the bogie would have a high value of vibration transmissibility compared to other tilting systems.

Hybrid Tilting System The hybrid tilt system (Fig. 3), developed by a research team uses a combination of two conventional tilting systems in order to gain extended tilt angles of up to 8°. The first system uses a conventional hydraulic actuator capable of initiating a tilt angle of 6°. The second uses the secondary air spring suspension system to initiate an additional tilt angle of 2°. The combined system would enable a tilting train achieve a speed of

+50 kmph as total tilt angle comes to 8° [7]. They also claim that the system would be safe and comfortable for passengers as feedback and preventive control systems would be put in place to ensure synchronisation of both tilting systems. Considering that this project aims to implement tilting systems in light rail or metro systems, this tilting mechanism will not be considered. A critical observation indicates that the system is still in its research phase and proper feasibility studies have yet to be undertaken. Motion sickness would be an issue and when one considers a vehicle with tilt capability of 6° only, it becomes slightly confusing why a need for an additional 2° would be needed to go faster.

Italian Pendulum Tilting System The Italian pendolino (Italian for pendulum) trains are quite common and is also currently used in the UK. This system shown in Fig. 4 is an active controlled system using hydraulic actuators which connects to swing links which are in turn connected to the tilting bogie. The system is safe, reliable and although slightly more complex than other tilting systems, it does have a higher positive feedback in terms of passenger comfort and motion sickness. This system would be a good choice to consider when considering implementation on the current metro system. Reason for this is its proven reliability and its already widespread patronage.

2.2 Control Systems

In tilting trains, naturally tilting systems hardly need any controls to initiate tilt. However, most naturally tilting systems can be upgraded with an actuator in order to improve their performance and give more control to operators and/or computer systems. Such are most actively tilting systems. As defined previously, actively tilting trains require some sort of device or component in other to initiate tilting.

Several control system types have been developed over the years with each having its merits and demerits. The three prominent control systems are

Fig. 3 Hybrid tilting bogie [7]

Fig. 4 Pendulum tilt system [12]

1. The body feedback control system
2. The bolster feedback control system
3. Reference transducer control system

The body feedback systems were first developed and worked by installation of an accelerometer in the body of the car which reads the total lateral force on the car body. The body accelerometer then sends a signal to the controller which then compensates with tilting at the actuator. This system had to be upgraded due to numerous stability problems arising due to low frequency movements in the secondary suspension. The bolster feedback system used accelerometers installed in the tilting bolster of the bogie. Although this system was much more efficient than the previous version, it still had to be upgraded. The main reason was because the tilting train when approaching a curve could be overly compensated or under compensated. The reference transducer system where a transducer is installed on the bogie receives signals from another transducer installed on the tracks. The signals include information of the curve, permissible speed and required tilt all of which are important factors which leads to an efficient negotiation of the curve. The reference transducer system is often called predictive tilt control system. Rightly so it ensures the train control systems possess the necessary information before it reaches the curve. Having this information beforehand, the system has proven to be a more stable and enjoyable system. It has also been found to have a positive impact on motion sickness which turns out to be the most challenging factor in terms of passenger comfort.

Figure 5 is a schematic diagram which shows how control systems have evolved over time and it is easily obvious why the reference transducer system is considered more efficient than the other systems. An example of tilting train systems which use the reference transducer control system is the Japanese tilting trains. These trains use wayside information to improve the performance of tilt. These systems combine an automatic train protection

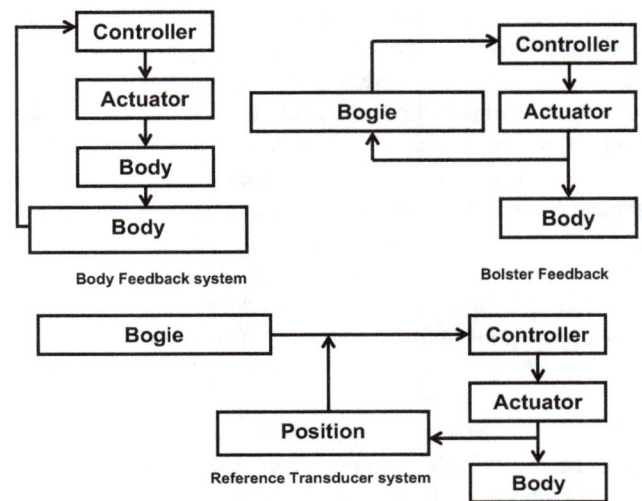

Fig. 5 Layout of Tilt control systems modified from [10]

(ATP) system, where track information is sent from a wayside element to the ATP system in the train and this is then transmitted to the tilt controller. There are debates on whether system track data should be sent to the train via wayside elements or if the data can be stored beforehand in the train [13]. Experts have debated on which is the best with the former having a higher vote of confidence. The reason why is because should there be any track changes, it could be easily registered unlike the latter system which could pose potential safety concerns if there are any track changes like a damaged track. However, there is a possible flaw in terms of cost to the wayside system. If there are sensors and transducers that have to be put in place before every track curve, it may become more expensive to enact compared to if there is a single system located in the train where all track data are stored. It may turn out that the system which is best may differ in pertaining to different train systems and their requirements. A compromise

Fig. 6 Illustration of predictive tilt control system [15]

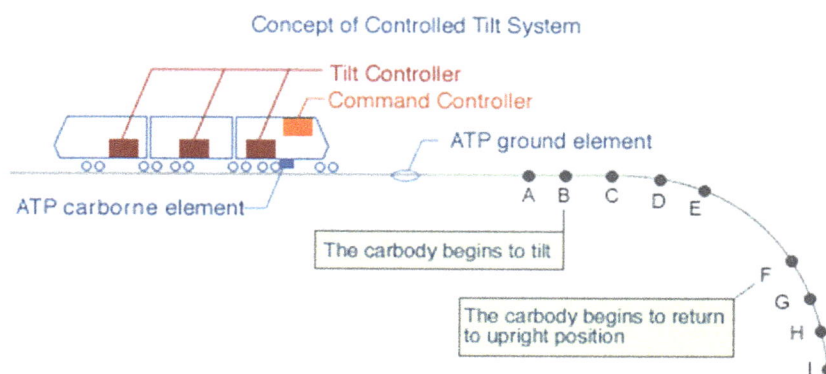

between system requirements and cost may ultimately lead to the best decision. There have also been claims that the Spanish supplier Construcciones y Auxiliar de Ferrocarriles (CAF) has in the series R-598 showing that a system based on stored track data can work without wayside input [14]. According to the claim, the stored track data can either be based on a track data register or on a train measured track data (Fig. 6). A positioning system would be required to pick the right data at the right time.

Important design parameters for a control system according to [14] include

- Filtration of input signals and optimization of response time delays
- Good tilt compensation ratios
- Designing control loops to minimise undesired car body behaviours ultimately optimising fast response.

Frostberg [12] went further to state that "control loops have to be designed together with the car body vehicle suspension dynamics so as to minimise any interference with the tilt system".

2.3 Tyne and Wear (Newcastle) Metro

The Tyne and Wear metro is a light rail service system located in the north east of the UK. It serves over 60 stations spanning the majority of North East city centers all the way from Newcastle to Sunderland. Having been in operation since 1980, several refurbishments and overhaul of the train fleet has been done over the years. However, with the fleet currently over 30 years old, and set to remain in service for the remainder of the next decade, there is a growing need for modernization.

Table 1 shows the specifications of the metro car being used in on the Tyne and Wear network. Power is obtained from 1500 V overhead lines through a pantograph which can be designed in a manner so as not to be affected during tilting, a weight of about 39 tonnes when fully laden and it currently seats 68 people sitting and 232 standing. The

Table 1 Metro car specifications

Construction	1974–1984
Beginning of service	1980
Train set formation	2 cars
Train set capacity	68 sitting, 232 standing
Maximum speed	50 mph (80 km/h)
Weight	39 metric tonnes
Power system	1500 V DC overhead lines
Current collection method	Pantograph
Track gauge	1435 mm (4 ft 8½ in)

train has the capabilities of reaching a maximum speed of 80 km/h.

2.3.1 Rolling Stock

The rolling stock of the Tyne and Wear has remained unchanged ever since beginning of operations, which puts them at over 30 years old. With several refurbishment schemes in place since 2010, the main goal is to extend their service life till the next 10–15 years after which they would be replaced by a new fleet set. The current fleet is made up of a total of 90 train sets in a two car per unit formation which can also be coupled together in pairs to form a four car unit formation. The metro also has in operation a three battery run electric locomotives constructed by Hunslet which are used to pull engineering trains around the network [17]. The metro train sets are each around 56 m long, using articulated six-axle twin vehicles, with 68 passenger seats and spacing for up to 200 standing passengers. All 90 of the metro fleet are in use the system; uses a power supply of 1500 V DC via overhead lines. A pantograph is used to transmit the current from the overhead lines to the train Fig. 7. The trains are also equipped with an air braking system and also an electro-magnetic emergency braking system.

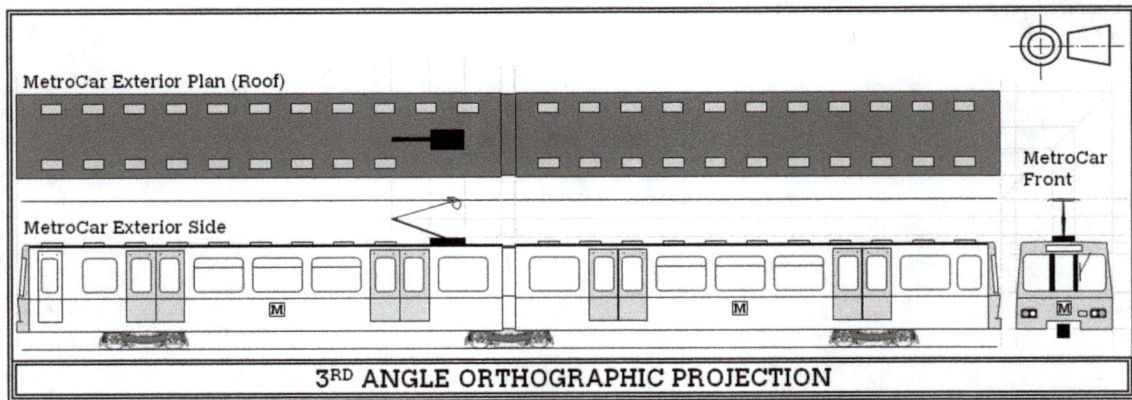

Fig. 7 Schematic Illustration of Tyne and Wear Metro Car [16]

Fig. 8 Current Tyne and Wear Metro Network [16]

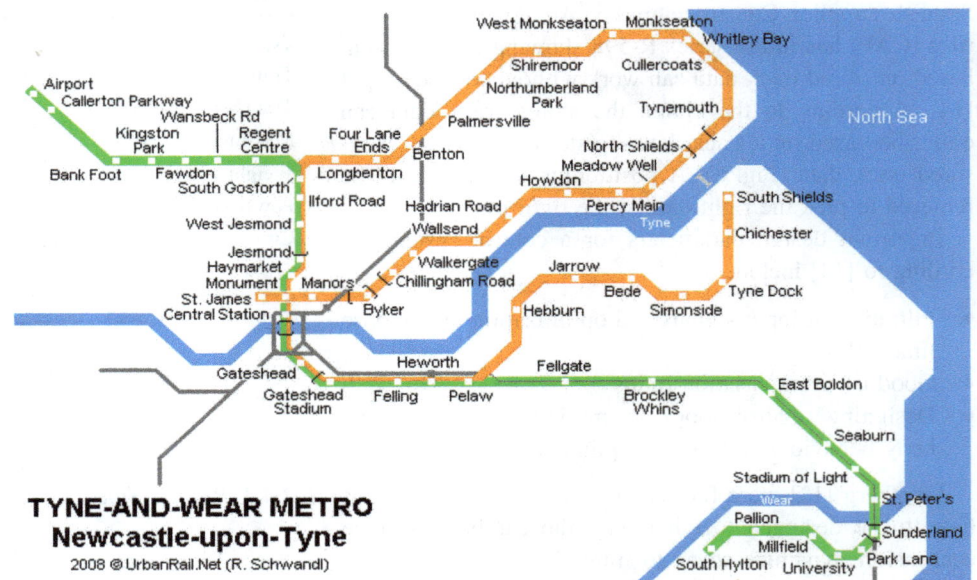

TYNE-AND-WEAR METRO
Newcastle-upon-Tyne
2008 @ UrbanRail.Net (R. Schwandl)

2.3.2 The Tyne and Wear Network

The Tyne and Wear network consists of two lines, called the green and yellow lines. As shown in Fig. 8, the green line begins at Newcastle Airport and runs through Newcastle City centre all the way to South Hylton in Sunderland. The yellow line begins at St James' Stadium, goes all the way to Four Lanes End and loops on itself before crossing the Tyne River heading to South Shields.

Figure 8 shows the Tyne and Wear Metro system including its green and yellow lines encompassing its 60 stations. It is easily obvious that the relative distances between most of the stations like Jesmond and Haymarket are quite small. However, one could also notice a much larger distance between some other stations like Pelaw and Hebburn or Brock whins and East Boldon. Since considerations are also going to be made for speed, it would be important to note the relative distances between each station and identify sections where the metro reaches its maximum speed. This would help identify possible sections where the metro may be capable of surpassing its maximum speed.

2.3.3 Future Plans

With the current metro fleet showing signs of old age as the come to a 35 year mark, plans have been put in place by Nexus for a full replacement by 2030. In a plan called the Metro strategy, Nexus outlines its future plans and how it would be making its case to the United Kingdom Government in hopes of the realisation of its visions and also the expansion of the current metro network including the possibilities of tram services in areas not covered by the metro. As shown in the network map below, several route expansions have been proposed. It is worth noting that the decisions for expansion would only be made after careful considerations of cost, investment benefits and also the general effect on the economy. In a survey conducted by Nexus, it was discovered that an investment of one pound in the metro yields a return of eight pounds [18]. This gives a sort of incentive on the values of the metro despite its being highly subsidised by the government. The network expansions being proposed include Silverlink in North Tyneside, Team Valley in Gateshead, The MetroCentre and Newcastle's West End [19].

3 Customer Satisfaction, Ride Comfort and Motion Sickness

Lauriks et al. [21] define comfort as "the wellbeing of a person or absence of mechanical disturbance in relation to the induced environment" [20]. It also stipulates that such wellbeing is achieved or disturbed based on a number of different factors which could be physiological in nature, for example, individual expectations and individual sensitivity and stress, or physical in nature like vehicle motions, noise, temperature and seating characteristics [20]. When looking at a situation combing some or all of these factors, the perception of comfort based on similar vibration values may be judged as uncomfortable in a given environment and comfortable in another environment [20]; it also defines ride quality as an entity representing a passenger's judgement of comfort level of a specific journey based on all experiences encompassing all factors. In every railway system, maintaining a healthy customer base is an essential part to every network's future. A sustained and/or increasing number of passengers ensure that the business grows which in turn provides avenues for service improvements, network extensions and ultimately increasing the growth of the immediate economy. The only way in which such strong, stable and sustained customer base can be achieved is through endeavours made to ensure that the overall customer satisfaction factors are made the highest priority. A satisfied customer is a happy customer; thus, if railway systems put measures in place to ensure customer satisfaction, then overall business will experience growth.

Over the years, several literatures have been studied in order to identify the factors which most influence customer satisfaction. As much as each and every factor remains absolutely important, some are consistently rated over time as being more important than others [21]. According to surveys conducted by different sources, identified customer satisfaction factors include

- Punctuality
- Reliability
- Station and train cleanliness

- Comfort level of train
- Train cleanliness
- Station condition
- Staff availability
- Passenger behaviour
- Train running information
- Safety and security
- Transport integration

With different authorities on the subject having different views on which of these factors have the most influence on overall satisfaction of passengers, The Transport for Greater Manchester amongst others claims that the key factors which would be the best predictors of overall passenger satisfaction in a train journey are

1. Cleanliness inside the trains
2. Comfort of the seating area(ride quality)
3. Punctuality and reliability

These findings coincide properly with other authorities like passenger-focus or Nexus thus making their findings increasingly accurate.

The current metro system scores very well on the subject of punctuality and reliability as there are hardly any breakdowns of the trains due to quality and preventive routine maintenance. The cleanliness of the train is also very good as the train vehicle is consistently cleaned before embarking on new journeys. However, the overall comfort of the train vehicle is not so good. Passengers frequently complain of vibrations felt during the journey and these can become concerning especially for passengers more susceptible to motion sickness. The current network of the metro covers a total distance of 77 km with average distances between stations around 2 km. On a short journey, say from south Gosforth to central station, passengers may not feel at all bothered by the vibrations of the train vehicle. This could be because they might be more energetic and would not feel any impacts from the vibrations. However, if we look at a case of someone who has travelled a couple of hours by flight and has to use the metro from the airport to Fellgate, it becomes understandable that such a person would be impacted by the amount of vibrations experienced. During a personal survey, it was observed that taking a journey in the morning, after a full breakfast and ready to start the day, there was little thought as to the magnitude of vibrations felt in the train vehicle. However, when taking the journey back home at the end of the day, the magnitude of vibration was almost unbearable as the train could not arrive at the destination station soon enough. On subsequent days, the experience was the same. This leads to a conclusion that when it comes to passengers who have already had a long day, the comfort level of the current metro system scores very low.

3.1 Motion Sickness

Motion sickness is said to occur when a person or group of people feel dizzy, nauseated or sick specifically due to conflicting body responses to real, perceived or anticipated movement. Mostly experienced by individuals in moving vehicles, for example ships, airplanes, automobiles or trains, it is caused when the human brain does not understand signals sent to it by the body's natural sensory organs (eyes, ears and more) which it makes to assume that the body may be host to foreign objects which may need to be expelled. This leads to several impulses where the individual feels like dizziness, vomiting, increased temperatures, increased salivation and many other symptoms. Reported cases of motion sickness date as far back as 5th century BC when Hippocrates stated that sailing on the sea creates disorder in the body [3]. Observations have shown that drivers hardly experience motion sickness leading to a general understanding that having forehand information significantly reduces the likelihood of experiencing these symptoms.

3.1.1 Motion Sickness in Trains

Tilting trains have been claimed to cause more motion sickness cases than non-tilting trains although these claims remain debatable. Reasons for debate range from knowledge about cases of motion sickness that are generally reported in both train types. In many surveys that have been made, a stark contrast in results show that possibly motion sickness may also be affected on an individual and geographical basis. In a survey done in 1964, it was reported that 0.13 % out of 370,000 individuals suffered symptoms related to motion sickness on non-tilting trains in America, whilst on a similar test in Tokyo, 18 % of passengers experienced motion sickness symptoms [3]. Conflicting test reports have also been noticed in reports by Frostberg on tests in Sweden (Swedish X2000 tilting train) and Norway where the Swedish report claimed that there was a clear correlation between tilt compensation and motion sickness, whilst the Norway report claimed that there was no significant correlation between tilt compensation and motion sickness [10, 22]. Persson [10] also claims that the extent of reported cases in Europe seems to be less than those reported in Japan. These reports tend to show a pattern of differences pointing to individual, geographical and psychological reasons. Also during a test done by this project, it was noticed that when taking a train ride in the morning, thoughts of being uncomfortable were virtually non-existent compared to taking a train at mid-day or worse, at the end of the day. It can be said that depending on many different factors including

- State of mind
- Individual susceptibility
- Sensitivity to motion
- Daily work stress
- Travel fatigue
- Geographical area
- Track conditions
- Train characteristics
- Track gauge
- Magnitude of tilt compensation

Some individuals could show symptoms of motion sickness in some cases or situations and not show any in other situations even when other people show these symptoms. A typical example is if an individual, who has been on a 6 h flight into Newcastle, takes a train ride and is exposed to several train movements, he/she may experience motion sickness compared to another individual who simply went to the airport as an escort or family member to pick them up. Table 2 shows the ranges of symptoms of motion sickness ranging from gastro-related issues to deeply psychological issues. Though not an exhaustive list, it can be seen that symptoms occur in line with individual states. An individual who may have had a heavy meal in the morning would be more likely to experience vomiting compared to an individual who probably just lost his livelihood. A woman who is pregnant may experience nausea compared to a heavily weighted woman who may experience high blood pressure.

4 Evaluation of Ride Comfort

In order to make an effective evaluation of the current metro ride comfort, it is important to understand the sort of forces or accelerations experienced by passengers. When an individual sits in a train, the individual is subjected to several forces due to reactions by the movement of the train. For example, when a train brakes, the passenger feels a pulling force. When the train accelerates, the passengers react as though a force pushes him or her unto the seats. These reactive forces experienced acts in a direction of six degrees of freedom. Each action of forces on the train car body results in a corresponding reaction on the passenger in mostly three out of the six degrees of freedom. As a train commutes from one point to the next, vibrations can be felt by passengers, especially when the vehicle suspension system has a high value of transmissibility.

4.1 Ride Comfort

Ride comfort in a train vehicle is mostly evaluated using the mean comfort standard scale method or ISO 2631:1997

Table 2 Common symptoms caused by motion sickness [3]

Gastro-related	Somatic	Objective	Emotional
Stomach Awareness	Dizziness	Skin humidity	Anxiety
Nausea	Exhaustion	Pulse rate	Nervousness
Sickness	Increased salivation	Blood Pressure	Scared
Retching	Drowsiness	Respiration rate	Apathy
Vomiting	Cold sweating	High body temperature	Tense

or the Sperling ride index (Werzungzahl) methods [23]. When these methods are considered as not being sufficiently accurate, the running r.m.s. method or fourth power vibration methods are used [23]. These different methods use measured r.m.s. weighted acceleration values got through the use of sensors or accelerometers. The weighted accelerations is defined by the equation below; however, these values were measured for three degrees of freedom, namely

- Vertical vibration
- Longitudinal vibration
- Lateral vibration

4.2 ISO 2631:1997

According to ISO-2631, the basic evaluation method uses frequency weighted r.m.s. accelerations defined by the equation below [23]

$$a_{\mathrm{w}} = \left[\frac{1}{T} \int_0^T a_{\mathrm{w}}^2(t) \mathrm{d}t \right]^{\frac{1}{2}},$$

where a_{w} is the r.m.s. acceleration values (as a function of time) and T is the time in seconds.

Likely reactions to various magnitudes of overall vibrations in a public transport vehicle are shown in Table 3.

From Table 3, we see that when the overall vibration in the car body is below 0.3 m/s^2, it is considered that the overall perception of ride comfort of the passenger is that the ride is comfortable. This means that any recorded value above 0.3 m/s^2 would be uncomfortable for the passengers ranging from being slightly uncomfortable to being extremely uncomfortable.

4.3 Sperling Ride Index (Wz)

This method is commonly called the Werzungzahl (Wz) method which is a frequency weighted r.m.s. value of accelerations evaluated over given time intervals or over a defined track section [23]. The Wz formula is given below as

Table 3 R.M.S vibration levels and passenger perception [24]

R.M.S. vibration level	Passenger perception
<0.315 m/s^2	Not comfortable
0.315–0.63 m/s^2	A little uncomfortable
0.5–1 m/s^2	Fairly uncomfortable
0.8–1.6 m/s^2	Uncomfortable
1.25–2.5 m/s^2	Very uncomfortable
>2 m/s^2	Extremely uncomfortable

Table 4 Ride index (Wz) and passenger perception [23]

Ride index Wz	Vibration sensitivity
1	Just noticeable
2	Clearly noticeable
2.5	More pronounced but not unpleasant
3	Strong, irregular, but still tolerable
3.25	Very irregular
3.5	Extremely irregular, unpleasant
4	Extremely unpleasant: prolonged exposure harmful

$$\mathrm{Wz} = 4.41 \left(a^{\mathrm{wrms}} \right)^{0.3},$$

where a^{wrms} is the r.m.s. value of frequency weighted acceleration a_{w} in m/s^2.

With the Sperling ride index, each direction has to be determined separately. This makes it less accurate compared to ISO-2631 and also more disadvantageous. Howbeit, it does make for a much simpler read and understanding as its final values are whole numbers. The use of Sperling ride index is most handy when comparing two or more situations. From Table 4, it is seen that with ride index of less than 2.5, the comfort level ranges from being very comfortable to being at a not so pleasant level. Between 2.5 and 3.5, the comfort level ranges from being slightly uncomfortable but tolerable to being very uncomfortable. Beyond this point, passengers become increasingly frustrated, nauseated and they cannot wait for the journey to be over. Any higher than 3.5, prolonged exposure becomes harmful and the train could be considered as a hazardous environment.

4.4 Mean Comfort Standard Method N_{MV}

There is a third method for quantifying customer comfort called the mean comfort standard method [25]. This method, similar to the ISO 2631 method uses values of acceleration from the three translational directions converted into a comfort evaluation scale [26]. This method is not considered as a very reliable and accurate evaluation method due to several reasons including

$$N_{MV} = 6 \times \sqrt{(a_{XP95}^{W_d})^2 + (a_{YP95}^{W_d})^2 + (a_{ZP95}^{W_d})^2}$$

- It is only valid for straight lines and cannot be used in curves.
- It only uses the 95th percentile of frequency weighted acceleration values.
- Final values cannot be correlated to track conditions as critical lateral and vertical values may be kilometres apart.
- There is loss of information as it does not utilise all data values.

As seen in Table 5, the mean comfort standard scale seems quite similar with the Sperling ride index scale. Mean comfort values less than 1.5 are considered to be very comfortable with values between 1.5 and 4.5 ranging from being comfortable to being uncomfortable. Any high than 4.5, the ride comfort becomes very uncomfortable.

Table 5 Mean comfort standard scale [26]

$N_{MV} < 1.5$	Very comfortable
$1.5 \leq N_{MV} \geq 2.5$	Comfortable
$2.5 \leq N_{MV} \geq 3.5$	Medium
$3.5 \leq N_{MV} \geq 4.5$	Uncomfortable
$N_{MV} \geq 4.5$	Very uncomfortable

4.5 Vibrations

The vibration on passengers in the metro was measured using "Vibration Monitoring". Vibration Monitoring is a simple app in android app stores which uses accelerometers in a smart phone and measures weighted acceleration changes about a given point of rest of the phone.

4.5.1 Test Setup

The vibration data were taken over the following five chosen stations on the metro network shown in Fig. 9:

- Pelaw
- Fellgate
- Brockley Whins
- East Boldon
- Seaburn

Looking specifically at customer comfort, these stations were chosen because they sit in a section where the current metro car reaches its maximum speed of 80 km/h. Being able to measure the sort of vibrations experienced by passengers as the metro starts from rest to maximum speed, this section of the track proved to be the best choice for the evaluation. The journey was taken starting from Seaburn to Pelaw and back during which speed and vibration data were taken.

The simple test was set up when the phone is placed on a passenger's body (thighs or strapped to the chest) to measure changes in body motion due to vibration experienced. As the train moves on the tracks, every sudden change in direction of the car body results in a corresponding change in the direction of the passenger's body which is recorded by the software. At the same time, speed view uses the phone's internal GPS system, connects to as many satellites as cell reception would allow and records changes in speed of the train as shown in Fig. 10. This process is done in the background. Figure 11 shows the relatively easy setup of vibration testing. As previously explained, vibration monitoring is used as the vibration software, most especially, because of its ability to collect recorded acceleration data from the accelerometers in the phone and this data can be read using Microsoft excel. This

Fig. 9 Analysed section of the metro network

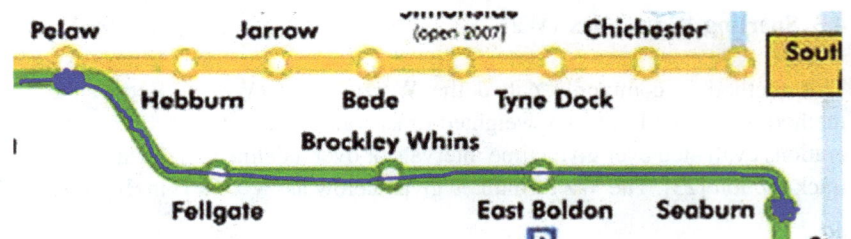

Fig. 10 Speedview setup

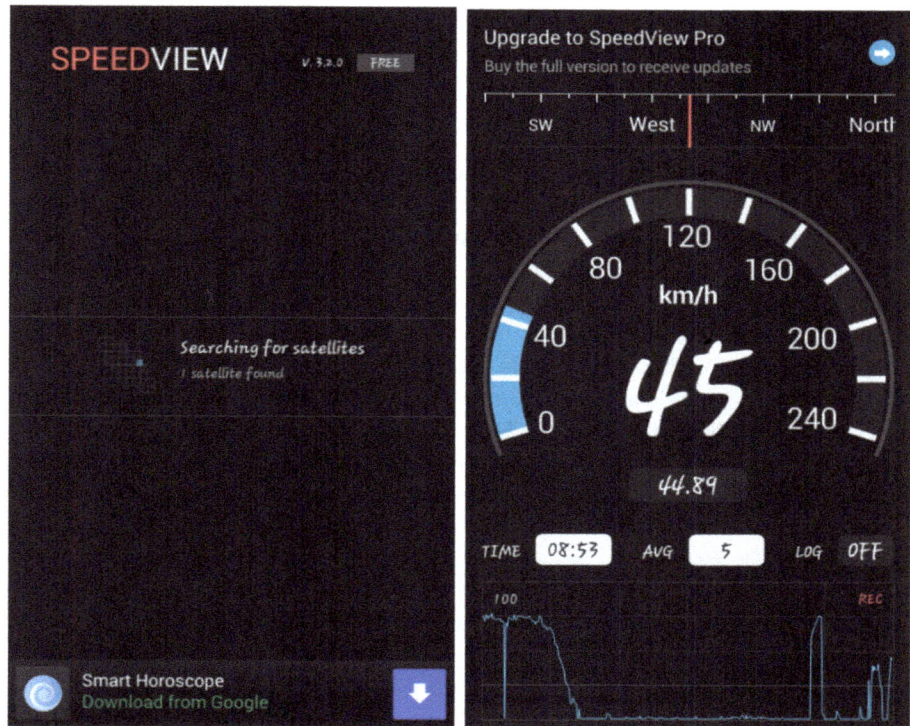

Fig. 11 Vibration monitoring setup

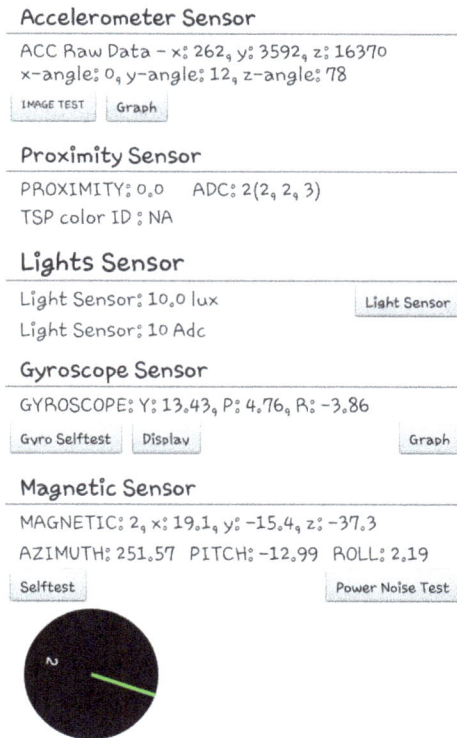

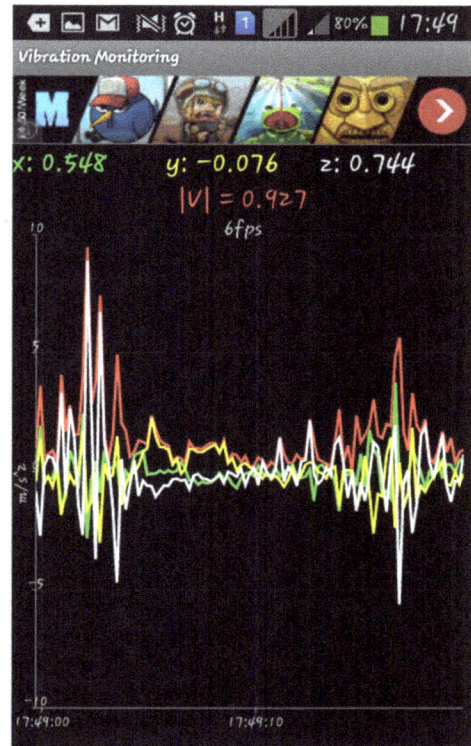

process was better favourable as it saves the need for procurement of accelerometers. As seen in Fig. 11, the measurements are made for all three tri-axial directions and values are shown in a graphical format in real time. *X*, *Y* and *Z* directions (lateral, longitudinal and vertical directions, respectively) are indicated by the green, yellow and white lines respectively, whilst the red line indicates overall vibration.

4.5.2 Nature of Data

The graph is plotted as acceleration against time with acceleration values rising and falling in conjunction with recorded vibration. The green line indicates the comfort threshold below which acceleration values would essentially be comfortable for passengers, whilst the red line is the point above which any acceleration value would be extremely uncomfortable. These lines are based on ISO-2631 which puts its scale at 0.3 for comfort and around 0.8 for extreme disturbances in comfort. The zone between 0.3 and 0.8 (green and red lines) ranges from being slightly uncomfortable to being extremely uncomfortable.

4.5.3 Seaburn to Pelaw

Figure 12 shows the speed/time performance graph of the metro from seaburn to pelaw. As expected, recorded values show that the metro reaches its maximum speed of 80 km/h in all four track sections. The graph also shows some quick changes in speed which is seen as the train accelerates from rest and maintains a relatively constant speed for a specific time frame after which it decelerates to a stop at the next station. A critical analysis of the graph shows that there are also many slight and sudden changes in speed as the vehicle moves along. This is especially seen in the journey

from Seaburn to East Boldon and between Fellgate and Pelaw. This is caused by a combination of a not so smooth speed change system in the driver's cabin, poor track conditions and vibrations of the car body which causes slight and sudden losses in momentum. Further analysis of the graph shows that the commuting time between stations follows the advised timetable with the train taking approximately 3 min from Seaburn to pelaw, 3 min from East Boldon to Brockley Whins, 2 min from Brockley Whins to Fellgate and finally 6 min from Fellgate to Pelaw.

From Fig. 13, it is seen that from the start of each journey, the vibration level starts well below the comfort line at the stations. Acceleration levels at the stations are between 0.03 and 0.1 m/s^2 and these values were expected although the car body is at rest as there are vibrations due to the car engine and also natural whole vibrating bodies. The amount of vibration increases as the train begins to move away from the station; in the journey between each station, a large amount of vibrations falls between the green and red lines. This indicates that when ever the train is at rest, passengers would feel very minuted vibrations which would be comfortable. However when the train leaves the station, vibrations increase and passengers may feel uncomfortable. Looking at these values in the Sperling ride index, we see similar results with the majority of values falling between the comfortable and extremely uncomfortable lines. A point worth of note is that with the Sperling ride index, 35 % of Wz values fall well above the red line, indicating that in 35 % of the time, passengers

Fig. 12 Speed/time graph of the metro from Seaburn to Pelaw

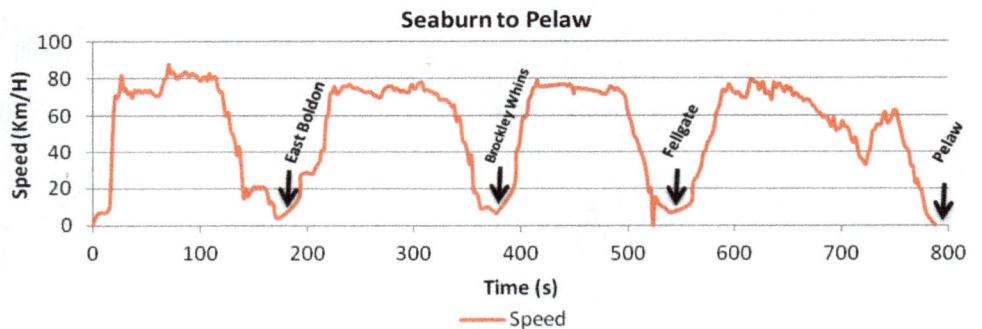

Fig. 13 Seaburn–Pelaw vibration spectrum (time domain)

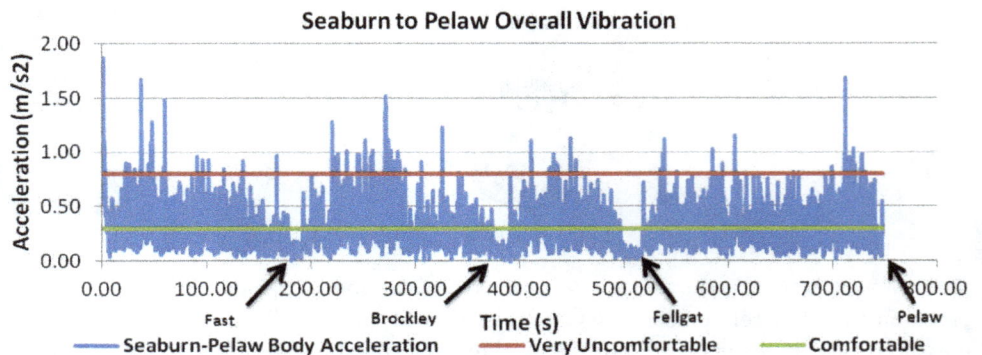

Fig. 14 Seaburn–Pelaw
Sperling ride index spectrum

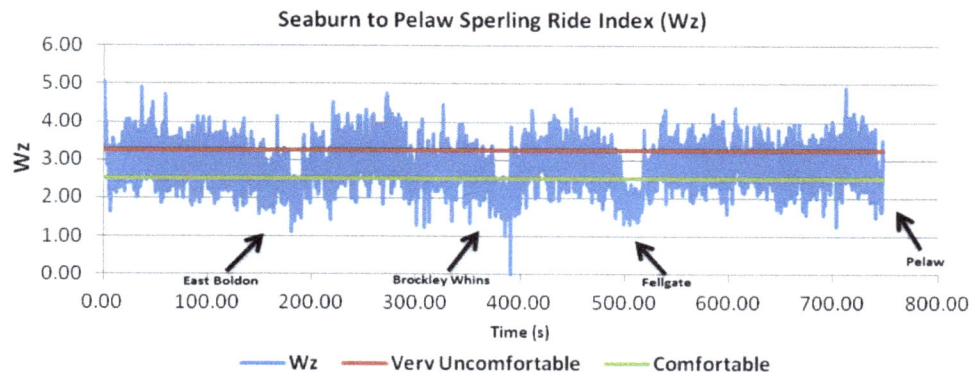

Fig. 15 Pelaw–Seaburn speed
graph

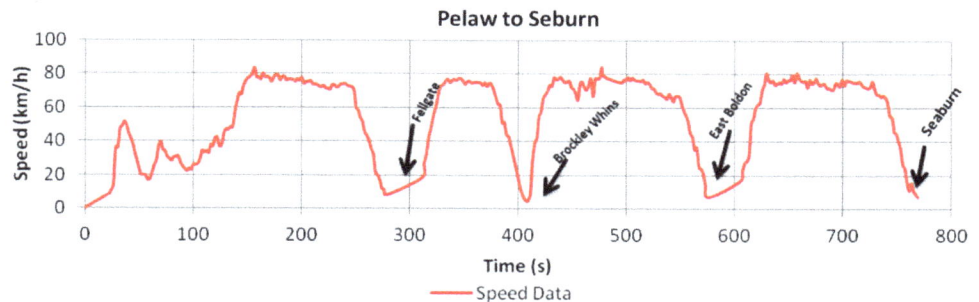

experience very irregular vibrations however it is notable that the Sperling ride index is not as accurate as ISO-2631 [23].

In conclusion, Figs. 13 and 14 indicate that passengers who spent 70 % of their time on the metro will feel vibrations ranging from being slightly uncomfortable to being very uncomfortable. Relating these findings to real situations, it can be predicted that for the current metro system, there may not be extreme cases of vibrations which could make passengers make complaints but there are certainly high magnitude vibrations which would make passengers feel uncomfortable.

4.5.4 Pelaw to Seaburn

On the return journey, a different metro car was used in which comparisons can be made on the values of speed and vibrations from different metro cars. It is expected that considering there are 90 cars among the metro fleet, no two car vibrations would be exactly the same. The dip in speed seen in the beginning of the graph in Fig. 15 between Pelaw and Fellgate is due to a combination of a gradient and curve where the train slows down to about 25 km/h in order to successfully negotiate the curve whilst climbing the gradient. At this speed, vibration stays normal because the speed is relatively low. Despite the change in the metro car, a smooth speed graph could still not be observed as the same sudden changes in speed are seen all through the line.

This goes to show that even with different metro cars, speed is still affected by poor tracks. Further analysis of Fig. 16 shows that the timings were relatively the same with Fig. 12. This observation supports claims of the metro's high reliability.

The trend of acceleration values is very similar to previous results with over 70 % of recorded values above the comfort threshold line. Notice the increase in vibrations between Brockley Whins and East Boldon. This section of the tracks is generally made up of straight line tracks, where better comfort would be expected; however, there is a definite increase in vibration due to below par suspension systems and poor track conditions. When the overall vibration and speed graphs are overlapped, it is noticed that the train vehicle is travelling at a maximum speed but the speed changes suddenly at the same time. This goes to show how the speed is affected by the vibrations of the car body and vice versa. Figure 17 (Sperling ride index) shows similar results with Fig. 14. The train starts from rest, accelerates to about 80 km/h during which point the vibration values rise above the comfort line. Accelerating values fall well below the line as the train decelerates and comes to a stop at each station.

4.5.5 Conclusion on Comfort Test Results

After the analysis of test results, it is clearly seen that the current metro system cannot be considered as a

Fig. 16 Pelaw–Seaburn overall
vibration spectrum (time
domain)

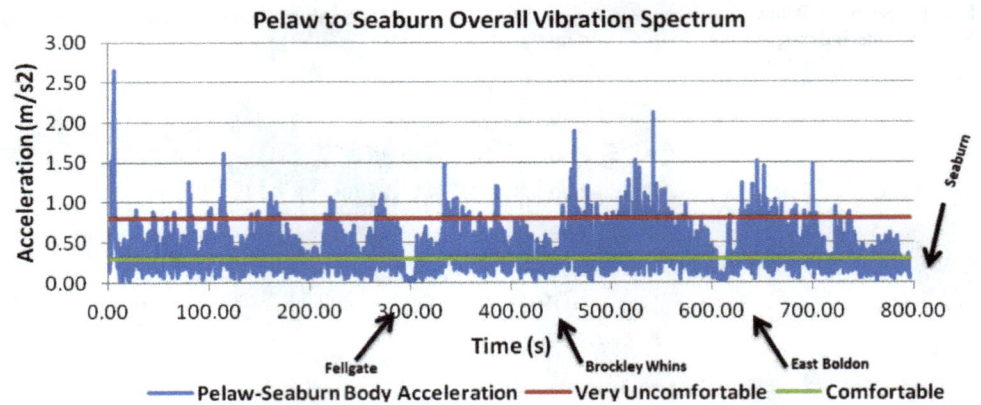

Fig. 16 Pelaw–Seaburn overall vibration spectrum (time domain)

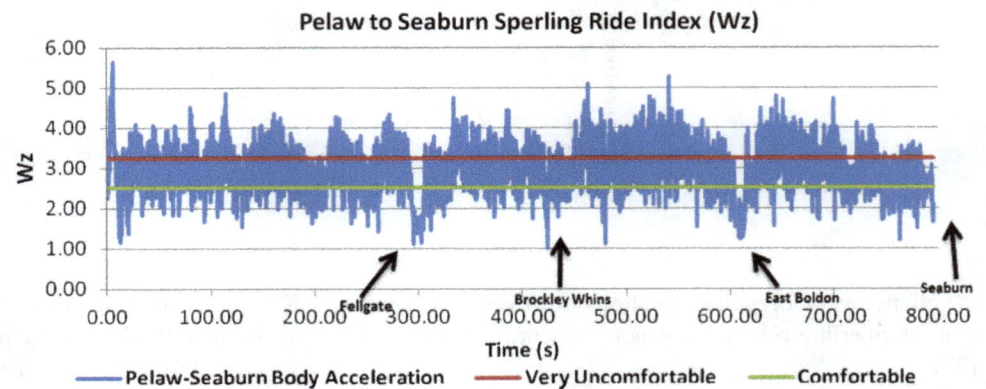

Fig. 17 Pelaw–Seaburn Sperling ride index

comfortable system. This result is quite expected as the system was designed in the 1980s and has only undergone over the years light refurbishments on its performance, cleanliness and seating. However, more has to be done in order to improve its services in terms of customer comfort. The current suspension systems are stiff and almost non-existent as virtually every bump on the rail track is felt by passengers. Better suspension systems would go a long way to providing the required level of comfort which customer really need. When considering cases of passengers who commute a relatively short distance, comfort may not be such an important topic, but when considering passengers who take longer journeys, for example from the airport to south shields or from Haymarket to Sunderland, passenger comfort becomes more of an issue.

5 OpenTrack Speed Simulation

OpenTrack is an object-oriented modelling software which simulates the movement of trains in rail networks. It started as a research project in the mid-1990s at the Swiss Federal Institute of Technology. The aim of the project was then to

produce an object-oriented modelling system for railways which could be used to find practical economic solutions to complex railway technology problems [27]. The software is now used as an established railway simulation tool in the railway industries, consultancies and universities in many different countries [27]. OpenTrack allows users the capability to model, simulate, analyse and make decision in various subject areas including

- High-speed rail systems
- Heavy rail
- Intercity rail
- Commuter rail systems
- Heavy haul freight systems
- Mining rail systems
- Metro/subway or underground systems
- People mover systems
- Rack railways and mountain railways
- Maglev systems

The following types of tasks can be achieved using OpenTrack:

- Viewing the capacity of lines and stations
- Rolling stock studies (future requirements)

- Understanding the requirements for a rail networks infrastructure
- Running time calculation
- Timetable construction
- Evaluation and designing of various signalling systems
- Analysing the effects of system failures

5.1 OpenTrack Data

OpenTrack is a simulation software which uses a combination of various data inputs from users and also data already stored in its database to come up with simulation data [27]. User data would normally include but are not limited to

- Tractive effort values
- Rolling resistance values
- Load and weight of the train
- Length of train
- Length of the tracks
- Nature of tracks (for example bad or good adhesion)
- Power

The trains being simulated run according to the timetable on the selected railway network. During simulation, train movements are calculated, and shown in an animated format under constraints imposed by the user through signalling systems and timetable data input. After the simulation, users can display resulting data and analyse these data in the form on diagrams, train graphs, occupation diagrams and statistics.

5.2 Network Setup

The first step to running the simulation on OpenTrack is by first building the network to be simulated [28]. Having already chosen five initial stations (Fig. 18), building the network simply requires using the build line tool to build lines connecting each designated station and setting a specific distance between each of them.

Determining distances between each station was obtained using Google maps as shown in Fig. 19 and values were subsequently logged into OpenTrack database. Values observed from the Google maps are as follows:

- Seaburn–Brockley Whins = 2.9 km or 2900 m
- Brockley Whins–East Boldon = 2.7 km or 2700 m
- East Boldon–Fellgate = 1.8 km or 1800 m
- Fellgate–Pelaw = 3.7 km or 3700 m

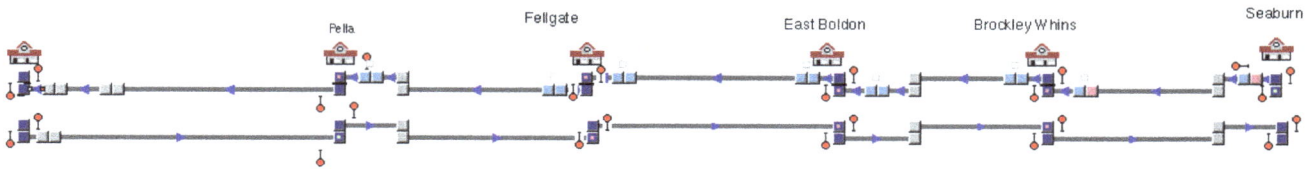

Fig. 18 Built network line on OpenTrack

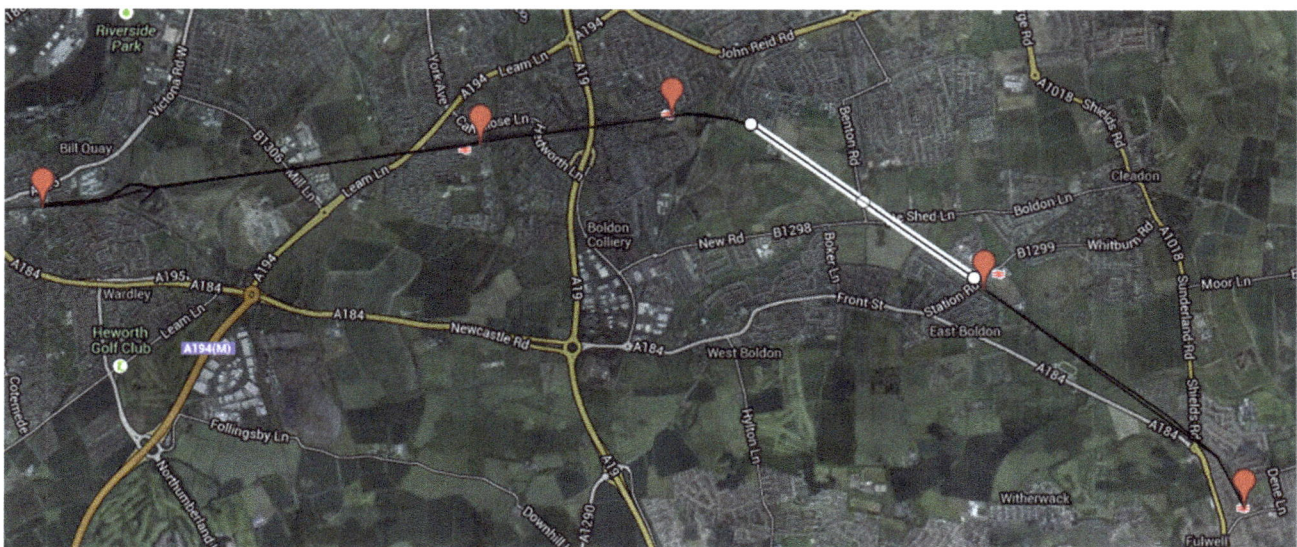

Fig. 19 Google map indications of station location [29]

Unfortunately, all track gradients and curve radius could not be accurately ascertained from available tools or materials thus these values were roughly estimated and put into the database.

5.3 Engine Setup

Input of engine characteristics was done using the engine tool. OpenTrack allows the user to set maximum speed limit by simply defining the necessary tractive effort graph which automatically allocates necessary parameters into the tool [28]. Weight and adhesion values of a fully loaded metro fleet are known as 39 tonnes and length of the train is also known as 27 m. Input of these values allows replication of the current metro system; thus, any change to these values can be easily noticed and compared.

5.4 Train Setup

After setup of the engine is completed, the next step is making sure the engine runs as a train. The train tool helps us achieve this through a selection of trains already in the database or if none are found to be useful [28], the user can create a new train and attach engine parameters to it. Two simulations were made. The first being a replication of the current metro system, whilst the second is the replication of a tilting train. The train type chosen was a commuter train with a deceleration rate of 0.68 m/s^2. This was used because it comes closest to the current metro system. Engine parameters are subsequently copied unto the train tool thus making the setup complete.

5.5 Timetable Setup

During the simulation, the train tries to obey an already set timetable. Timetable data have already been analysed from the current metro timetables and also measured time taken between stations. Approximate time between each stations is given below:

- Seaburn–Brockley Whins = 3 min
- Brockley Whins–East Boldon = 3 min
- East Boldon–Fellgate = 2 min
- Fellgate–Pelaw = 5 min

These values were put into the system's timetable tool whilst strict checkbox is unchecked. This allows the simulation to run correctly whilst not necessarily adhering to the input timetable. This setting allows the viewing of any differences between real situations and ideal situations. In a real situation, extra times are added to actual times to allow for possible delays of unforeseen circumstances. This simulation would show the metro car behaviour in an ideal situation and total time taken between trains. Dwelling time value of 30 s is used to allocate the required amount of time the system would take to stop at each station. This would replicate the metro train stopping and starting at each station which allows passengers to embark or disembark from the vehicle.

5.6 Simulation

After all required routes, paths, parameters and system characteristics have been built, simulation is started. The simulation tool is used which shows the progress of the train as it commutes from each station to the next. Two simulations were made using this system with the first aimed at replication of the present situation of the metro and the second done with slight change of system parameter in order to simulate a tilting system as closely as possible. The simulation runs from Seaburn station to Pelaw station on the following assumptions:

- It is a clear sunny day, thus adhesion between track and wheel is good.
- There are no tunnels between all five stations thus tunnel adhesion is good.
- The train vehicle is in impeccable condition thus performance is 100 %.
- Track quality is good.

5.6.1 Metro Simulation

Figure 20 shows how OpenTrack simulates the train. The simulation tool is used and at the click of the start button, simulation is started. A green label is seen moving on the built network indicating that the built network is valid and trains would be able to commute. Time stamps show the progress being made by the train as it accelerates and

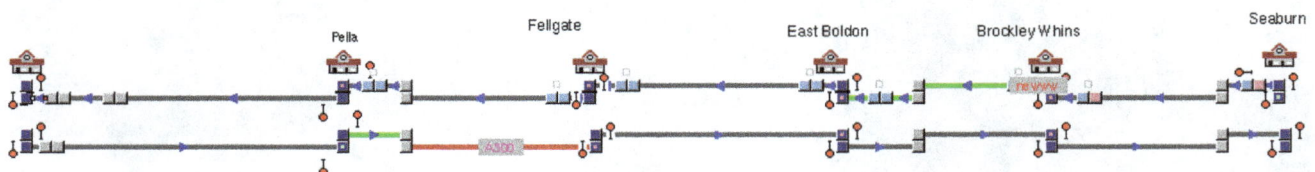

Fig. 20 Simulation of metro

decelerates at each station. With this simulation, the train takes approximately 11 min and 44 s (706 min) to move from Seaburn to Pelaw. The real situation shown in Fig. 13 takes approximately 13+ min. Slight difference can be explained by the dwelling time of the train at each station which can vary widely depending if train drivers wait for longer or shorter times for passengers in cases where the train arrives earlier or later than scheduled time. Approximate dwelling time for this train simulation was set at 30 s for each station.

5.6.2 Tilting Simulation

Several parameters like performance and timetable constrains were changed in the system so as to closely mimic tilting technology systems. Scientific research claims that tilting technology reduces commuting time by as much as 15 % [12]; however, since the current metro actually reaches its maximum speed of 80 km/h, a new metro engine has to be incorporated and this is done in this simulation by increasing engine parameters by approximately 15 %. After simulation, it was observed that the train takes approximately 9 min 48 s to make the journey. When compared with the previous simulation, it is found that a 17 % reduction in commuting time was achieved. If such is the case, it can be claimed that a 17 % increase in engine power leads to a 17 % decrease in commuting time. This information would come in handy in managerial decisions which would need to be made in the near future.

5.7 Results

Results achieved confirmed the notion that tilting technology can reduce commuting time in any given journey. Both simulations gave an output similar to real data and this is evident when they are plotted on the same graph. In Fig. 21, it is seen that when compared to real situations, travel time is reduced by as much as 10 % with a 15 % increase in maximum achievable speed. It is noted that the time difference between the real and non-tilt simulated

situations is about 40–50 s. This can be explained by differences in dwelling times in both situations. The differences of the sum of each station dwelling time would give the difference in time shown on the graph. For example, if in real situations, average dwelling time for each station is 40 s, the sum would be 2 min. The average dwelling time chosen for simulation is 30 s with total dwelling time summed up as 1 min and 30 s. Difference between both dwelling times gives the lag time between real and simulated time values. Other results observed include

- Sudden changes in speed due to poor track conditions and vibrations are virtually eliminated or non-existent as the state of poor track conditions were set as good.
- The dip in speed normally seen between Fellgate and Pelaw is still seen; however, the simulated trains seem to be able to negotiate this curve at a much higher speed. The tilting train simulation travels at 55 km/h at this point whilst the non-tilting train travels at 40 km/h. That means the non-tilting train travels 27 % slower than the tilting train at this curve.
- The maximum speed of the tilting train as expected is 90+ km/h and it maintains this speed for about 1 min. An interesting observation is seen with the old metro which maintains its maximum seep of 80 km/h for almost 2 min. It is expected that the new metro would be faster at this section but it is unexpected that it would be almost 50 % faster. A future research could be made to ascertain why this is.

5.7.1 Accuracy and Validation

Both vibration and speed analysis done on the metro were done in a manner which could produce as accurate a result as possible; however, several parameters could reduce or undermine the maximum accuracy of the tests. Possible problems which could undermine the maximum achievable accuracy could include

- A strapped phone to the thighs or chest may have some inaccurate readings arising from normal human body

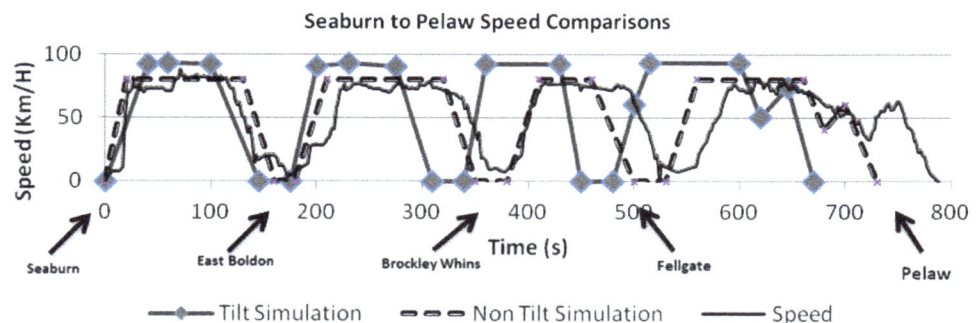

Fig. 21 Speed comparisons between real and simulated speed graphs

movement (movements not induced by train motion) although conscious efforts were made to limit this factor as much as possible. It can be stated with 100 % confidence that at least 95 % of data recorded in all vibration tests were due to train movements.

- Phone app data may not be as accurate as accelerometers, however it can be predicted with confidence that actually accelerometer values would be slightly higher than phone app data values.
- The tests only confirmed that the current metro system does require an increase in ride comfort, it does not determine the effects of tilting technology. Unfortunately, accurately determining tilting technology effects on the metro cannot be done without a properly modelled system either with prototype train assemblies or a simple similar test done on a working tilting train. All options are limited by time allocated for this project.
- Using a phone app as opposed to actual accelerometer devices which could give better readings of vibration.
- Speed graphs were done also using a phone app which may or may not have been affected by cell reception at the time. However during testing, it can be stated with reasonable confidence that this is not the case.
- Replication of the Tyne and Wear Metro network was not 100 % complete due to insufficient curve data of the network although every endeavour was made to input major curve data on the network.

All of these factors may reduce the overall level of accuracy of these tests but it can also be confidently stated that the tests were successful in reasonably identifying the level of comfort currently experienced by the passengers and the level reduced travel time should tilting technology be incorporated in the current metro fleet.

6 Conclusion and Discussion

A review of test results shows that tilting technology does have the merits of being a suitable choice of technology which could improve both performance, commuting time and ride comfort. This project has shown that in the current metro network

- An increase to a top speed of 93 km/h is achievable on the metro network through the use of tilting technology.
- The current metro system may not be overly uncomfortable but could do with significant improvements to ride comfort which can be provided by tilting technology.
- Commuting time can be increased by as much as 10 % for every 15 % increase in speed which can be provided by tilting technology.

- With the current metro fleet scoring high in majority of customer satisfaction factors, an increase in comfort would significantly increase passenger patronization.

In conclusion, from a management point of view, these benefits may not be enough to properly pursue implementation of tilting technology in the current metro system. A cost/benefit analysis would likely be the deciding factor should such a decision be made. Considerations on the amount of infrastructural upgrading of each train fleet or whole track network makes for a much less optimistic view of tilting technology systems being used in increasing customer satisfaction, speed or even customer comfort. With the current metro system really only in need of a comfort upgrade, the amount of upgrades required to implement tilting technology just does not in the author's opinion seem like a wise business decision.

6.1 Recommendations

A more reasonable solution could be the implementation of a system which does not necessarily improve speed but definitely improves ride comfort without the vast need for system upgrades. Obviously such a system would be mainly focused on the vehicle suspension. One such system being researched with such promise of improving ride comfort is the active or semi-active lateral suspension (ALS) system researched by FIAT [30]. This is a system which basically works by introducing fast controlled hydraulic dampers placed both vertically and longitudinally on the bogie frame. The system can also be combined with tilting technology systems and has been claimed to have the capabilities of increasing comfort by over 37 % [30, p. 10]. Figure 22 shows the possible impacts of a 37 % decrease in overall vibrations of the current metro system and it immediately becomes apparent that the impacts could be quite positive. The level of comfort falls well

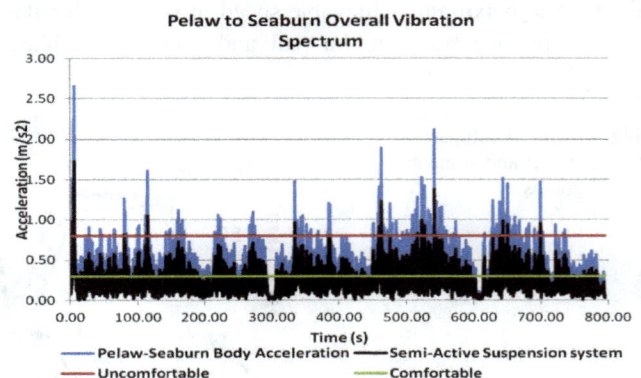

Fig. 22 Data showing possible impacts of implemented active suspension systems on metro

below the red line with more than 95 % of acceleration values below this line. Considering the availability of such a system, it may be more logical or economic to implement an active suspension system as opposed to a tilting technology system.

6.2 Future Study

Future studies on this topic should focus on ALS systems and its suitability to ride comfort. If the current metro system is to be improved, emphasis should be placed on ride comfort. As this study has confirmed that tilting technology may be beneficial to improving speed and comfort, it has noted that expenditures required for such an upgrade far surpasses the required benefits. ALS systems have been proven to work impeccably well in the automotive industry [31, 32] and is being fully researched for train applications [33–35]. It can be predicted that considering speed is not important in the current metro network, an implementation of active suspension systems can meet the required comfort levels which would warrant highly increased customer satisfaction without the need for an "over the top" overhaul of the entire metro fleet and network. Happy customers make for increased revenues and considering that the current metro scores highly in many customer satisfaction factors, growth appears to be limited by its current level of ride comfort.

References

1. Stout KL (2013) Deadly high speed train crash. CNN, Hong Kong
2. Kim NP, Park TW (1999) A study on the dynamic performance of the 200 km/h Korean tilting train by means of roller rig test. J Mech Sci Technol 23:910–913
3. Persson R (2011) Tilting trains: enhanced benefits and strategies for less motion sickness. KTH Royal Institute of Technology, Stockholm
4. Forstberg J, Andersson E, Ledin T (1988) Influence of different conditions for tilt compensation on symptoms of motion sickness in tilting trains. Brain Res Bull 47(5):525–535
5. Barnett R (1992) Tilting trains: the Italian ETR and the Swedish X-2000, vol 113. California High Speed Rail Series, California
6. Yashushi N, Yoshi S, Shoji N, Hiroaki N, Toshiaki H (1997) Tilting control system for railway vehicle using long-stroke air springs
7. Shikimura A, Inaba T, Kakinuma H, Sato I, Sato Y, Sasaki K, Hirayama M Development of Next-generation Tilting Train by Hybrid Tilt System. s.l.: Hokkaido Railway Company, Sapporo; Railway Technical Research Institute, Kokubunji
8. Zhou R, Zolotas A, Goodall R (2011) Integrated tilt with active lateral secondary suspension control for high speed railway vehicles. Mechatronics 21:1108–1122
9. Nedall B-L (1998) The experience of the SJ X2000 tilting train and its effect on the market. 103, In: Proceedings of the institution of mechanical engineers. J Rail Rapid Transit Part F 212:103–108.
10. Persson R (2008) Tilting trains: technology, benefits and motion sickness. Stockholm Network, Stockholm
11. Hitachi-rail.com (2014) Pneumatic mechanism. http://www.hitachi-rail.com/products/rolling_stock/tilting/feature04.html. Accessed 17 April 2014
12. Frostberg J (2000) Ride comfort and motion sickness in tilting trains: human responses to motion experiments and simulator experiments (TRITA-FKT). Adtranz, Sweden
13. Pearson JT, Goodall RM, Pratt I (1998) Control system studies of an active anti-roll bar tilt system for railway vehicles. J Rail Rapid Transit 43:212
14. Persson R (2007) Tilting trains: description and analysis of the present situation. KTH Rail Vehicles, Stockholm
15. Hitachirail.com (2013) Predictive tilt control system. http://www.hitachi-rail.com/products/rolling_stock/tilting/feature02.html. Accessed 06 Jan 2014
16. Urbanrail.net (2014) Newcastle-upon-Tyne. http://www.urbanrail.net/eu/uk/new/newcstle.htm. Accessed 25 Aug 2014
17. Railway-technology.com (2013) Tyne & Wear Metro, United Kingdom. http://www.railway-technology.com/projects/tyne/. Accessed 20 July 2014
18. NEXUS (2014) Metro Strategy 2030: background information. http://www.nexus.org.uk/sites/default/files/Metro%20Strategy%202030%20summary%20document.pdf. Accessed 30 April 2014
19. Anna H (2014) Plans to expand the Tyne and wear metro. http://www.capitalfm.com/northeast/on-air/news-travel/local-news/plans-to-expand-the-tyne-and-wear-metro/. Accessed 28 June 2014
20. Lauriks G, Evan J, Förstberg J, Balli M and Barron de Angoiti I (2003) UIC comfort test. http://www.vti.se/en/publications/pdf/uic-comfort-tests-investigation-of-ride-comfort-and-comfort-disturbance-on-transition-and-circular-curves.pdf. Accessed 26 April 2014
21. Greater-Manchester-Commitee (2014) Transport for customer satisfaction survey: rail passenger survey autumn 2013.Transport for greater Manchester committee report for information, Manchester
22. Frostberg J (2000) Ride comfort and motion sickness in tilting trains: human responses to motion environments in train and simulator experiments. s.l.: TRITA-FKT
23. Ramasamy N, Shafiquzzaman K, Mats B,Virendra Kumar G, Huzur Saran V, Harsha SP (2014) Determination of activity comfort in Swedish passenger trains. http://www.uic.org/cdrom/2008/11_wcrr2008/pdf/R.2.4.3.2.pdf. Accessed 12 April 2014
24. Sathishkumar P, Jancirani J, Dennie J (2014) Reducing the seat vibration of vehicle by semi active force control technique 2. J Mech Sci Technol 28:473–479
25. Annelli O (2010) Methods for reducing vertical carbody Vibrations of a rail vehicle. KTH Engineering Sciences, Sweden
26. Kufver, B (2014) EN standard 12299 for evaluation of ride comfort for rail passengers. http://www.railwaygroup.kth.se/polopoly_fs/1.347076!/Menu/general/column-content/attachment/Bj%C3%B6rn%20Kufver%201.pdf. Accessed 08 Dec 2014
27. OpenTrack.ch (2014). OpenTrack railway simulation. http://www.opentrack.ch/opentrack/opentrack_e/opentrack_e.html. Accessed 08 April 2014
28. Huerlimann D, Nash AB (2014) OpenTrak manual, simulation of railway networks. s.l.: OpenTrack Railway Technology Ltd and ETH Zurich Institute for Transport Planning and Systems
29. Google (2014) Seaburn-Pelaw track map. s.l.: Google maps
30. Montiglio M, Stefanini A (1999) Development of a semi-active lateral suspension for a new tilting train. s.l.: Fiat Research Centre

31. van der Sandea TPJ, Gysen BLJ, Besselink IJM, Paulides JJH, Monolova EA (2013) Robust control of an electromagnetic active suspension system: simulations and measurements 2. Mechatronics 23:204–212
32. Abduljabbar ZS, ElMadany MM (2000) Optimal active suspension with preview for a quarter-car model incorporating integral constraint and vibration absorber. Cairo-Egypt: international MDP conference
33. Sasaki K (2000) A lateral semi-active suspension of tilting trains. 1, QR of RTRI, vol. 41, Ken-Yusha, Tokyo
34. Foo E, Goodall RM (2000) Active suspension control of flexible-bodied railway vehicles using electro-hydraulic and electro-magnetic actuators 5. Control Eng Prac 8:507–518
35. Tanifujia K, Koizumi S, Shimamune R (2002) Mechatronics in Japanese rail vehicles: active and semi-active suspensions 9. Control Eng Prac 10:999–1004

18

Testing the Efficacy of Platform and Train Passenger Boarding, Alighting and Dispersal Through Innovative 3D Agent-Based Modelling Techniques

Selby Coxon[1] · Tom Chandler[2] · Elliott Wilson[2]

Abstract Suburban railways around the world are experiencing a rapid increase in patronage. Higher passenger densities, particularly during peak times of the day, have implications for train punctuality, crowding, accessibility and passenger comfort. Research indicates that the design of the train carriage and the impediments of platform furniture all have an influence on accessibility and passenger dispersal, with consequences for service punctuality and network capacity. Building new concepts in train and station design are expensive undertakings and carry with the investment a high level of risk. Computational simulation methods such as agent-based modelling (ABM) can mitigate this risk at much lower cost. Many contemporary ABM modellers represent passenger flow at a macroscale, often in a single plan view and with agents travelling at same speeds and represented crudely as dots on a flat plane. This paper discusses a body of work concerning the building of a boarding and alighting simulator at a more detailed scale where a deeper and richer experience of crowd behaviour has been modelled using 3D animated figures. The primary benefit of these methods of evaluation is that they take away the expense and lack of realism present in experiments with full-size mock-ups. The outcomes of this work have resulted in sophisticated imagery, underpinned by technical accuracy that provides a tool for

the development of station infrastructure, train carriage design with implications on timetabling and network planning.

Keywords Dwell time · Agent-based modelling · 3D

1 Introduction

Rail is an important contributor to the movement of people and goods in many of the world's large cities. Suburban, metro and subway systems are very efficient in terms of the number of people moved relative to land use. Rail is a popular means of transport and becoming more so as urban populations increase. In the latter part of the 19th century, when the London Underground opened, only 10 % of the world's population lived in cities. Now in the early 21st century, over 50 % of the world's population live in a city [1]. In terms of transit use, 80 % of the population of Tokyo uses the subway, making some 2930 million passenger journeys per year (2009 figure) (Ibid), the highest level of patronage anywhere in the world.

Trains are independent of congested road traffic conditions and therefore have the potential to be faster at delivering passengers into city centres. Automation and advances in signalling reduce the impediments to a smooth and timely rail system. The growth in city populations has fuelled increased rail patronage with the consequence that many train networks can struggle to be punctual. The most significant variable in the journey of a train is the time it will take paused at each station. This 'dwell' time is at the mercy of how long it takes passengers to board, alight and disperse within the train carriage or across the platform. At peak periods, dwell times can become extended as passengers jostle to board or alight. It is

✉ Selby Coxon
  selby.coxon@monash.edu

[1] Faculty of Art Design & Architecture, Monash University, 900 Dandenong Road, Caulfield, Melbourne 3145, Australia

[2] Faculty of Information Technology, Monash University, 900 Dandenong Road, Caulfield, Melbourne 3145, Australia

Editor: Baoming Han

general practice that timetables have built-in 'recovery' time and attempt to predict extensions of dwell time during peak periods. However, with increased patronage, the predictability of dwell times becomes more difficult [4]. Delayed trains create a number of implications beyond poor punctuality, including the extension of headways (the time gap between services). This extension is especially onerous if the lines are shared with express services and freight trains. Extended dwell times reduce network capacity leading to less services and more late services, ultimately impacting upon an operator's revenue and contributing to poor passenger perceptions of the mode.

Dwell time predictability is important in the creation of service timetables. To this end, operators subdivide the dwell time to better understand where problems lie. Current timetable orthodoxy determines dwell times by mathematical means. While there are variations to the formula, they all in essence treat boarding and alighting as a linear period of time multiplied by a coefficient representative of how much passengers have been slowed down by the circumstances of other passengers, width of the doors and if they are carrying belongings [5]. Accurate calculation of these dwell times will inform operators of the predicted capacity of the network and so drive timetables with some accuracy.

However, while building mathematical models might simplify determining dwell times as they may be, they also mask the intricate composition of the causes of extended dwells. Studies show [2] that there is a wide range of qualitative variables that impact upon passenger behaviour while boarding and alighting. These factors include the prevailing culture of the passengers, their age, relative athleticism, the gap between the platform and the train, the level of the occlusion at the door and their motivations once within the train to finding a seat. These human factor variables are difficult to be determined quantitatively, but they do relate strongly to the interface between the passenger and carriage. Figure 1 shows the points between predictable timing with where the unpredictable variation in dwell is located. Figure 2 encapsulates, as a flowchart, each of the 'factors' that affect the efficacy of the passenger

to board or alight from a train and by implication impact upon the dwell time variability of the service.

Extended dwell times imply difficulty in passenger boarding and alighting at anyone or more of the stages outlined above. With significant increases in patronage, particularly during peak time, crowding itself is the significant determinant of extended dwell times. While passengers may not be particularly aware of the wider implications of delays at the station during boarding and alighting, crowding (the cause of the delay) tends to have a greater impact especially upon the perception of comfort.

2 Measurement and Evaluation Methods: The Role of Computational Modelling

Historically, methods of determining boarding and alighting performance have been confined to the building of full-size mock-ups and inviting a sample group of passengers to enter and alight from the interior and a platform structure. This method has varying degrees of success as a method of dwell time calibration. Documenting the experimental process through, for example, video aids the evaluation and decision making of the manufacturer. However, this method is time consuming and costly and to some extent flawed by the unreal nature of the setting [2].

Agent-Based Modelling seeks to direct digitally animated 'agents' by way of a series of algorithms originally derived empirically. ABM interactions exhibit the following two properties:

(1) The interactions are composed of individuals (agents) with a designated set of characteristics.
(2) The system in which these interactions take place exhibits emergent properties, that is, new properties arising from the interactions of the agents that cannot be deduced simply by aggregating the combined properties of the agents.

The primary benefit of these methods of evaluation is that they take away the expense and lack of realism present

Fig. 1 Linear diagram of dwell time structure

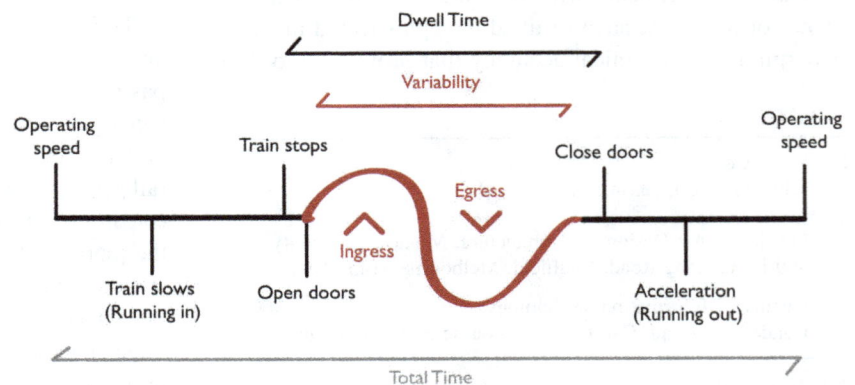

Alighting

Preparation to alight ─────

Awareness of location ─────

Access to exit

Cultural behaviour ◄── Negotiating passing other passengers

Doorway configuration

Boarding

Emergence onto platform ─────

Dispersal along platform ◄── Extent of patronage ◄── Coverage from weather elements

Queueing behaviour ◄── Time of day ◄── Time pressure

Locating door to carriage

Ingress / Egress

Physical ability ──→ Disability ──→ Age ──→ Gender

Doorway configuration

Threshold ─────

Step(s) ─────

Gap between platform and train ──→ Holding of doors

Negotiating passing other passengers ──→ Cultural behaviour

Personal belongings ──→ Carried / Stowed

Dispersal within carriage / out onto platform

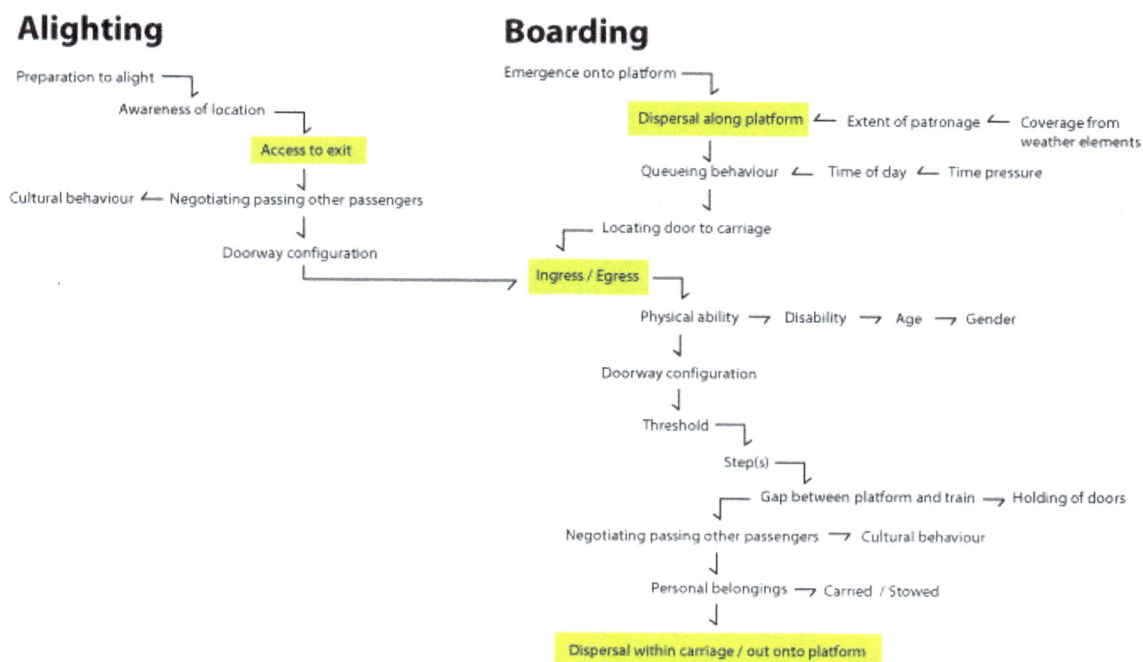

Fig. 2 Flow chart of boarding and alighting influences

in physical experiments with full-size mock-ups. In computer simulation, animated passengers are programmed to undertake simple tasks with directed goals, e.g. board and find the nearest free seat. This is done irrespective of any sense of urgency that might be present at a real boarding or lack of urgency at a static mock-up.

ABM begins with assumptions about the agents (passengers) and their interactions and then uses computer simulation to reveal the dynamic consequences of these assumptions. ABM researchers can investigate how large-scale effects arise from the micro-processes of interactions among many agents. Large-scale effects of interacting agents can be surprising because it can be hard to anticipate the full consequences of even simple forms of interaction. For problems such as determining the ebb and flow of large groups of train passengers where predicting the effects of individuals on each other is difficult, ABM techniques have great potential. What is difficult to determine is how accurate and representative the salient aspects of the agents are of the travelling public. In highly sophisticated simulations, it is possible to equip the agents with the ability to learn and develop over time. The key issue here is the extent to which the resulting outcomes are orderly within the environment where they have been placed.

3 Developments in Computational ABM

While the simulation of 3D environments is a long established practice in the computer-aided design and engineering disciplines, the inclusion of 3D human models with

animated behaviours into crowd modelling spaces remains comparatively rare. Some of the reasons for the absence of human models can be attributed not only to the unique challenges involved in animating 3D figures, but also to the absence of animators and computer game researchers involved in the simulation design process. Recent advances in game engine technology and computer processing power have enabled human 'agents' to be realistically graphically depicted in simulations, walk cycles, idle cycles and a range of poses and animated behaviours that can enhance the interpretability of the simulation (Fig. 3).

4 Creating the 3D Characters

Fundamental to the authenticity of animated passengers is their walk cycle. For the 3D characters designed for the simulation discussed here, different animated walk cycles were designed to communicate whether an agent was male or female, young or old. Sped into a run, a walk cycle can also convey the urgency of the agent's objectives. When at rest, such as waiting for a train or as a passenger on board the animated characters function in an idle cycle, which is an animated loop of small and almost imperceptible movements that characters make when they are sitting or standing. Characters may, for example, shift their balance slowly from one foot to another or turn their head slightly in each direction to look about. The 3D characters in this research were modelled as Melbourne city commuters; they were realistic enough that their facial features could

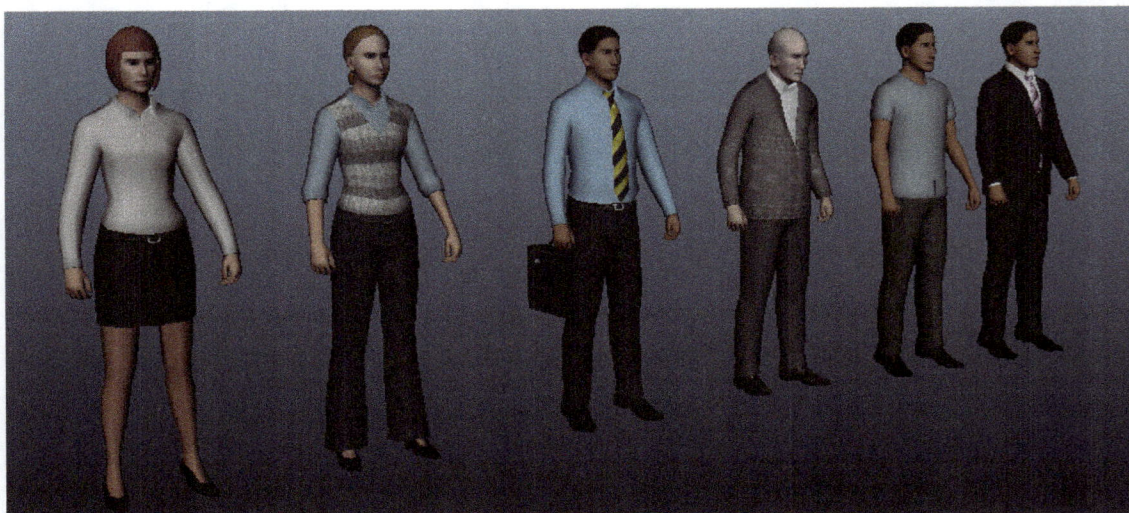

Fig. 3 Selection of 3D commuter characters: the visual and animated embodiments of the agents in the simulation. Created by Chandara Ung

be discerned and attire to indicate gender or replicate the dominant dress styles of commuters during peak periods. Their clothes were to some degree interchangeable and appropriate for Melbourne's temperate climate, though not for any particular season. The male and female figures ranged in age from high school students to retirees and accommodated a number of body shapes and ethnicities.

As the simulation was developed, the interplay of character typologies, attire and idle cycles presented challenges. For example, the sitting posture in the idle cycle differs for women and men reflecting their physiology. Creating a stance that could be perceived as relaxed and not tense required subtleties in the building of each iteration of character model. The building in of commonplace distractions such as the inspecting of mobile phones and the shifting of body weight all added to the appearance of a realistic environment (Fig. 4).

5 Outline of Simulation Logic and Rendering

The *Unity* game engine was created in 2005 by Unity Technologies. The software allows for the creation of interactive 3D content with minimal technical effort. Though mostly used in video game production, it has also become a popular visualisation and scientific research tool in a number of varied disciplines. In this project, the Authors used *Unity* to develop and test the efficacy of carriage interior seat and impediment arrangements. *Unity* has a pathfinding engine built-in which allows the author of the simulation to define which areas of the scene are "walkable" and the different kinds of "agents" that will move through the scene Fig. 5. Once these have been defined, the author can write code that will give the agent a goal position. The agent will then calculate the shortest path towards this goal through the "walkable" area thus avoiding issues such as walking through walls and other designated solid objects. The path is updated in real time to avoid walking into or through other agents within the simulation. In order to determine this goal position, each 3D character was programmed to follow a simple logic tree and make decisions depending on their current situation. For example, an agent who has walked into the train carriage will 'look' for a free seat (deciding on the closest free seat near them) and then walk towards that seat. As it is walking the agent is constantly rechecking if the seat is still free (another agent could have taken it). If the seat is free and they are close enough to sit in it, they will sit down. Once the agent is seated, the simulation considers them 'complete' and no further action is required. A flowchart outlining the decisions that each agent makes can be seen below in Fig. 6.

6 Application of this Technique with an Alternative Design of Carriage

To test the efficacy of the modelling simulation, example carriage interior designs were prepared. This was part of a long-running larger project concerning dwell time and passenger dispersal behaviour. A model of the existing rolling stock was designed to create a baseline performance of the system. A concept design whereby the interior was re-arranged to accommodate a different seating arrangement and an alternative set of doors was devised. The alternative test design concept attempted to manipulate passenger flow with the following strategies:

- re-arranging seating into centrally mounted clusters to open dual corridors through the carriage;

Fig. 4 A 3D character in 'walk cycle' and an idle loop for standing and sitting. Created by Chandara Ung

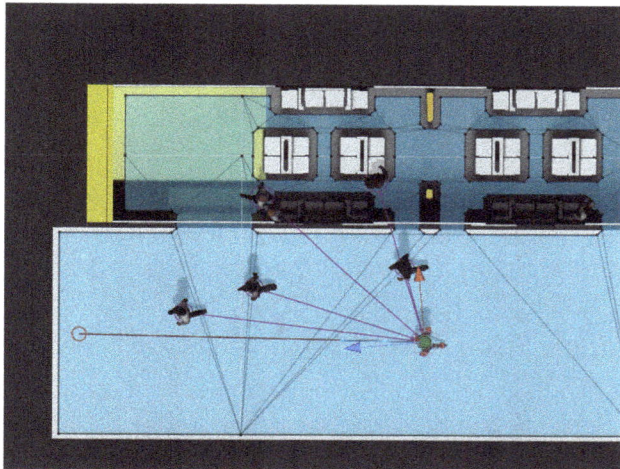

Fig. 5 Plan view of the mapping of agent movement in Unity. Areas permissible for agents to walk in are outlined in *blue*

- door arrangements that dictate passenger flow either by dint of their width (greater than 2 m) or by the implementation of 'rules' to impose passenger behaviours, particularly in determining the direction of movement to and away from ingress and egress;
- creating the largest possible seating capacity.

The relative performances were then compared using the simulation with variable crowding figures. The conceptual design contained a central arrangement of seating clusters with dual corridors running along the length of the carriage and high numbers of folding and perch seats. The central innovation offered here is the concept of the 'peak door'. In essence, the authors are speculating that the three-door arrangement as utilised on contemporary trains remains in place and that extra boarding and alighting capacity is only required at certain times of day and at those times an extra two doors per side become operational. These peak doors would be relatively discreet during the off-peak period and indeed folding seats would be located across the temporary vestibule. These seats would fold into the door framework at designated times on the early shoulder of the peak period, locking into place as the doors become operative. Seating would be lost, but standing space is increased (Fig. 7).

The operational mechanism by which these peak doors are implemented would be aligned with the start and finish of services. A change in the prevailing culture of passenger expectation would be required for patrons to come to expect extra doors to be operative at certain times of the day. This concept opens up the notion of internal space being flexible beyond the use of folding seating. For most of the day, the services can manage a dispersed patronage through the carriage, and, with the exceptions of some accessibility issues, only at peak times do crowding, poor dispersal and lengthened dwell times reflect negatively on the carriage design. The use of peak doors, utilised only temporarily, overcomes the issue of the loss of seats due to multiple vestibule spaces.

7 Results

The simulation serves to validate, via experimentation, an improved boarding and alighting time. The validity of the conceptual carriage interior as an arbiter of improved passenger exchange and stabilised dwell times is seen most keenly by applying the simulation to the highest loads.

Boarding Agent

Alighting Agent

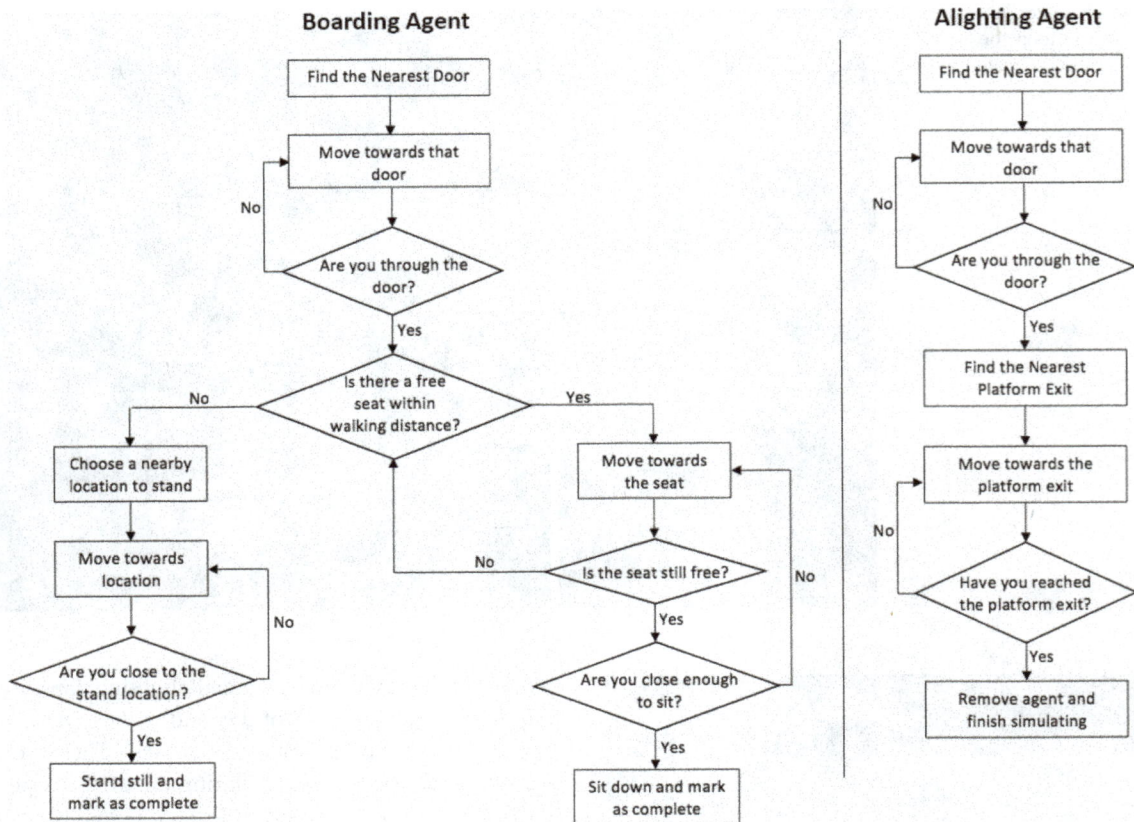

Fig. 6 Flowchart decision tree for agents moving through the simulation

Fig. 7 Plan view of the simulation as applied to the concept interior where Peak Doors have been applied and there is a heavy patronage

The worst case situation for dwell time delays is caused when all the seats in the carriage are taken and the excess in capacity is beginning to build as standees cluster around the door vestibules. When such a train arrives at a significant interchange, where passengers need to alight and significant numbers need to board, then delays in dwell time occur.

In these simulations, both the existing rolling stock (Comeng) and the new concept carriage are populated to the capacity of the simulation with 250 passengers seated and standing. There are no datasets indicating the exact numbers of passenger exchanges, boarding, alighting and standees, so a number of simulations were created, incrementally increasing the percentage of passengers alighting and boarding. The percentage increments are based upon an existing study originally contracted by the Melbourne Transport Operating Company, in which Weston's formula was used to calculate a graphical distribution of anticipated dwell times. In each simulation, the number of passengers either boarding or alighting is increased by 5 %. The distribution of these passengers, both within the carriage and on the platform, is randomly driven by the *Unity* software and so no two simulations can be exactly repeated. For each of the conditions, both carriages fall within the generally accepted dwell time of 20 s; however, the peak door two-corridor solution (Fig. 7) gives consistently quicker passenger exchanges. Only for very small numbers (5 % equating to 14 patrons) were exchange times roughly comparable. As the number of passengers in the boarding and alighting exchange increased, the extra door and corridor space had an observable impact on dwell time reduction.

As an example of an extreme exchange, simulations were run based on the very high passenger loads, with 36 % above the intended carriage capacity. This figure is based on the recorded loading counts at Clifton Hill, Melbourne during May 2011. In this scenario, the train was carrying an average, assuming an even distribution, of 182 passengers per carriage. The test aimed to determine, at this high loading, the results if all the seats remained occupied and the remaining standing passengers all alighted from the carriage to be replaced by the same number boarding. Over 100 repeated simulations, on average the concept train carriage had a 33% shorter dwell time, although it was noted that neither the existing train design nor the concept achieved the desirable standard 20s dwell time at these very high loads (Fig. 8).

8 Future Work

These initial simulations are encouraging in terms of the data they are able to generate and the insights gained therein to how passenger crowds might board, alight and disperse in various carriage arrangements. Extending the range of behaviours that 3D agents exhibit when reacting to other agents poses a number of interesting possibilities. If a crowd consists of passengers mostly travelling alone, the simulation could account for people's general preference in not only their spatial separation but also for seats facing forward and next to a window. The Authors' current simulation reflects an environment most like a peak time commute, whereby most passengers are not associated with a group of other people. What if our agents were travelling in the company of friends and companions? Fine tuning the

simulation to account for groups of agents gathering together near the doors or in adjacent seats would entail a consideration for how other agents navigate around these groups. A large crowd of vocal high school students boarding at a station, for example, brings about a new range of agent behaviours and movement dynamics. We discussed above the possibility of agents moving at very different speeds, but how might agents be convincingly programmed to consider other agents, for example, to give up their seat or make way for elderly or disabled passengers? And finally, how might other agents react to groups of disruptive passengers or even aggressive and intoxicated ones? Refinements of agent behaviours such as these would clearly benefit from the collaboration of behavioural psychologists.

9 Conclusion

Commuter rail is experiencing growth in patronage with higher passenger densities and the effects of crowding, accessibility and extended dwell times. Research indicates that the design of the train carriage and the impediments of platform furniture all have an influence on dwell time performance and therefore network capacity. Building new train and platform infrastructure concepts are expensive undertakings and carry a high level of risk. The Authors have demonstrated that contemporary high-level game engine software can create authentic simulations of crowd behaviour and dispersal to the extent that designs can be tested in advance of implementation.

In these final simulations, it can be seen that multiple doors, dual-flow passenger exchange and dual corridors made for consistently shorter dwell times for the same numbers of patrons as in the current existing rolling stock. Certain assumptions have been made. Motivations for agents to seek a seat and obey certain cultural norms have been assumed and while observable evidence would suggest these assumptions have validity [3], this does not negate all possible behaviours that might be encountered at stations. Equally, passenger motivations discussed in the literature but not applied in these simulations due to technical difficulty include sitting next to known people or away from strangers, sitting in the direction of travel and next to windows, bunching at certain doors, as in rainy or hot sunny conditions and when some patrons circumnavigate control conventions and so work against prevailing crowd movements. Contemporary software such as the one used by the Authors enables these refinements to be undertaken in the next iteration of this simulation tool.

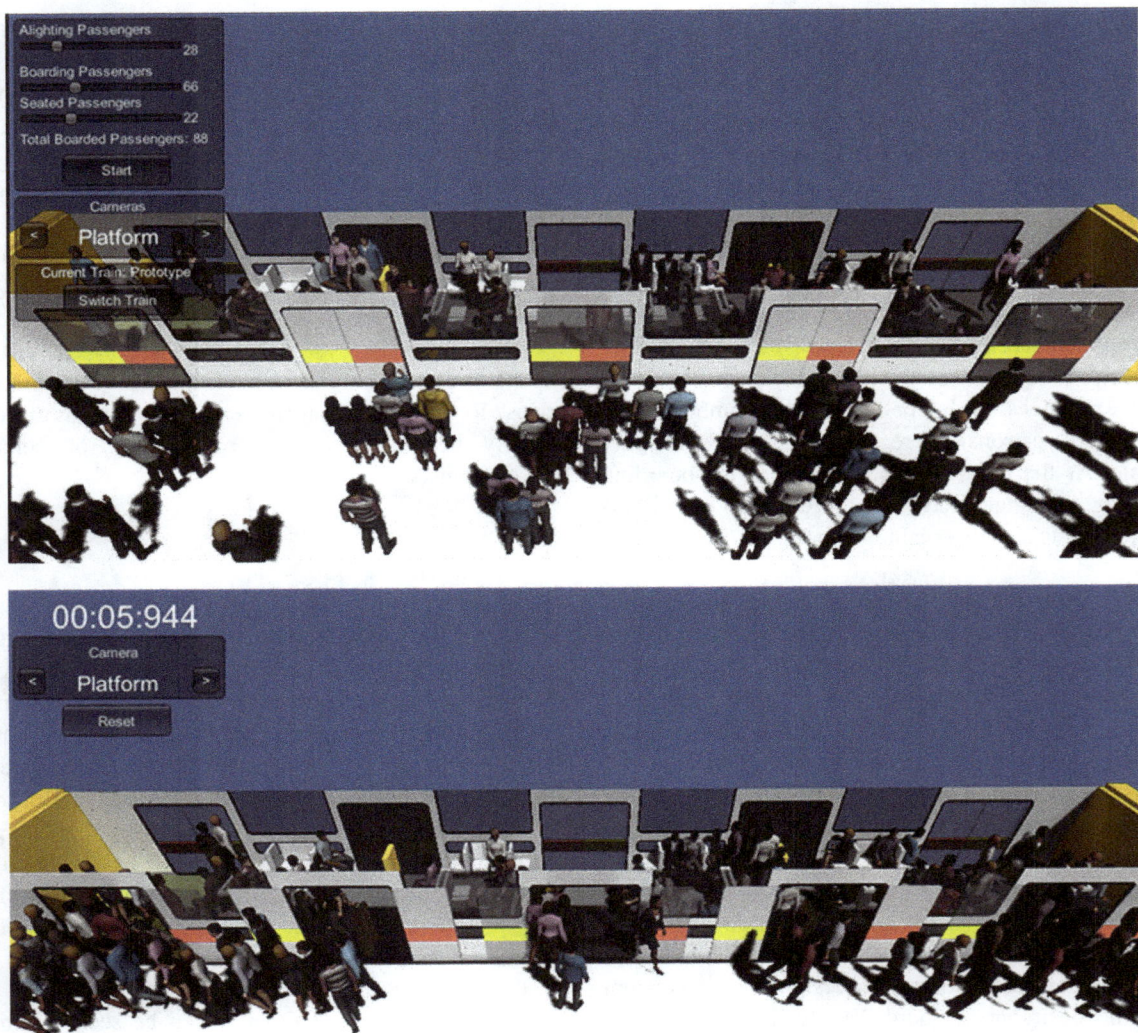

Fig. 8 A perspective view of the simulation demonstrating a realistic three-dimensional view of the passengers boarding and alighting. Note also the user interface in the *top left corner* of the screen where loading patterns of passengers is determined

References

1. Burdett R, Sudjic D (2009) The endless city—the urban age project, London School of Economics & Herrhausen Society. Phaidon, London

2. Daamen W, Lee Y, Wiggenraad P (2008) Boarding and alighting behaviour of public transport passengers. J Trans Res Board 2042:71–81. doi:10.3141/2042-08
3. Hirsch L, Thompson K (2011) I can sit but I'd rather stand: Commuter's experience of crowdedness and fellow passenger behaviour in carriages on Australian metropolitan trains. Australian Transport Research Forum 2011 Proceedings, Adelaide, Australia, 28–30 Sept 2011
4. Ruger B, Tuna D (2008) Influence of railway interiors on dwell time and punctuality. Railway Interiors International, UKIP Media Events, Dorking, Surrey
5. Wiggenraad P (2001) Alighting and boarding times of passengers at Dutch railway stations; analysis of data collected at seven stations in October 2000. TRAIL Research School. Delft University of Technology, Delft

19

Public Perception of Driverless Trains

Anna Fraszczyk[1] · Philip Brown[1] · Suyi Duan[1]

Abstract The global trend for rail automation is increasing but there are very few publications on public perception of the ongoing changes in the railways. In order to fill this gap and to better understand people's perception of driverless trains, the paper focuses on automation of metro systems with a particular interest in unattended train operation (UTO). A survey seeking a public opinion on UTO was conducted, and the results show that 93 % of female and 72 % of male respondents think that a "fake" driver room should be present on a driverless train. In terms of human error, a great majority of respondents expressed no worries about a train design or maintenance issues. However, staff communication, selected by 36 % males and 43 % females, and a technical failure, highlighted by 50 % of males and 43 % of females, were two issues that raised most safety concerns amongst the respondents. Other results related to passenger's safety, employment, advantages and limitations of the UTO, amongst other issues, are presented and discussed in the paper.

Keywords Metro · Automation · Driverless train · Attitudes

1 Introduction

There are 148 cities with metro systems around the world [15], and so far 32 of them adapted automated metro systems [14]. The global trend for automation is increasing

✉ Anna Fraszczyk
anna.fraszczyk@newcastle.ac.uk

[1] NewRail, Newcastle University, King's Gate, Newcastle upon Tyne NE1 7RU, UK

Editor: Baoming Han

with eight new systems being introduced into full operation between 2011 and 2013 [14]. Although the driverless technology is progressing quickly, public perception of unattended train operation (UTO) has not been researched much. With more UTO systems planned for operation by 2025, mainly in Australia, Asia and South America, this paper aims to highlight public perception of driverless trains, which, if taken into account, might help with better understanding of passengers' perspective on UTO and contribute to seamless implementations of the new systems around the globe.

A metro system, or a rapid transit system, is an urban transport system, which uses exclusive rails to run trains of high capacity without interruptions or contact with other transport systems or modes of transport [5]. Metro systems often involve some level of automation, from the most basic automatic train protection (e.g. automatic brakes application) to fully automated and driverless trains (e.g. Dubai Metro). There are four grades of train automation and the highest, with no staff on board, is referred to as UTO [14].

According to Karvonen et al. [7], there are three main reasons for automated train operation (ATO): cost effectiveness, high traffic frequency and flexibility. Moreover, these reasons are accompanied by a number of other advantages, such as punctuality and efficiency, which are widely highlighted by UTO enthusiasts (e.g. Observatory of Automated Metros). However, the UTO has a strong opposition in worker's unions and automation sceptics, who stress the safety issues of driverless trains and the drivers' loss of jobs [1].

Malla [8] argues that from a technical perspective, the debate on UTO having an advantage over conventional rail system is "almost over". However, from a passengers' perspective, the debate on advantages and disadvantages of

UTO continues and is often based on people's perception of driverless trains, in terms of safety, rather than a reality [9]. This paper contributes to the debate by presenting results of a survey on public attitudes to and perceptions of driverless trains. A better understanding of this human–system interaction is important in order to facilitate a smooth shift from conventional to automated metro systems, if this shift is going to happen on a greater scale in the future.

2 History

The debate on automated trains started over four decades ago when a number of publications on benefits of automated metro systems appeared. In 1973, Vuchic reviewed benefits of a driverless train, which included a high-frequency service and the flexible adjustment of train schedules [15]. Also in 1973, Berwell stated that "it is to be expected that the railway should be the first transport system to be automated" and listed a number of reasons for consideration [2]. The reasons, amongst others, included the same ownership and management of infrastructure and vehicles and the fact that automation was already in place with signalling or power control elements of the railway system, which would make a full train automation on a driverless train a natural step forward.

The first fully automated metro system was the SkyTrain introduced in Vancouver, Canada in 1985, which was originally built in time for Expo 1986. It is the oldest and one of the longest ATO systems in the world [10], with three lines and 47 stations in total. Since SkyTrain era, many other cities introduced automated metro systems with driverless technology with Everline in South Korea and Line 5 in Milan being amongst the most recent driverless systems implemented [14].

In 2013, there were 148 cities with metro systems around the world [14] and 32 of them used UTO [13]. The global trend for a full metro automation is increasing with

eight new systems being introduced into full operation between 2011 and 2013 [13].

3 Levels of Automation

There are four grades of train automation (GoA), and Table 1, based on [12], explains the GoAs in more details. In general, the number of the grade depends on staff involvement in basic functions of train operation. The four main automated functions are setting train in motion, stopping train, door closure and operation in event of disruption. In the first grade, GoA1, a driver is involved in all four functions listed in Table 1, but his/her involvement is gradually reduced to zero in GoA4 where a train is fully automatic. The difference between GoA3 and GoA4 is that the first employs a train attendant, whereas the latter grade offers an unattended and fully ATO. UTO means that a rail vehicle runs fully automatically without a train driver or other operating staff onboard. It is a driverless train; however, some operators prefer to put a driver or a member of staff on board (e.g. Beijing Subway's Airport Express operates with a driver in a cab).

4 Advantages of UTO

UITP [12] argues that UTO (GOA4) brings many benefits to all key players in the system: passengers, train operators, funding bodies and staff. The key benefits of the driverless trains are [9, 12, 15] train running time optimisation, average speed of the system increase, headways shortening and dwell time in stations reduction, which all together translate into the first great benefit of UTO which is increased network capacity. Secondly, the UTO enthusiasts [2, 14, 15] argue that by removing a driver from the train, the human-risk factor is reduced and overall safety and reliability of the system increases. Thirdly, in terms of operational costs of a driverless railway system, the

Table 1 The grades of train automation

Grade of automation (GoA)	Type of train operation	Setting train in motion	Stopping train	Door closure	Operation in event of disruption	Example
GoA 1	ATP with driver	Driver	Driver	Driver	Driver	London underground Victoria line
GoA 2	ATP and ATO with driver	Automatic	Automatic	Driver	Driver	Paris Métro line 3
GoA 3	DTO	Automatic	Automatic	Train attendant	Train attendant	Airport express Beijing subway
GoA 4	UTO	Automatic	Automatic	Automatic	Automatic	Dubai metro

ATP automatic train protection, *ATO* automatic train operation, *DTO* driverless train operation, *UTO* unattended train operation
Source Based on [12]

argument is that less train drivers equal cost savings [14], although more staff are recruited for other tasks. Fourthly, automated acceleration and deceleration patters help with energy recovery and savings contributing to environmentally friendly driving and cost savings [9]. Finally, from staff's perspective, drivers are no longer involved in monotonous tasks of driving a train, as their job profile changes, and they can be re-qualified and deployed along the line providing passengers with more customer service and staff–passenger interaction options [4, 14].

5 Disadvantages of UTO

Nevertheless the great number of advantages of UTO, the system has a number of disadvantages too when compared with a conventional system. Firstly, UTO requires a higher cost of implementation as it involves automation at the levels of rolling stock, signalling and platform [12]. Secondly, maintenance costs of the UTO system are higher as additional platform and track protection systems must be installed and maintained. Overall, the initial investment into UTO infrastructure and driverless vehicles is high [11].

Thirdly, from a human–system interaction perspective, as Karvonen et al. [7] argue, as the driver disappears from the train, the significant link between the passengers and the metro system becomes weaker or is lost. Karvonen et al. [7] studied Helsinki metro drivers' behaviour and identified 16 metro train drivers' sub-tasks (hidden roles), such as making announcements to the passengers, guiding passengers out, interpreting events in the environment or fixing small faults in exceptions. In the light of "hidden roles" of a driver, UTO might be a great disadvantage, especially in the case of emergencies happening in the field, as unattended train will no longer provide a driver in situ capable of fixing simple failures or informing the control centre about problems and current situation in the field [3]. Finally, UTO requires a highly qualified maintenance personnel in the field [15], but also in the control room, which leads to changes in driver's job profile and the need for new qualifications and training for staff. Rail trade unions around the globe are generally against UTO arguing that train automation raises safety concerns and causes job loses [1, 6, 11].

6 Methodology

This study used a paper-based survey as a data collection method. Although other methods of data collection, such as focus group, interviews or observations, were also considered, the questionnaire method was selected as best for

the project in terms of the shortest time scale needed for data collection and analysis and the lowest budget required.

6.1 Questionnaire Design

The questionnaire used in the study was designed by a Master student as part of her rail major project focused on passengers' perception of automatic trains. The questionnaire was divided into three parts, with the first and largest part being about attitudes to and perceptions of automated trains, the second requesting information about the respondent and the third part offering space for additional comments.

The questions included in the first part of the questionnaire could be divided into technical questions and questions of opinions and preferences, and examples of the questions are presented in Table 2.

The great majority of questions were of closed type and offered specific answers, and respondents were asked to mark one answer per question only. However, two open-ended questions were included as follow-up on reasons why people might be afraid of using driverless trains and opinions about driverless train technology in general. Overall, the survey included 21 questions on attitudes to and perceptions of driverless train technology and three personal questions asking for respondent's age, gender and country of origin.

6.2 Data Collection

The questionnaire was distributed in July 2014 and was answered by participants of a rail summer school, both students and professors. The student respondents included in the sample had some background knowledge on the complexity of the railway system before completing the

Table 2 Examples of questions and answers included in the survey

Question of opinion and preference	Answer options given
What do you think about driverless train technology?	a. Very good
	b. Good
	c. Neutral
	d. Bad
	e. Very bad
How would you rate the importance of a driver on a train?	a. Very important
	b. Important
	c. Neutral
	d. Not important
	e. Not at all important, not necessary
Do you think the driver room should be built on driverless trains?	a. Yes
	b. No

questionnaire; however, driverless trains and metro automation topics were not included in the summer school's curriculum.

7 Analysis of Results

7.1 Sample Size and Age

The questionnaire was answered by 50 people from 10 countries (see Fig. 1 for details). The age range within the sample was from 17 to 62 with 75 % under 30 years old, and a gender split was 36 men and 14 women. It is shown in Fig. 1 that majority of respondents represented six countries: Romania (8 respondents), Italy and the UK (7 respondents each), Portugal (6 respondents), Poland and Bulgaria (5 respondents each). Interestingly, Bulgaria was the only country with all respondents being females (no male Bulgarian students attended the summer school in 2014). The remaining four countries, namely Czech Republic, Turkey, China and Germany, had between two and four representatives within the sample.

7.2 Preferred Type of a Train

As the purpose of the survey was to investigate public perception of driverless trains, the respondents were asked for their preferred train when travelling and were given a choice of three answer options: "Driver train", "Driverless train" and "Any train", where the latter answer indicated no specific preference and could be interpreted in favour of driverless trains option.

The respondents were also asked whether they are worried about using a driverless train and were given three answer options: "Yes", "No" and "Not Sure". Responses to this question were combined with responses to the question on preferred type of train and both are presented in Fig. 2. The results displayed in Fig. 2 show that a great majority of the sample, up to 64 % of all respondents (middle of the graph with "No" answer option), is not worried about using driverless trains. Moreover, the answers are at a similar level for both genders, and over half of males (58 %) and females (57 %) do not worry about using a driverless train and could use a driverless train or whatever train. Only 11 % of males and 14 % of

Fig. 1 Sample size and countries represented (count)

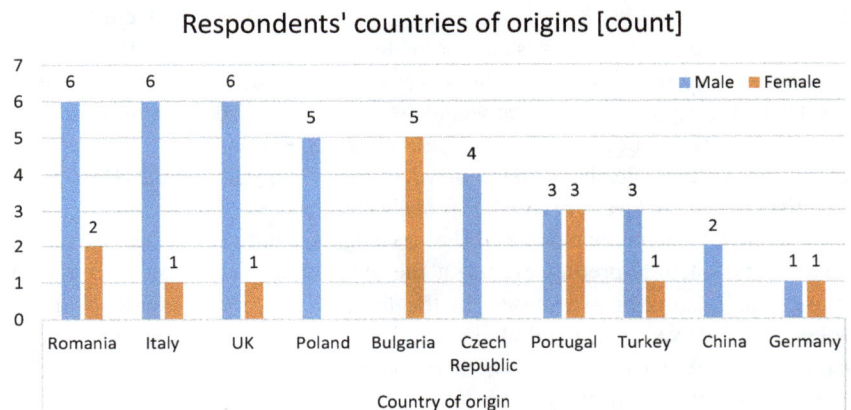

Fig. 2 Choice of a type of a train versus worry of using a driverless train (%)

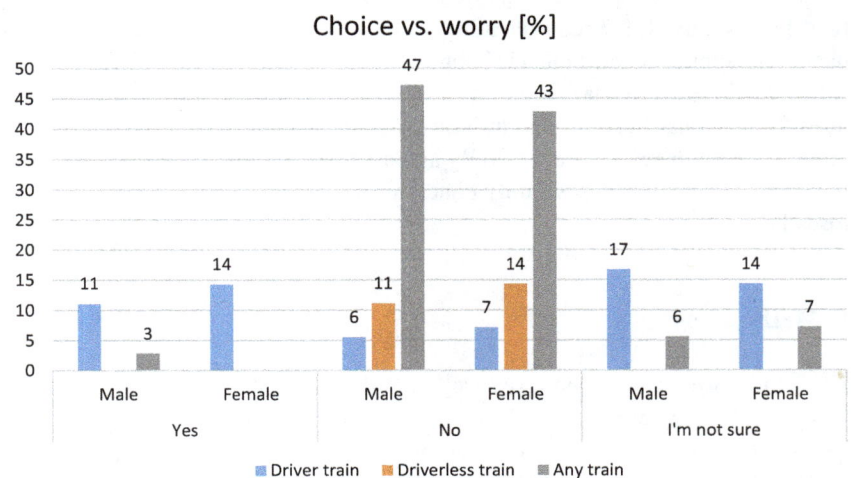

females are worried about using a driverless train and, if given a choice, would prefer to use a traditional train with a driver. This result shows that in general the respondents are not worried about being on a train without a driver and in fact nearly half of them is not event bothered about the train type ("Any train" option in Fig. 2).

7.3 Driverless Train Technology

Although more driverless trains are in operation worldwide and passengers are using them on a daily basis, peoples' opinions on the technology used on driverless trains are not publicised much. Therefore, the respondents have been asked to rate driverless train technology on a 5-point scale with "Very Good" being the highest rate and "Very Bad" being the lowest rate.

Figure 3 shows that 72 % of males and 93 % of females rated the driverless train technology as at least "Good". Although the majority of the sample is of a positive opinion, there are still 25 % of males and 7 % of females who are neutral and only 3 % of males with a negative view on a driverless train technology.

Overall, 78 % of the sample rated UTO as "Very Good" or "Good". Despite the fact that 34 % of the sample would prefer a train with a driver, majority of the respondents within this group still rated driverless trains as "Very Good" (4 %) or "Good" (16 %). The driverless train enthusiasts were in minority and formed 12 % of the sample only. Although none of the driverless train enthusiast rated this option negatively ("Bad" or "Very Bad"), the split between "Very Good" and "Good" was from 6 % to 4 %. The largest group of respondents, 60 % of the sample, selected "Any train" train as their preferred option showing that they could ride either a driver or a driverless train (Fig. 4).

7.4 Factors Influencing Preferences

According to UTO advocators, the driverless system brings a number of benefits to their users (see Sect. 4), mainly in terms of time and frequency of services. The respondents were asked to select reasons which would influence their preference for driverless train over a driver train. The list of options included reduced ticket price, extended running periods, increased train frequency and other.

Fig. 3 Respondents' opinions on driverless train technology (%)

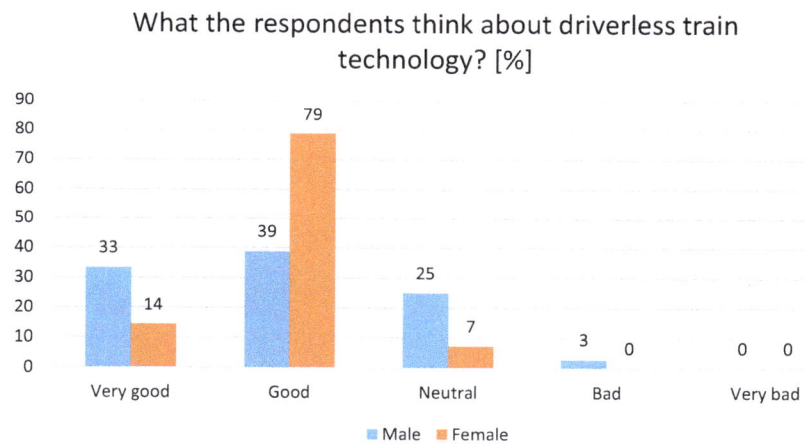

What the respondents think about driverless train technology? [%]

Fig. 4 Opinions on driverless trains versus choice of a type of a train (%)

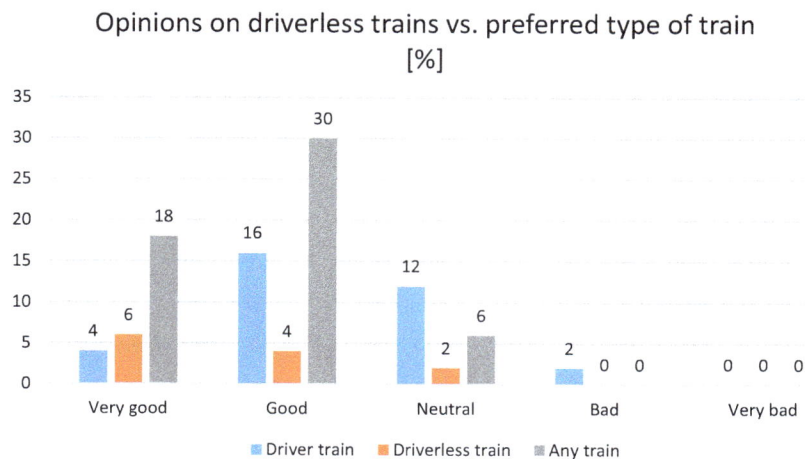

Opinions on driverless trains vs. preferred type of train [%]

Results displayed in Fig. 5 show that none of the options given would convince the majority of respondents to choose a driverless train as a preferred option. However, based on the answers given, it can be seen that it would be more difficult to convince females to use driverless trains as over half of the female sample stated that none of the three factors presented in Fig. 5 would influence their choices. Although male responses were similar, the split between "Yes" and "No" answers for "extended running periods" and "increased train frequency" was more equal (47 % for "Yes" vs. 53 % for "No" and 50 % for "Yes" vs. 50 % for "No", respectively). The results suggest that perhaps new or other measures, to these presented in the survey, should be used when campaigning for change in public's perception of driverless trains and the benefits the UTO systems offers as the benefits listed in Fig. 5 did not get a great respondent's support.

7.5 Importance of a Driver

Although UTO is designed to be fully operational without a member of staff on board, some operators choose to put staff on board (e.g. Budapest Metro Line M4, Airport Express Beijing Subway), especially at the early stages of system's implementation. In this light, the respondents were asked about the importance of a driver on a train, but also about a need for a driver room on a driverless train. The latter is obviously a "fake" room, but in principle its purpose is to help with a shift from a driver to a driverless system and accommodate a smooth change in users' acceptance of the new system.

Both male and female respondents agree that a driver room should be present on a driverless train; however, the issue seems to be much more important to females (93 % of females) than males (72 % of males). Moreover, majority of females who would like to see a driver room on a driverless train rated the presence of a driver on a train as "Very Important" or "Important" (14 and 50 %, respectively). This result shows that females within the sample are much more than males attached to the idea of a driver on a train as well as more comfortable with a train with a driver room installed. The gender differences in the responses presented in Fig. 6 highlight the fact that how

Fig. 5 Factors that would influence a driverless train as a preferred option (%)

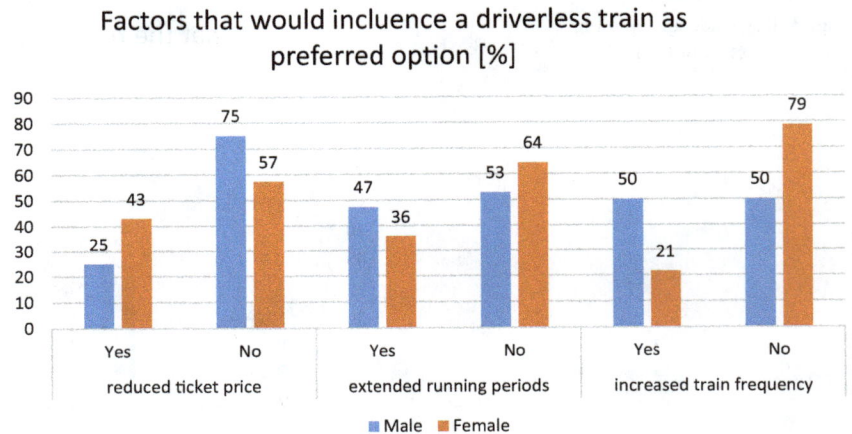

Factors that would incluence a driverless train as preferred option [%]

Fig. 6 Importance of a train driver versus a need for a driver room on a driverless train (%)

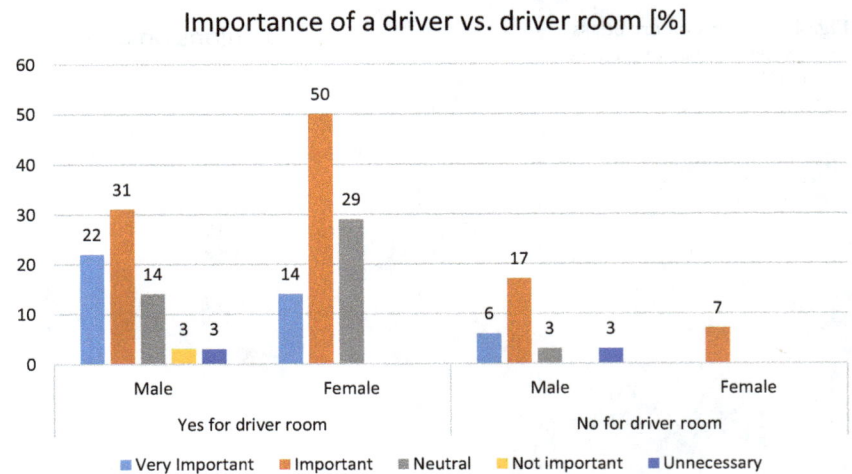

Importance of a driver vs. driver room [%]

both genders see the role of a driver on a train, and this issue requires further investigation as it can potentially lead to other issues and identification of other explanatory variables.

7.6 Human Error

Although UTO enthusiasts highlight the advantage of the automated systems where a human error is reduced or eliminated, the fact is that people, from train designers to control room staff, are still involved in the UTO system. The respondents therefore have been asked to select areas where, according to their opinion, a human error is likely to occur.

Results displayed in Fig. 7 show that a great majority of respondents expressed no worries about a train design or maintenance issues (only 11 % of males and 28 % of males versus 14 % of females, respectively). However, a communication between staff boosted the level of worried respondents to 36 % amongst males and 43 % amongst females. The results suggest that respondents see a staff communication as an area where human error is more likely to occur than in a design or a maintenance domain. Moreover, 50 % of males and 43 % of females were worried about a technical failure of UTO and, although this is only half of the sample or less, this issue strikes as an area of greatest concern amongst the respondents out of the four areas listed on Fig. 7.

7.7 Unemployment Issue

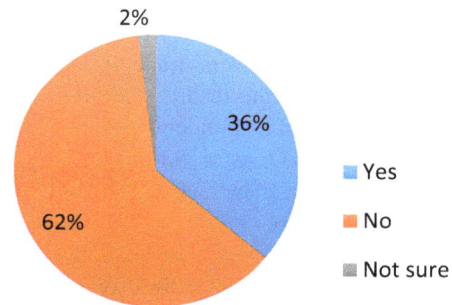

The position of train drivers' trade unions campaigning against driverless trains is well known as well as their argument of drivers losing jobs and contributing to higher unemployment rates. However, opinions of the public on this issue are unknown. Thus the respondents were asked

about their opinions on unemployment increase connected to driverless trains' implementations. Figure 8 displays clearly that 62 % of the respondents are not worried about drivers losing their jobs as they believe that the drivers could requalify and do other jobs. However, 36 % of the respondents agree that the unemployment rates will increase with implementations of driverless trains.

8 Conclusions

Although automated and driverless trains have been in operation for over three decades, there has not been many scientific research work published on public attitudes to and perceptions of UTO.

In order to contribute to a better understanding of people's perception of driverless trains, this paper presented results of a survey where 50 individuals were asked about their opinions on UTO. Although it might be argued that the sample was biased because all respondents were somehow interested in the railways, it must be highlighted

Fig. 8 Will implementations of driverless trains contribute to increase of drivers' unemployment? (%)

Fig. 7 Areas of human error in an UTO system (%)

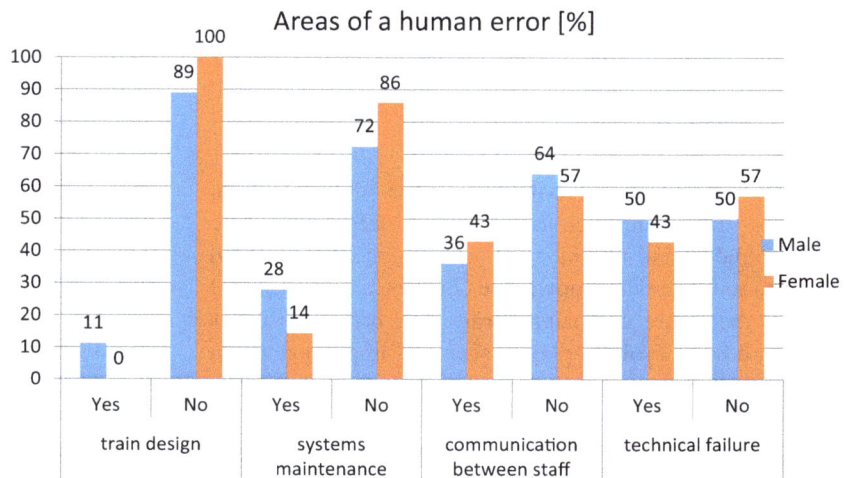

that they are also passengers with personal opinions about the railway system and the survey sought their individual opinions on UTO.

The results presented in the paper can be grouped into three thematic areas: train type preferences and opinions on driverless trains, importance of a driver on a train and the unemployment issues and a human error issue.

Firstly, only 11 % of males and 14 % of females stated that they would prefer to use a traditional train with a driver rather than a driverless train or whatever train. This result shows that the majority of the respondents is not bothered about the train type they are using. Moreover, opinions about the driverless technology are very positive and rated as a "Very Good" or "Good" technology by the overwhelming majority of 93 % females and 72 % of males within the sample. This shows that in general the respondents are keen on UTO and they do not have a problem to trust the technology.

Secondly, the importance of a driver on a train was rated as "Important" or "Very Important" by the majority of the respondents who highlight the perception of a driver as an important component of the system. Moreover, over 50 % of the sample agreed that there should be a driver's room on the train, which in the case of a driverless technology is obviously not necessary.

Thirdly, despite many drivers' trade unions campaigning against driverless trains, the results presented in the paper show that the majority of respondents (62 %) do not see the implementation of UTO as a thread to a driver's job security. However, as there was no follow-up of the unemployment question, it is difficult to understand respondents' reasons for being "for" or "against" the idea that driverless trains will affect train drivers' employment.

Fourthly, the results revealed that the respondents overall are not worried about human error occurring on a driverless train. However, when looked into more detail, it appears that a technical failure and a staff communication issues are the two main areas of concern in relation to a human error on UTO.

9 Further Research

Overall, this paper contributes to the discussion on driverless trains but much more research needs to be done to fully understand and monitor public perceptions of and attitudes to UTO. This knowledge could be a powerful tool used in the future campaigns promoting driverless trains and could have a role to play in seamless implementations of the new systems around the globe.

More specifically, a more detailed investigation of public level of understanding of technology behind UTO might help to examine the reasons why although respondents trust the technology they do not specifically go for it if given a choice between a driver train and a driverless train.

Next, the public perception of the role of a driver on a driverless train requires further investigation where links with issues such as safety and security and anti-social behaviour on a train could be explored.

To follow-up the employment issue, a further investigation into the reasons why the public perceives chances of the drivers to requalify and stay on the job quite high would be needed to better understand their motivations which could be used in the future promotion of the driverless trains to drivers' trade unions and the public.

Finally, in order to further investigate public opinions of areas where they fear a human error might occur, a more detailed study of perceptions and preferences on UTO would be needed. This could help to design public campaigns explaining how UTO system works and enforce technical strategies for overcoming the possibility of a human error to occur on UTO system.

Acknowledgments This paper is based on results collected by an MSc student Suyi Duan who investigated passengers' perception of driverless trains as part of her major project at the School of Mechanical and Systems Engineering at Newcastle University. Philip Brown, a college student on Nuffield Research Placements at Newcastle University, contributed to the analysis of results of the project.

References

1. BBC (2014) Driverless tube trains: Unions vow 'war' over plan. http://www.bbc.co.uk/news/uk-england-london-26381175. Accessed 9 Dec 2014
2. Berwell FT (1973) Automatic railways: automation and control in transport. Pergamon Press, Oxford, pp 177–191
3. Brown P (2014) Are driverless trains the future? Rail Technology Magazine, February/March 2014, p 19
4. Fisher E (2011) Justifying automation. In: Railway technology. http://www.railway-technology.com. Accessed 5 Dec 2014
5. Fraszczyk A, Magalhães da Silva J, Gwóźdź A, Vasileva G (2014) Metro as an example of an urban rail system. Four case studies from Europe. Transp Probl 9:101–107
6. Hasham N (2013) Driverless trains plan must overcome public scepticism. http://www.smh.com.au/nsw/driverless-trains-plan-must-overcome-public-scepticism-20130607-2nvjq.html. Accessed 9 Dec 2014
7. Karvonen H, Aaltonen I, Wahlström M, Salo L, Savioja P, Norros L (2011) Hidden roles of the train driver: a challenge for metro automation. Interact Comput 23:289–298
8. Malla R (2014) Automation sets a new benchmark. Metro report, May 2014

9. Rumsey A (2009) Communications based train control. IRSE seminar
10. TransLink (2014) SkyTrain. http://www.translink.ca/en/Schedules-and-Maps/SkyTrain.aspx. Accessed 9 Dec 2014
11. UIC (2014) Automatic train control. Energy efficiency technologies for railways. http://www.railway-energy.org/static/Automatic_train_control_79.php. Accessed 9 Dec 2014
12. UITP (2011) Media backgrounder. Metro automation facts, figures and trends. UITP, Brussels
13. UITP (2013) Metro automation in 2013. Observatory of Automated Metros World Atlas Report. UITP, Brussels
14. UITP (2014) Statistics brief. World metro figure. UITP, Brussels
15. Vuchic V (2014) Maintaining performance with full automation. Metro report international, March 2014, pp 36–39

Permissions

List of Contributors

Peter E. Timan
Bombardier Transportation, Kingston, ON, Canada

Xihe He
Chongqing Rail Transit Design and Research Institute, Chongqing 401122, China

Florin Codruţ Nemţanu, Dorin Laurenţiu Bureţea and Luigi Gabriel Obreja
Transport Faculty, University Politehnica of Bucharest, Bucharest, Romania

Zhili Zhang
School of Mechanical, Electronic and Control Engineering, Beijing Jiaotong University, Beijing 100044, China

ChunQiang Wang and Wenqiang Zhang
Transportation Administration of Beijing Municipal Commission of Transport, Beijing 100053, China

Teodora Stefanova, Christian Wullems, James Freeman and Andry Rakotonirainy
Centre for Accident Research and Road Safety – Queensland, Queensland University of Technology, 130 Victoria Park Road, Kelvin Grove 4059, Australia

Patricia Delhomme and Jean-Marie Burkhardt
IFSTTAR, AME, LPC, 78000 Versailles, France

Aleksandrs Rjabovs and Roberto Palacin
NewRail – Centre for Railway Research, School of Mechanical and Systems Engineering, Newcastle University, Newcastle upon Tyne NE17RU, UK

Grégoire S. Larue
Centre for Accident Research and Road Safety - Queensland, Queensland University of Technology, Brisbane 4000, Australia
Australasian Centre for Rail Innovation, Canberra 2600, Australia

Christian Wullems
Centre for Accident Research and Road Safety - Queensland, Queensland University of Technology, Brisbane 4000, Australia
Cooperative Research Centre for Rail Innovation, Brisbane 4000, Australia

Paul Batty and Roberto Palacin
School of Mechanical and Systems Engineering, NewRail – Centre for Railway Research, Newcastle University, Stephenson Building, Newcastle Upon Tyne NE1 7RU, UK

Dan Lu and Futian Wang
State Key Laboratory of Rail Traffic Control and Safety, Beijing Jiaotong University, Beijing, China

Suliang Chang
The Line Branch of Beijing Subway Operation Limited Company, Beijing, China

Anjum Naweed
Appleton Institute for Behavioural Science, Central Queensland University, 44 Greenhill Rd, Wayville, SA 5034, Australia

Helen Moody
Injury Prevention and Management, Corporate Health Group, Adelaide, SA, Australia

J. P. Powell and R. Palacín
NewRail - Centre for Railway Research, Newcastle University, Stephenson Building, Claremont Road, Newcastle upon Tyne NE1 7RU, UK

Binbin Liu and Stefano Bruni
Dipartimento di Meccanica, Politecnico di Milano, Via La Masa 1, 20156 Milan, Italy

Chun Zhang
School of Architecture and Design, Beijing Jiaotong University, Beijing, China

Joyce Man
School of Public and Environmental Affairs, Indiana University Bloomington, Bloomington, USA
Lincoln Institute Center for Urban Development and Land Policy, Beijing, China

Alex Dampier
Mechanical and Systems Engineering School, Newcastle University, Newcastle upon Tyne, UK

Marin Marinov
NewRail, Mechanical and Systems Engineering School, Newcastle University, Newcastle upon Tyne, UK

Daniel Brice
School of Mechanical and Systems Engineering, Newcastle University, Newcastle upon Tyne, UK

Marin Marinov
NewRail, School of Mechanical and Systems Engineering, Newcastle University, Newcastle upon Tyne, UK

Bernhard Rüger
TU-Wien, Wien, Austria

Jing Teng and Wang-Rui Liu
Key Laboratory of Road and Traffic Engineering, Ministry Education, Tongji University, 4800 Cao'An Road, JiaDing District, Shanghai, China

Marin Marinov
NewRail, Mechanical and Systems Engineering School, Newcastle University, 2nd Floor, Stephenson Building, Newcastle upon Tyne NE1 7RU, UK

Agajere Ovuezirie Darlton
Mechanical and Systems Engineering School, Newcastle University, Stephenson Building, Newcastle upon Tyne NE1 7RU, UK

Selby Coxon
Faculty of Art Design & Architecture, Monash University, 900 Dandenong Road, Caulfield, Melbourne 3145, Australia

Tom Chandler and Elliott Wilson
Faculty of Information Technology, Monash University, 900 Dandenong Road, Caulfield, Melbourne 3145, Australia

Anna Fraszczyk, Philip Brown and Suyi Duan
NewRail, Newcastle University, King's Gate, Newcastle upon Tyne NE1 7RU, UK

www.ingramcontent.com/pod-product-compliance
Lightning Source LLC
Chambersburg PA
CBHW070152240326
41458CB00126B/4443